# Selected Titles in This Series

90 **Christian Gérard and Izabella Łaba,** Multiparticle quantum scattering in constant magnetic fields, 2002

89 **Michel Ledoux,** The concentration of measure phenomenon, 2001

88 **Edward Frenkel and David Ben-Zvi,** Vertex algebras and algebraic curves, 2001

87 **Bruno Poizat,** Stable groups, 2001

86 **Stanley N. Burris,** Number theoretic density and logical limit laws, 2001

85 **V. A. Kozlov, V. G. Maz′ya, and J. Rossmann,** Spectral problems associated with corner singularities of solutions to elliptic equations, 2001

84 **László Fuchs and Luigi Salce,** Modules over non-Noetherian domains, 2001

83 **Sigurdur Helgason,** Groups and geometric analysis: Integral geometry, invariant differential operators, and spherical functions, 2000

82 **Goro Shimura,** Arithmeticity in the theory of automorphic forms, 2000

81 **Michael E. Taylor,** Tools for PDE: Pseudodifferential operators, paradifferential operators, and layer potentials, 2000

80 **Lindsay N. Childs,** Taming wild extensions: Hopf algebras and local Galois module theory, 2000

79 **Joseph A. Cima and William T. Ross,** The backward shift on the Hardy space, 2000

78 **Boris A. Kupershmidt,** KP or mKP: Noncommutative mathematics of Lagrangian, Hamiltonian, and integrable systems, 2000

77 **Fumio Hiai and Dénes Petz,** The semicircle law, free random variables and entropy, 2000

76 **Frederick P. Gardiner and Nikola Lakic,** Quasiconformal Teichmüller theory, 2000

75 **Greg Hjorth,** Classification and orbit equivalence relations, 2000

74 **Daniel W. Stroock,** An introduction to the analysis of paths on a Riemannian manifold, 2000

73 **John Locker,** Spectral theory of non-self-adjoint two-point differential operators, 2000

72 **Gerald Teschl,** Jacobi operators and completely integrable nonlinear lattices, 1999

71 **Lajos Pukánszky,** Characters of connected Lie groups, 1999

70 **Carmen Chicone and Yuri Latushkin,** Evolution semigroups in dynamical systems and differential equations, 1999

69 **C. T. C. Wall (A. A. Ranicki, Editor),** Surgery on compact manifolds, second edition, 1999

68 **David A. Cox and Sheldon Katz,** Mirror symmetry and algebraic geometry, 1999

67 **A. Borel and N. Wallach,** Continuous cohomology, discrete subgroups, and representations of reductive groups, second edition, 2000

66 **Yu. Ilyashenko and Weigu Li,** Nonlocal bifurcations, 1999

65 **Carl Faith,** Rings and things and a fine array of twentieth century associative algebra, 1999

64 **Rene A. Carmona and Boris Rozovskii, Editors,** Stochastic partial differential equations: Six perspectives, 1999

63 **Mark Hovey,** Model categories, 1999

62 **Vladimir I. Bogachev,** Gaussian measures, 1998

61 **W. Norrie Everitt and Lawrence Markus,** Boundary value problems and symplectic algebra for ordinary differential and quasi-differential operators, 1999

60 **Iain Raeburn and Dana P. Williams,** Morita equivalence and continuous-trace $C^*$-algebras, 1998

For a complete list of titles in this series, visit the AMS Bookstore at **www.ams.org/bookstore/**.

# Multiparticle Quantum Scattering in Constant Magnetic Fields

Mathematical
Surveys
and
Monographs

Volume 90

# Multiparticle Quantum Scattering in Constant Magnetic Fields

**Christian Gérard**

**Izabella Łaba**

**American Mathematical Society**

2000 *Mathematics Subject Classification.* Primary 35P25, 35Q40, 34L25, 47A40, 81U10.

ABSTRACT. This book is devoted to the scattering theory of systems of $N$ interacting quantum particles in an external constant magnetic field. Particular emphasis is placed on the development of the Mourre theory, applications of geometrical methods, and the proof of asymptotic completeness for a large class of systems.

---

**Library of Congress Cataloging-in-Publication Data**

Gérard, Christian, 1960–
Multiparticle quantum scattering in constant magnetic fields / Christian Gérard, Izabella Łaba.
p. cm. — (Mathematical surveys and monographs ; ISSN 0076-5376 ; v. 90)
Includes bibliographical references and index.
ISBN 0-8218-2919-X (alk. paper)
1. Scattering (Physics) 2. Quantum theory. 3. Few-body problem. 4. Magnetic fields. I. Łaba, Izabella, 1966– II. Title. III. Series.

QC20.7.S3 G47 2001
539.7′58—dc21 2001053521

---



∞ The paper used in this book is acid-free and falls within the guidelines
established to ensure permanence and durability.
Visit the AMS home page at URL: http://www.ams.org/

10 9 8 7 6 5 4 3 2 1 06 05 04 03 02 01

# Contents

# Preface

This monograph is devoted to the spectral and scattering theory of quantum Hamiltonians describing systems of $N$ interacting particles in an external constant magnetic field. Most of it consists of the results obtained by the authors from 1993 to 1999.

Quantum scattering theory is the subfield of quantum mechanics which deals with the large-time asymptotics of the solutions of the Schrödinger equation and with the structure of the continuous spectrum of the corresponding Schrödinger operator. One of its main problems is to prove (or disprove) *asymptotic completeness*, which roughly speaking, is a statement that all solutions of the Schrödinger equation under consideration must follow asymptotically certain prescribed patterns. (The precise mathematical formulation of this will be given in Chapter 6.) There is a vast body of literature on this and other aspects of 2-particle scattering, see e.g., [**RS**] or [**Hö**, vols. II, IV] for an overview. For $N \geq 3$ particles, the problem becomes much more complicated. It was only in the last 20 years that the $N$-body scattering theory underwent a period of rapid development, beginning with the work of Enss [**E2**], [**E3**], and culminating in the proof of $N$-body asymptotic completeness by Sigal-Soffer [**SS1**] and Dereziński [**De1**], with significant contributions by many other authors, see e.g., [**M1**], [**PSS**], [**FH1**], [**Gr**], [**Y**]. We refer the reader to [**DG1**] for a more detailed account of that story and for a self-contained presentation of the results obtained in the 1980's and 90's.

Our work was largely inspired by these developments: we set out to extend the new results on asymptotic completeness to the case of $N$-body systems in a constant magnetic field. Such systems are of considerable interest in quantum physics. There is a large body of research on the *quantum Hall effect*; most of it assumes that there are no interactions between the particles save for the Pauli exclusion principle, but it is possible that at some point the scattering effects will have to be taken into account. In astrophysics, there is some evidence that strong magnetic fields exist on the surfaces of neutron stars and white dwarfs. "Quantum dots" are a prime example of quantum systems which can be significantly affected by magnetic fields of strength comparable to what can actually be achieved in existing laboratories. Physicists have also been studying highly excited (Rydberg) atoms in magnetic fields, which offer an opportunity to study the phenomena of "quantum chaos". See e.g., [**RWHG**] for a survey of some of the recent (theoretical and experimental) work on the subject. Furthermore, there is a growing interest in magnetic Hamiltonians among mathematicians and mathematical physicists. In particular, questions such as the stability of matter [**Li**], [**LSS**], [**Fe**], eigenvalue or resonance asymptotics [**Iv1**], [**Iv2**], [**FW1**], [**FW2**], [**FW3**], and decay of eigenfunctions [**Er**], [**N**], [**So**] were recently addressed in the literature.

The purpose of this book is twofold. Firstly, in Chapter 1 we provide a general introduction to the spectral theory of $N$-body magnetic Hamiltonians, aimed at a wider audience of mathematical physicists. Secondly, we present a proof of asymptotic completeness for wide classes of magnetic Hamiltonians, namely for generic 3-body systems and for $N$-body systems whose all proper subsystems have nonzero total electric charge. The proof requires much more than simply applying the known methods in a slightly different situation; this book focuses on the new methods and techniques that are specific to the magnetic case. In particular, this includes an extension of the Mourre theory to "dispersive" Hamiltonians with a rather complicated structure (Chapter 3) and a geometrical analysis of the propagation of charged systems (Chapter 5). Our goal was to give a clear and reasonably self-contained presentation of the subject and to provide a solid foundation for further research.

The book is addressed mostly to researchers and graduate students in mathematical physics. We do expect the reader to be familiar with quantum mechanics, functional analysis, and modern PDE theory (especially with pseudodifferential calculus). A background in $N$-body scattering and abstract Mourre theory will be useful, but not indispensable. To the readers who wish to acquire such background we recommend the monographs [**DG1**] and [**ABG**]. However, anyone willing to accept without proof the results of [**DG1**] and [**ABG**] that we will invoke should also be able to follow all of our arguments. In fact parts of this book (especially Chapters 1 and 2) may serve as an introduction to the $N$-body theory. We emphasize that no previous exposure to magnetic Schrödinger operators is required.

Some of the results presented here were first published in [GŁ1–3]. However much of the material, including all of our results in the 2−dimensional case and a large part of the geometrical analysis of Chapter 5, is published here for the first time. The Mourre theory for magnetic Hamiltonians (Chapter 3) has been completely reworked and rewritten, especially in the case we call "dispersive".

# Notation

For the reader's convenience, we collect and explain here the notation used throughout this book.

**Spaces:** $\mathrm{X}, \mathrm{Y}, \mathrm{Z}, X, Y, Z$, often subscripted or superscripted, will be Euclidean spaces (isomorphic to $\mathbb{R}^m$ for some $m$). The coordinates in these spaces will be denoted by $\mathrm{x}$, $\mathrm{y}$, $\mathrm{z}$, $x$, $y$, $z$, with appropriate subscripts or superscripts. The Roman letters X, Y, Z, will be used to denote the configuration spaces before the center of mass separation, and the italic $X$, $Y$, $Z$ denote the same spaces after the center of mass separation. (Most of the time we will only work with the latter spaces.) A similar convention will be used for the coordinates $\mathrm{x}$, $x$, etc. Since we will actually separate the center of mass only in one direction, we will have $\mathrm{Y} = Y$ and $\mathrm{y} = y$.

**Dual spaces:** The spaces dual to $\mathrm{X}, \mathrm{Y}, \mathrm{Z}, X, Y, Z$ will be denoted by $\mathrm{X}', \mathrm{Y}'$, etc. The duality will be denoted by $\langle \cdot, \cdot \rangle$. If $\mathcal{X}$ is one of the configuration spaces as above, $T^*\mathcal{X} := \mathcal{X} \times \mathcal{X}'$ will be the cotangent bundle of $\mathcal{X}$.

**Derivatives:** The symbol $\partial_x f$ or $\frac{\partial f}{\partial x}$ will mean the partial derivative of $f$ in $x$ if $x$ is a variable in $\mathbb{R}$, and $\nabla_x f$ if $x$ is a vector variable. We will also write $D_x f = -\mathrm{i}\partial_x f$. We will often omit the subscripts $x$ if the choice of variables is clear from context, and abbreviate $\partial_{x_j}$ to $\partial_j$, $D_{x_j}$ to $D_j$, etc.

For $\alpha = (\alpha_1, \ldots, \alpha_n)$, we use the standard notation

$$\partial^\alpha f = \partial_{x_1}^{\alpha_1} \ldots \partial_{x_n}^{\alpha_n} f, \ |\alpha| = \alpha_1 + \cdots + \alpha_n.$$

For $0 < \epsilon < 1$, we will denote:

$$C^\epsilon(\mathbb{R}) = \{f \in C(\mathbb{R}) : \ |f(x+y) - f(y)| \le c(x)|y|^\epsilon \text{ for all } x \in \mathbb{R}, \ |y| \le 1\},$$

$$C^{k+\epsilon}(\mathbb{R}) = \{f \in C^k(\mathbb{R}) : \ f^{(k)} \in C^\epsilon(\mathbb{R})\}.$$

**Classes of symbols:** For an Euclidean space $\mathcal{X}$ (usually equal to one of the configuration spaces above), $\mathcal{S}(\mathcal{X})$ is the Schwartz class of functions on $\mathcal{X}$, and $\mathcal{S}'(\mathcal{X})$ is the space of tempered distributions on $\mathcal{X}$. We will also work with the following classes of symbols:

$$S^0(\mathcal{X}) = \{f \in C^\infty(\mathcal{X}) | \ |\partial^\alpha f(x)| \le c_\alpha, \ x \in \mathcal{X}, \ |\alpha| \ge 0\},$$

and, for $\epsilon \in \mathbb{R}$,

$$S^\epsilon_{\mathrm{cl}}(\mathcal{X}) = \{f \in C^\infty(\mathcal{X}) | \ |D^\alpha f(x)| \le c_\alpha \langle x \rangle^{\epsilon - |\alpha|}, \ x \in \mathcal{X}, \ |\alpha| \ge 0\}.$$

We will use the convention that whenever an estimate is stated for all functions $f_1, f_2, \ldots$ in a specified class of symbols $S$, the constants in the estimate are understood to depend only on the seminorms of $f_i$ in $S$, and that this dependence will be linear for each $f_i$.

**Operators:** $\mathcal{H}$, $\mathcal{H}'$, etc., often subscripted or superscripted, will be Hilbert spaces, usually equal to $L^2(\mathcal{X})$, where $\mathcal{X}$ is one of the configuration spaces above. The inner product on $\mathcal{H}$ will be denoted by $(\cdot, \cdot)$, and will be linear in the second variable and antilinear in the first one.

$A, B, H$, etc. will be linear operators on $\mathcal{H}$. For Hamiltonians of the $N$-body systems under consideration, we adopt the convention that the Roman H and the italic $H$ are, respectively, the Hamiltonians of the system before and after the center of mass separation.

The identity operator on $\mathcal{H}$ will be denoted by $\mathbf{1}_{\mathcal{H}}$, or simply by $\mathbf{1}$ if it is clear from context what $\mathcal{H}$ is.

$B(\mathcal{H}_1, \mathcal{H}_2)$ is the algebra of bounded linear operators from $\mathcal{H}_1$ to $\mathcal{H}_2$; we will abbreviate $B(\mathcal{H}, \mathcal{H}) =: B(\mathcal{H})$.

We will use s$-$ lim and w$-$ lim to denote the limits in the strong and weak topology, respectively.

$\mathcal{D}(H)$ is the domain of $H$; it will always be either specified explicitly or clear from context whether the form domain or the operator domain of $H$ is meant. The graph norm on $\mathcal{D}(H)$ is

$$\|u\|_{\mathcal{D}(H)} = \|u\| + \|Hu\|.$$

The symbols $\sigma(H)$, $\sigma_{\rm pp}(H)$, $\sigma_{\rm cont}(H)$ will denote the spectrum of $H$, the point spectrum of $H$, and the continuous spectrum of $H$, respectively. We will also use $\mathcal{H}_{\rm pp}(H)$, $\mathcal{H}_{\rm cont}(H)$, $\mathcal{H}_{\rm ac}(H)$ to denote the pure point, continuous, and absolutely continuous spectral subspaces of $H$.

$\mathbf{1}_\Omega(H)$ is the spectral projection of the self-adjoint operator $H$ on a set $\Omega \subset \mathbb{R}$.

If $A$ is a symmetric quadratic form and $B$ is a symmetric operator on $\mathcal{H}$, the phrase "$A$ is bounded on the domain of $B$" means that

$$|A(u, u)| \leq C\|Bu\|^2, \quad u \in \mathcal{D}(B).$$

(In particular, it implies the inclusion $\mathcal{D}(A) \subset \mathcal{D}(B^2)$ between the form domains of $A$ and $B^2$.)

If $A$ is an operator, the phrase "$A$ preserves the domain of $B$" means that if $u \in \mathcal{D}(B) \cap \mathcal{D}(A)$, then $Au \in \mathcal{D}(B)$ and

$$\|BAu\| \leq C(\|u\| + \|Bu\|).$$

A mapping $M \ni k \to A(k)$, where $M$ is a metric space (usually a subset of $\mathbb{R}^n$) and $A(k)$ are self-adjoint operators on $\mathcal{H}$, is said to be *continuous in the norm resolvent sense* if the mapping

$$M \ni k \to (A(k) + z)^{-1} \in B(\mathcal{H})$$

is norm continuous in $k$ for any $z$ with $\operatorname{Im} z \neq 0$, and *analytic in the norm resolvent sense* if the above mapping is analytic in $k$ for any $z$ with $\operatorname{Im} z \neq 0$.

We will use "$A + \text{h.c.}$" to denote "$A + A^*$; typically, this notation will be employed when $A$ is a long expression of the form $A = A_1 \dots A_k$, in which case the "h.c." stands for $A_k^* \dots A_1^*$.

We define $\operatorname{ad}_A(H) := [H, \mathrm{i}A]$ and $\operatorname{ad}_A^k(H) := [\operatorname{ad}_A^{k-1}(H), A]$.

By "undoing a commutator" we will mean writing $[A, B]$ as $AB - BA$ and estimating each term separately.

If $A(t)$ is a family of operators on $\mathcal{H}$ depending on a parameter $t$, we will write $A(t) = O(t^\alpha)$ if $A(t) \in B(\mathcal{H})$ for $t$ large enough and $\|A(t)\| = O(t^\alpha)$. A similar convention will be used for $A(t) = o(t^\alpha)$.

The notation "$A \geq_* B$ at $H = \lambda$", used heavily in Chapter 3, is explained at the beginning of Section 3.1.

The classes of operators $C^s(A)$, where $A$ is a self-adjoint operator on $\mathcal{H}$, are defined in Section 3.2.

**Cut-off functions:** $\mathbf{1}_\Omega$ will be the characteristic function of the set $\Omega$. $F(x \in \Omega)$ is a "smoothed out" characteristic function of $\Omega$: it is a $C^\infty$ function equal to 0 outside $\Omega$ and to 1 in a slightly smaller set.

**Miscellanous:** $\langle x \rangle$ is a function in $C^\infty(\mathcal{X})$, greater than $1/2$ for all $x \in \mathcal{X}$ and equal to $|x|$ for $|x| \geq 1$.

CHAPTER 1

# Fundamentals

The results of this chapter are of a very basic nature: they will provide a foundation for all of the further work presented in this book, and will likely be of independent interest to a wider group of researchers.

Section 1.1 is a non-technical introduction to magnetic Hamiltonians, written mostly for the newcomers to the subject. Section 1.2 is devoted to preliminaries and bookkeeping. We define the Hamiltonians under investigation, state and discuss the assumptions on the pair interactions, address the question of self-adjointness, and prove a few related technical results. We also review certain mathematical tools which will be used throughout the book. Most of this section consists of well known results, borrowed from the references cited in the text.

It is well known that magnetic Hamiltonians do not admit center of mass separation; the goal of Section 1.3 is to find a suitable substitute for it. We use a constant of motion called *total pseudomomentum* to define *reducing unitary transformations*, which yield non-trivial direct integral representations of the Hamiltonians in question. We then introduce our "preferred" reducing transformations (with additional good properties), and find the associated direct integral representations. The general theory is illustrated by explicit computations for one- and two-particle systems.

Finally, in Section 1.4 we define the *bound states* of $N$-body systems and study their external motion. Without a magnetic field these questions would be trivial; here, however, we must proceed much more carefully. As it turns out, the external motion of a bound state will depend on its total electric charge. Charged systems behave essentially in the same manner as charged particles: they may escape to infinity only in the direction of the magnetic field. Neutral systems, however, may travel in all directions, and their external and internal dynamics are interdependent.

The main results of Sections 1.3 and 1.4 were first proved in [**GŁ1**]. The importance of the pseudomomentum in the study of $N$-body magnetic Hamiltonians had been recognized earlier in [**AHS2**]: the main results of that paper were, in the language of this book, the existence of reducing transformations, an explicit computation of the reduced Hamiltonians of a neutral pair and an investigation of their spectral properties (see also the comments at the beginning of Chapter 3), and an HVZ-type theorem. The calculations in Subsection 1.3.7 follow [**Ł3**]. Physicists have been aware of the dependence of the external motion of a bound state on its total electric charge and pseudomomentum (see [**RWHG**, Section 2.2], for a brief discussion and a quantum-mechanical analogue of our Example 2 in Section 1.4); however, the precise statements of Propositions 1.4.3, 1.4.6 do not appear to have been known prior to our work, even in the 2-body case.

## 1.1. Introduction

The purpose of this section is to explain some of the main concepts and ideas of scattering theory for $N$-particle magnetic Hamiltonians in a non-technical way and, whenever possible, in a setting simplified to the extreme. We will thus ignore many important but technical points such as selfadjointness or domain questions, and will not state rigorously our main results (including asymptotic completeness). The details are of course included in the main text of this book. Moreover, each chapter begins with an introduction which further explains the main ideas involved and includes brief historical notes and bibliographical references.

**1.1.1. One-particle Hamiltonian.** We first consider a single quantum particle in a constant magnetic field $\vec{B}$. We introduce coordinates $x = (y^{(1)}, y^{(2)}, z)^T$ in $\mathbb{R}^3$ so that the magnetic field is parallel to the third axis: $\vec{B} = (0, 0, 2b)$, $b > 0$. We also denote $D = -\mathrm{i}\partial$. For simplicity, we assume that the particle has mass $m = 1$. Then its wave function $\phi(x, t)$ solves the Schrödinger equation $\mathrm{i}\partial_t \phi = H_0 \phi$, where $H_0$ is the free magnetic Hamiltonian

$$H_0 = \frac{1}{2}(D - qJx)^2,$$

$J$ is the vector potential associated with the magnetic field, and $q$ is the electric charge of the particle. The vector potential is not unique and is defined only up to *gauge transformations*; we choose to work in the *transversal gauge* in which

$$J = \begin{pmatrix} 0 & -b & 0 \\ b & 0 & 0 \\ 0 & 0 & 0 \end{pmatrix}.$$

How the magnetic field affects a particle depends mostly on its electric charge. If the particle is *neutral* ($q = 0$), it travels freely throughout the space, as $H_0$ is simply the free Hamiltonian $H_0 = \frac{1}{2}D^2$. However, for *charged* particles ($q \neq 0$) the picture is quite different. To see this, we rewrite $H_0$ as

$$H_0 = H_0^{\perp} + \frac{1}{2}D_z^2,$$

where

$$H_0^{\perp} = \frac{1}{2}(D_{y^{(1)}} + qby^{(2)})^2 + \frac{1}{2}(D_{y^{(2)}} - qby^{(1)})^2.$$

The dynamics of the particle in the $(y_1, y_2)$ plane and along the $z$ axis are thus described by Hamiltonians acting on different variables, hence independent. The motion in the $z$ direction is free and requires no further explanation. We therefore turn to the (more complicated) dynamics in the transversal plane, governed by $H_0^\perp$.

The spectrum of $H_0^\perp$ is well known to consist of the *Landau levels*

$$\Lambda_n = bq(2n+1), \ \ n = 0, 1, 2, \dots;$$

each Landau level is infinitely degenerate. (See Subsection 1.3.8.) To understand this better, we observe that $H_0$ commutes with the *pseudomomentum*, defined as

$$K = D + qJx = \begin{pmatrix} D_{y^{(1)}} - qby^{(2)} \\ D_{y^{(2)}} + qby^{(1)} \\ D_z \end{pmatrix}.$$

The first two components of $K$ have an interesting physical interpretation. Let $y = (y^{(1)}, y^{(2)})^T$, and define

$$\mathbf{C} = \frac{1}{2}(qJ^\perp)^{-1}(D_y + qJ^\perp y),$$

where $J^\perp = \begin{pmatrix} 0 & -b \\ b & 0 \end{pmatrix}$ is the "non-trivial part" of $J$. Thus $\mathbf{C}$ is essentially a multiple of the first two components of $K$, therefore a constant of motion. In classical mechanics, the trajectory of a charged particle in a constant magnetic field is a helix obtained by superposing circular motion in the directions transversal to the field (denoted here by $y$) with free motion in the direction of the field. The line $y = \mathbf{C}$ is the central axis of this helix. (See Subsection 5.1.1.) We therefore call $\mathbf{C}$ the *center of orbit*. We note also that the difference

$$\mathbf{C} - y = \frac{1}{2}(qJ^\perp)^{-1}(D_y - qJ^\perp y)$$

is bounded along the evolution by the total energy $H_0^\perp$ of the motion in transversal directions. (In classical mechanics, $|C - y|$ is the radius of the helix just mentioned.)

Both $K$ and $\mathbf{C}$ will play important roles later on. For now, we return to the analysis of the two-dimensional dynamics generated by $H_0^\perp$ and note that each Landau level $\Lambda_n$ has infinitely many eigenfunctions, each one "centered" at a different $\mathbf{C}$. (This corresponds to the invariance of the Hamiltonian under *magnetic translations*, see Subsection 1.3.2.) We further observe that the two components of the quantum-mechanical observable $\mathbf{C}$ do not commute and cannot be localized simultaneously. Later in Chapter 1 we will construct a unitary transformation which maps the first two components of $K$ to a multiple of $(\eta, D_\eta)$ for a suitable $\eta$, and the Hamiltonian $H_0^\perp$ to a one-dimensional harmonic oscillator (Subsection 1.3.8).

It follows from the above discussion that $H_0$ has essential spectrum $[\Lambda_0, \infty)$, where $\Lambda_0$ is the lowest Landau level. The charged particle moves freely in the $z$ direction and remains in a bound state in the transversal plane; in particular, its $y$ coordinates remain bounded uniformly in time.

**1.1.2. Basic concepts of scattering theory.** Suppose now that we perturb the Hamiltonian $H_0$ by adding an external (scalar) potential $V(x)$. (This might describe e.g. the interaction with a second particle whose position is fixed.) For simplicity, we will assume here that $V(x)$ is a real-valued smooth function satisfying

$$|D^\alpha V(x)| \le C_\alpha(1+|x|)^{-\alpha-\mu} \tag{1.1.1}$$

for some $\mu > 0$ and all multiindices $\alpha$. (The actual assumptions on the potentials that we use in the rest of this book are much weaker.)

Let $H = H_0 + V$. Typically, $H$ will have *bound states* (i.e. $L^2$ eigenfunctions). If $u_0$ is such an eigenfunction with an eigenvalue $\lambda$, then the solution of the Schrödinger equation $\mathrm{i}\partial_t\phi = H\phi$ with the initial condition $\phi(x,0) = u_0(x)$ is simply given by

$$\phi(x,t) = \mathrm{e}^{-itH}u_0(x) = \mathrm{e}^{-it\lambda}u_0(x).$$

In particular, all physical observables (such as the position or velocity) remain unchanged.

The goal of scattering theory is to analyze the *scattering states* – the states orthogonal to all eigenfunctions of $H$. Such states are expected to have large-time asymptotics similar to the solutions of the free magnetic Schrödinger equation, the heuristic being that far from the origin the potential $V(x)$ is small and should not affect the dynamics significantly. To make this precise, we define the *wave operators*:

$$\Omega^\pm := \mathrm{s-}\lim_{t\to\pm\infty} \mathrm{e}^{\mathrm{i}tH}\mathrm{e}^{-\mathrm{i}tH_0}. \tag{1.1.2}$$

Assuming that $V$ is a *short range* potential, i.e., (1.1.1) holds with $\mu > 1$, we can prove that (1.1.2) exist and are *asymptotically complete.*

Suppose that $\Omega^+u^+ = u$ (the case $t \to -\infty$ is similar). Then

$$\lim_{t\to\infty} \|\mathrm{e}^{-\mathrm{i}tH}u - \mathrm{e}^{-\mathrm{i}tH_0}u^+\| = 0. \tag{1.1.3}$$

The existence of the wave operators (1.1.2) means that for every $u^+ \in L^2(\mathbb{R}^3)$ there is a $u \in L^2(\mathbb{R}^3)$ such that (1.1.3) holds. In other words, for every free trajectory $\mathrm{e}^{-\mathrm{i}tH_0}u^+$ there is a trajectory $\mathrm{e}^{-\mathrm{i}tH}u$ which has the same asymptotic behaviour as $t \to \infty$. *Asymptotic completeness* is the property that the range of $\Omega^\pm$ is equal to the continuous spectral subspace of $H$:

$$\operatorname{Ran}\Omega^\pm = \mathcal{H}_{\mathrm{cont}}(H) = (\mathcal{H}_{\mathrm{pp}}(H))^\perp.$$

This means that every trajectory $\mathrm{e}^{-\mathrm{i}tH}u$, where $u \in \mathcal{H}_{\mathrm{cont}}(H)$, is asymptotic to a free trajectory $\mathrm{e}^{-\mathrm{i}tH_0}u^+$ as $t \to \infty$ and to another free trajectory $\mathrm{e}^{-\mathrm{i}tH_0}u^-$ as $t \to -\infty$. (The mapping $u^- \to u^+$ is known as the *scattering operator.*) For one-particle magnetic Hamiltonians as above, we obtain that trajectories $\mathrm{e}^{-\mathrm{i}tH}u$ with $u \in \mathcal{H}_{\mathrm{cont}}(H)$ have the same behavior as the free trajectories discussed in the previous subsection: the motion of the particle is asymptotically free in the direction of the magnetic field and bounded in the transversal directions.

If $V$ is long-range ((1.1.1) holds with $0 < \mu \leq 1$), the wave operators defined as above do not usually exist. Instead, one considers the *modified wave operators*, given by (1.1.2) with $e^{-itH_0}$ replaced by a modified free evolution $U_0(t)$. The latter is constructed using (exact or approximate) solutions of the corresponding equations of classical mechanics. In the case under consideration – a single charged particle in a constant magnetic field and an external potential vanishing at infinity – the heuristic picture remains essentially the same as for short-range interactions, except that the dynamics in the $z$ direction is not completely free as the effects of the long-range potential are not quite negligible at infinity.

The modified comparison evolution $U_0(t)$ is not canonically defined and there are several (non-equivalent) well known standard constructions. We will not go into the details (see however Section 6.1); instead, we will explain briefly why long-range potentials are different from short-range and why the critical rate of decay is $\mu = 1$. A crucial intermediate result in the proof of asymptotic completeness is that for any scattering state the *asymptotic velocity* in the $z$ direction, defined roughly as the limit of $|z|/t$ as $t \to \infty$, exists and is non-zero. In particular, the wave function $\phi(x,t) = e^{-itH}u$, $u \in \mathcal{H}_{\text{cont}}(H)$, is (up to small errors) localized in a region $|z| > c|t|$ for some $c > 0$. In that region we have

$$|V(x)| \leq C(1+|x|)^{-\mu} \leq C'(1+|t|)^{-\mu},$$

and the right side is integrable in $t$ if and only if $\mu > 1$. This is precisely why the wave operators (1.1.2) work for $\mu > 1$ and have to be modified if $\mu \leq 1$.

**1.1.3. Two-particle Hamiltonians.** We now consider a system of two interacting particles with masses $m_i$ and electric charges $q_i \neq 0$. The corresponding Hamiltonian is

$$H = \sum_{i=1}^{2} \frac{1}{2m_i}(D_i - q_i J x_i)^2 + V(x_1 - x_2), \tag{1.1.4}$$

where $J$ is as before and the notation $D_i, x_i$ is self-explanatory. Again, for simplicity we will assume that (1.1.1) holds.

In the absence of a magnetic field ($b = 0$), a standard procedure (both in classical and quantum mechanics) is to subtract the free motion of the *center of mass* of the pair,

$$\frac{m_1x_1 + m_2x_2}{m_1 + m_2},$$

and consider only the relative motion of the particles which is described by a one-body Hamiltonian with a potential vanishing at infinity. This cannot be done if $b \neq 0$, basically because the magnetic Hamiltonian $H$ does not commute with translations. Instead, we have to use the more complicated reduction procedure described in Section 1.3. We do not explain it here as it is quite technical (for a single charged particle and a neutral pair, the details are worked out in Subsections 1.3.8 and 1.3.9). It is important, however, to mention that this procedure is based on the fact that the *total pseudomomentum*

$$K_{\text{cm}} = \sum_{i=1}^{2} (D_i + q_i J x_i) \tag{1.1.5}$$

is a constant of motion. We remark that all components of $K$ commute if and only if the total electric charge $Q = q_1 + q_2$ is zero, and that the details of our reduction procedure will depend on whether $Q = 0$ or $Q \neq 0$. This will be a recurring theme in our work.

It follows from the above that the bound and scattering states cannot be defined simply as the eigenfunctions and states from the continuous spectral subspace, respectively, of the Hamiltonian with the center of mass motion subtracted. Rather, we use the reduction procedure just mentioned to define "bound states" satisfying the common-sense requirement that $|x_1 - x_2|$ be bounded uniformly in time, and "scattering states" orthogonal to the bound states (Section 1.4).

We begin with a brief discussion of the scattering states, which are simpler. The main point is that for such states the two particles must separate in the $z$ direction with non-zero relative asymptotic velocity, hence the distance between them grows linearly in time. Eventually the two particles become independent, each of them behaving similarly to the single particle of Subsection 1.1.2 – moving freely in the direction of the field and performing bounded motion in the $y$ plane.

The behaviour of the bound states, on the other hand, depends on the total electric charge $Q$. We first consider the *charged* case $Q \neq 0$. In this case the bound states also perform only bounded motion in the transversal directions and move freely along the field. (In a way, the bound state can be viewed as a single charged object similar to a charged particle.) The following simple argument illustrates this. We noted above that $K_{\text{cm}}$, given by (1.1.5), is a constant of motion. On the other hand, we see from (1.1.4) that the expression

$$\sum_{i=1}^{2} (D_i - q_i J x_i)$$

is dominated by the total energy $H$. Subtracting this from (1.1.5), we see that

$$q_1 J x_1 + q_2 J x_2$$

is bounded along the evolution, hence so is $q_1 y_1 + q_2 y_2$. But the definition of a bound state implies that $y_1 - y_2$ is also bounded, and, since we assume that $q_1 \neq -q_2$, $q_1 y_1 + q_2 y_2$ is not a multiple of $y_1 - y_2$. It follows that $y_1$ and $y_2$ are bounded.

In Section 1.4 we actually use a different argument, based directly on the (rigorous) definition of bound states. Here we chose to sketch the above reasoning instead, not only because it is non-technical and does not require the reduction procedure of Section 1.3, but also because it illustrates ideas which will be important later on. The following interpretation in particular may seem now to be unnecessarily complicated, but it will be crucial in our analysis of the long-range "charged" $N$-body case in Chapter 5. Since $Q = q_1 + q_2 \neq 0$, we may define the *center of charge* of the pair:

$$\frac{q_1 x_1 + q_2 x_2}{Q}.$$

Suppose that the pair is in a bound state so that $x_1 - x_2$ is bounded uniformly in time. To prove that the transversal coordinates $y$ of both particles remain bounded, it suffices to prove that so do the $y$ coordinates of the center of charge. Define the *center of orbit* $\mathbf{C}_i$ of each particle as in Subsection 1.1.1; since $\mathbf{C}_i - y_i$ are bounded, we may replace $y_i$ by $\mathbf{C}_i$. But

$$\frac{q_1 \mathbf{C}_1 + q_2 \mathbf{C}_2}{Q} = (2QJ^{\perp})^{-1} \sum_{i=1}^{2} (D_{y_i} + q_i J^{\perp} y_i).$$

This is a multiple of the first two components of the total pseudomomentum, hence a constant of motion.

Assume now that the pair is *neutral*, i.e., $Q = q_1 + q_2 = 0$. It is clear that the above argument does not work in this case. Consider further that if the two particles were actually held together by an infinitely strong attractive force, they would become a single object of zero electric charge and as such would not be affected at all by the magnetic field. If on the other hand the particles did not interact, each of them would behave as described above in Subsection 1.1.1. The physical interactions of interest in scattering theory are somewhere in between: they cannot be regarded as infinitely strong, but they can hold the particles together in a stable bound state. It turns out that such bound states behave like neutral objects in that they can travel in all directions, not just in the direction of the magnetic field. However, that motion depends on the internal structure of the bound state in a non-trivial way. The mathematical description of this is roughly as follows. Assuming that $Q = 0$, we may write the Hamiltonian (1.1.4) as a direct integral

$$H = \int^{\oplus} H(k) dk,$$

where $k$ is the total pseudomomentum (note that all of its components commute), and each $H(k)$ acts on the "internal" variables. (For a neutral pair, $H(k)$ are computed explicitly in Subsection 1.3.9, and their spectrum is described in Section 3.5.) The bound states are direct integrals $\int^{\oplus} u(k) dk$, where $u(k)$ are eigenfunctions of $H(k)$ corresponding to eigenvalues $E(k)$. We prove in Section 1.4 that (under suitable regularity assumptions) the average velocity of such a bound state is $\nabla E(k)$. This yields the usual free motion in the $z$ direction, but the dynamics in the $y$ plane can be much more complicated. For further discussion and explicit examples from classical mechanics, see the introduction to Section 1.4.

**1.1.4. N-particle Hamiltonians: an overview.** Finally, we consider the more general $N$-body magnetic Hamiltonians

$$H = \sum_{i=1}^{N} \frac{1}{2m_i}(D_i - q_i J x_i)^2 + \sum_{i<j} V_{ij}(x_i - x_j). \tag{1.1.6}$$

We continue to assume that $q_i \neq 0$ for all $i$ and that the pair potentials $V_{ij}$ satisfy (1.1.1).

The first step is again to separate the bound states (those with all relative distances $|x_i - x_j|$ bounded uniformly in time) from scattering states. As for two-particle systems, this is done via the reduction procedure of Section 1.3, and the properties of the bound states depend on the total electric charge $Q = \sum q_i$ of the system; see Section 1.4 for details.

We now turn to the analysis of the scattering states. In addition to the free *scattering channel*, in which all particles eventually become asymptotically independent, there are also many other possible scenarios of the scattering process. For instance, in 3-particle scattering the particles 1 and 2 might form a bound state, while particle 3 breaks away and escapes to infinity on its own. In general, we expect the $N$-body system to break up into stable and asymptotically independent subsystems called *clusters*. We will formulate this rigorously in Chapter 6 in terms of the *channel wave operators*. For short-range potentials (with a decay rate $\mu > 1$), they are

$$\Omega_a^{\pm} = \text{s-}\lim_{t \to \pm\infty} \mathrm{e}^{itH} \mathrm{e}^{-itH_a}, \tag{1.1.7}$$

where $H_a$ is a channel Hamiltonian describing the evolution of independent clusters (more precisely, it is obtained from $H$ by removing the interactions between particles belonging to different clusters). If the pair potentials are long-range, additional corrections similar to those indicated at the end of Subsection 1.1.2 are needed.

The scattering theory of $N$-body magnetic Hamiltonians depends mostly on whether any of the possible clusters can have total electric charge 0. If the system contains no neutral subsystems, i.e.,

$$\sum_{i \in S} q_i \neq 0$$

for any proper subset $S$ of $\{1, 2, \ldots, N\}$, then clearly all clusters must be charged. We therefore expect that the particles – regardless of the cluster configuration – must perform bounded motion in the $y$ plane, and that there can be scattering only in the direction of the magnetic field. We are indeed able to prove this under fairly weak assumptions on the pair potentials. The main ingredients of the proof of asymptotic completeness in the "charged" case are:

- the *Mourre estimate* (Section 3.3), which is similar to its non-magnetic analogue except that the *conjugate operator* acts only in the $z$ direction and that we have to use the reduction procedure of Section 1.3;
- the existence of the *asymptotic velocity* in the $z$ direction, again similar to the $N$-body non-magnetic case (Chapter 4);
- a geometrical analysis involving the dynamics of the centers of orbit (Chapter 5), which generalizes the arguments in Subsection 1.1.3 but is much more complicated.

The general case turns out to be considerably harder. We saw in the last subsection that the dynamics of a bound state of a neutral cluster depends non-trivially on its internal structure. In particular, its average velocity is given by an expression $\nabla E(k)$, which usually cannot be computed explicitly (even for 2-particle Hamiltonians) and can be almost impossible to control. It seems difficult to understand the interaction between such neutral clusters and the remaining particles. Furthermore, there is the possibility of the system oscillating between two or more cluster configurations containing different neutral clusters. All this gives rise to a multitude of problems, most of which are too technical to be discussed here (we will say more about it later on).

In this book we will only prove asymptotic completeness for 3-particle systems which may contain neutral pairs; even so, the proof is rather complicated (see Section 3.6 in particular). Moreover, we need stronger assumptions on the pair potentials than in the "charged" case, and for 3-dimensional systems we also need an implicit assumption on the spectra of the reduced Hamiltonians. The precise statement of asymptotic completeness is again given in terms of the wave operators such as (1.1.7). This time, however, the qualitative picture is different: if a neutral pair becomes asymptotically independent from the third particle, it may travel in all directions as described in Subsection 1.1.3.

Proving asymptotic completeness in the general $N$-body case remains an open (and probably very difficult) problem.

## 1.2. Hamiltonians

**1.2.1. Three-dimensional systems.** The Hamiltonian of a system of $N$ interacting spinless particles of masses $m_i$ and electric charges $q_i$ in a constant magnetic field $\vec{B} = (0, 0, 2b)$ in $\mathbb{R}^3$ has the form

$$\mathrm{H} := \sum_{1}^{N} \frac{1}{2m_i}(D_{\mathrm{x}_i} - q_i J\mathrm{x}_i)^2 + \sum_{i<j} V_{ij}(\mathrm{x}_i - \mathrm{x}_j), \tag{1.2.1}$$

acting on $L^2(\mathbb{R}^{3N})$. Here $J\mathrm{x}_i$ is the vector potential resulting from the action of the magnetic field on the $i$-th particle, and $V_{ij}$ is the interaction potential between particles $i$ and $j$. From the point of view of physics, the most important case is that of *Coulomb potentials*

$$V_{ij}(r) = \frac{q_i q_j}{|r|}.$$

For spin-$\frac{1}{2}$ particles, instead of (1.2.1) one considers the *Pauli Hamiltonian*

$$\mathrm{H_P} = \sum_1^N \frac{1}{2m_i}(D_{\mathrm{x}_i} - q_i J\mathrm{x}_i)^2 - \frac{q_i}{2m_i}\sigma_i \cdot \vec{B} + \sum_{i<j} V_{ij}(\mathrm{x}_i - \mathrm{x}_j),$$

acting on $L^2(\mathbb{R}^{3N}, \mathbb{C}^{2N})$, where $\sigma_i = (\sigma_{i,k}), 1 \le k \le 3$ is the spin of the $i$-th particle. Since the magnetic field $\vec{B}$ is constant, $\sum_i \frac{q_i}{2m_i}\sigma_i \cdot \vec{B}$ commutes with $\mathrm{H_P}$, hence all of our results for the spinless case extend trivially to $\mathrm{H_P}$.

Note also that

$$\overline{\mathrm{e}^{-\mathrm{i}tH}u} = \mathrm{e}^{\mathrm{i}t\bar{H}}\bar{u}, \tag{1.2.2}$$

where $\bar{H}$ is obtained from $H$ replacing $q_i$ by $-q_i$.

Without loss of generality we may assume that $b > 0$. We will work in the transversal gauge, in which $J$ is the skew-symmetric matrix

$$J = \begin{pmatrix} 0 & -b & 0 \\ b & 0 & 0 \\ 0 & 0 & 0 \end{pmatrix}. \tag{1.2.3}$$

We consider systems such that all of the particles are charged, i.e.,

$$q_i \ne 0, \ 1 \le i \le N. \tag{$Q1$}$$

The configuration space of the system is $\mathrm{X} = \mathbb{R}^{3N}$. The coordinates in X will be denoted by:

$$\mathrm{x} = (\mathrm{x}_1, \ldots, \mathrm{x}_N), \ \mathrm{x}_i \in \mathbb{R}^3.$$

We equip X with the inner product

$$g(\mathrm{x}, \bar{\mathrm{x}}) = \sum_1^N m_i \mathrm{x}_i \bar{\mathrm{x}}_i.$$

We will denote the coordinates of the particles by

$$\mathrm{x}_i = (\mathrm{y}_i, \mathrm{z}_i),$$

where $\mathrm{y}_i = (\mathrm{y}_i^{(1)}, \mathrm{y}_i^{(2)}) \in \mathbb{R}^2$ and $\mathrm{z}_i \in \mathbb{R}$ are the coordinates of the $i$-th particle in the plane transversal to $\vec{B}$ and in the direction of $\vec{B}$ respectively. The reason for this notation is that the magnetic field $\vec{B}$ "does not act" in the z direction (i.e., $J(0,0,\mathrm{z}_i)^T = 0$), hence the y and z variables will often play very different roles.

Let

$$\mathrm{Y} = \{\mathrm{x} \in \mathrm{X} \mid \mathrm{z}_i = 0, \ 1 \le i \le N\},$$
$$\mathrm{Z} = \{\mathrm{x} \in \mathrm{X} \mid \mathrm{y}_i = 0, \ 1 \le i \le N\}.$$

Then

$$X = Y \oplus^{\perp} Z.$$

The projections of a vector $x \in X$ onto Y and Z will be denoted, respectively, by y and z. The dual spaces to X, Y, Z will be called $X'$, $Y'$, $Z'$. The duality is given by

$$\langle \xi, x \rangle = \sum_{1}^{N} \langle \xi_i, x_i \rangle,$$

and the inner product on $X'$ is

$$g'(\xi, \bar{\xi}) = \sum_{1}^{N} m_i^{-1} \xi_i \bar{\xi}_i.$$

We will also write $|x|^2$ and $|\xi|^2$ instead of $g(x, x)$ and $g'(\xi, \xi)$.

The $N$-particle vector potential is defined by

$$\mathbf{A}x = (q_1 J x_1, \ldots, q_N J x_N).$$

$\mathbf{A}$ is an antisymmetric operator $\mathbf{A} : X \to X'$, i.e.,

$$\langle Ax, x' \rangle = -\langle Ax', x \rangle.$$

$\mathbf{A}$ may also be considered as an antisymmetric bilinear form on $X \times X$. Note that

$$Z = \{x \in X |\ \mathbf{A}x = 0\}.$$

In this notation, H can be rewritten more concisely as

$$\text{H} = \frac{1}{2}(D_x - \mathbf{A}x)^2 + V(x) = \frac{1}{2}D_z^2 + \frac{1}{2}(D_y - \mathbf{A}y)^2 + V(x), \tag{1.2.4}$$

where

$$V(x) := \sum_{i<j} V_{ij}(x_i - x_j).$$

**1.2.2. Two-dimensional systems.** We will also consider systems similar to those described in Subsection 1.2.1, with the additional constraint that all particles are confined to a 2-dimensional plane perpendicular to $\vec{B}$. For physical systems, this may be realized using an external electric potential (as in the quantum Hall effect). Most of what follows is simply a repetition of the previous subsection with the z variables omitted; we will keep it concise and only explain the necessary changes in the notation.

The configuration space of such a system is $Y = \mathbb{R}^{2N}$, with coordinates

$$y = (y_1, \ldots, y_N),\ y_i = (y_i^{(1)}, y_i^{(2)}) \in \mathbb{R}^2.$$

The scalar product in Y is given by

$$g(y, \bar{y}) = \sum_{1}^{N} m_i y_i \bar{y}_i,$$

the dual space to Y is Y′ with the scalar product

$$g'(\eta,\bar\eta)=\sum_1^N m_i^{-1}\eta_i\bar\eta_i,$$

and the duality is $\langle\eta,\mathrm{y}\rangle=\sum_1^N\langle\eta_i,\mathrm{y}_i\rangle$. The Hamiltonian of the system is

$$\mathrm{H}:=\sum_1^N\frac{1}{2m_i}(D_{\mathrm{y}_i}-q_iJ\mathrm{y}_i)^2+\sum_{i<j}V_{ij}(\mathrm{y}_i-\mathrm{y}_j), \tag{1.2.5}$$

where $m_i$ and $q_i$ are the mass and the electric charge of the $i$-th particle respectively (as before, we assume that $q_i\neq 0$ for $1\le i\le N$), and

$$J=\begin{pmatrix}0 & -b\\ b & 0\end{pmatrix}.$$

As in Subsection 1.2.1, we introduce the $N$-particle vector potential:

$$\mathbf{A}\mathrm{y}=(q_1J\mathrm{y}_1,\dots,q_NJ\mathrm{y}_N);$$

then

$$\mathrm{H}=\frac{1}{2}(D_{\mathrm{y}}-\mathbf{A}\mathrm{y})^2+V(\mathrm{y}), \tag{1.2.6}$$

where

$$V(\mathrm{y}):=\sum_{i<j}V_{ij}(\mathrm{y}_i-\mathrm{y}_j).$$

**1.2.3. Assumptions on the Hamiltonians.** We collect here the assumptions on the Hamiltonians which will be invoked throughout the book, in both the 3−dimensional and the 2−dimensional case. We begin with the assumptions on the electric charges of the particles.

*(Q1)* $q_i\neq 0,\ i=1,\dots,N$.

*(Q2)* $q_{i_1}+q_{i_2}+\cdots+q_{i_n}\neq 0$ for any $\{i_1,\dots,i_n\}\subset\{1,\dots,N\},\ n<N$.

The following classes of magnetic Hamiltonians will be considered in this book.

- *The general case*: $N\ge 1$ is arbitrary, *(Q1)* holds (this will be assumed everywhere throughout the book).
- *The strongly charged case*: $N\ge 1$ is arbitrary, *(Q2)* holds (this class includes atoms, positive ions, and systems of identical particles).
- *The 3-body dispersive case*: $N=3$, *(Q1)* holds, *(Q2)* fails.

Some of our results (in particular, those of Chapters 1 and 2) are valid in the general case; however, we will be able to prove the Mourre estimate and asymptotic completeness only for the latter two classes of Hamiltonians. Note that the case $N=2$ is always strongly charged, since then *(Q1)* implies *(Q2)*.

Next, we state our assumptions on the pair potentials. We will use $r$ to denote a variable in $\mathbb{R}^3$ or $\mathbb{R}^2$, depending on which case is under consideration, and $z$ will denote the component of $r$ parallel to the direction of $\vec{B}$; those of the hypotheses below which contain $z$ or $D_z$ will not apply to the 2−dimensional case.

*(V1)* $V_{ij}$ is a multiplication operator which is $-\Delta_r$-bounded with relative bound 0;

*(V2)* $\|F(\frac{|r|}{R} \geq 1)V_{ij}(-\Delta_r + 1)^{-1}\| = o(1),\ R \to \infty;$

*(V3)* $\|\langle r\rangle^{\mu} V_{ij}(-\Delta_r + 1)^{-1}\| < \infty$ for some $\mu > 0$;

*(V4)* $V_{ij} = V_{ij}^{\mathrm{s}} + V_{ij}^{\mathrm{l}}$, where:

$$\||V_{ij}^{\mathrm{s}}(r)|\langle r\rangle^2(-\Delta_r + 1)^{-\frac{1}{2}}\| < \infty,$$

$$|\partial_r^{\alpha} V_{ij}^{\mathrm{l}}(r)| \leq C_{\alpha}\langle r\rangle^{-\mu_{\mathrm{l}}-|\alpha|},\ 0 < \mu_{\mathrm{l}} \leq 2, |\alpha| \geq 0;$$

*(M1)* $z\nabla_z V_{ij}(r)(-\Delta_r + 1)^{-1}$ is compact;

*(SC)* $V_{ij}(r) = V_{ij}^{\mathrm{s}} + V_{ij}^{\mathrm{l}}$, where:

$$\|F(\tfrac{|r|}{R} \geq 1)V_{ij}^{\mathrm{s}}(-\Delta_r + 1)^{-1}\| = O(R^{-\mu_{\mathrm{s}}}),\ \mu_{\mathrm{s}} > 1,$$

$$\|F(\tfrac{|r|}{R} \geq 1)\partial_r^{\alpha} V_{ij}^{\mathrm{l}}\| \leq CR^{-|\alpha|-\mu_{\mathrm{l}}},\ \mu_{\mathrm{l}} > 0,\ |\alpha| \leq 2;$$

*(C)* $V_{ij}$ are of *Coulomb type*: the decomposition $V_{ij} = V_{ij}^{\mathrm{s}} + V_{ij}^{\mathrm{l}}$ in *(SC)* may be chosen so that

$$V_{ij}^{\mathrm{l}}(r) = q_i q_j V^{\mathrm{l}}(r),$$

where $V^{\mathrm{l}}(r)$ is the same for all $i, j$.

Assumptions *(V1)*, *(V2)*, *(V3)* are fairly weak conditions under which we can define H as a self-adjoint operator and develop a basic spectral theory (such as HVZ-type theorems, see Section 6.5). *(V4)* is a stronger condition which will only be used in Section 3.7. Assumption *(M1)* is needed in the Mourre theory (Chapter 3). *(SC)* will be used in the proofs of propagation estimates and asymptotic completeness in Chapters 4, 5, 6. *(C)* will only be used in the proof of asymptotic completeness in the 3-body dispersive case in Sections 6.4, 6.6.

A simple set of assumptions on the potentials under which most of the results of this book hold is *(V1)*, *(V4)*, *(M1)*, *(C)*: these four conditions imply all of the other assumptions listed above. Indeed, it is clear that *(V1)*, *(V4)* imply *(V2)*, *(V3)*, and *(SC)*. Moreover, *(M1)* implies that

$$\|F(\frac{|r|}{R} \geq 1)z\nabla_z V_{ij}(-\Delta_r + 1)^{-1}\| = o(1),\ R \to \infty,$$

since $F(\frac{|r|}{R} \geq 1)$ tends strongly to 0 as $R \to \infty$.

We remark that the Coulomb potentials $V_{ij} = q_i q_j |\mathrm{x}_i - \mathrm{x}_j|^{-1}$ in $\mathbb{R}^3$ satisfy *(V1)*, *(V4)*, *(M1)*, *(C)*. Indeed, by Lemma A.1 in the Appendix $|r|^{-1}(\Delta_r + 1)^{-\frac{1}{2}}$ is bounded on $L^2(\mathbb{R}^3)$. *(V1)* follows immediately, and *(V4)*, *(M1)* follow upon rewriting $|r|^{-1}$ as $|r|^{-1}F(|r| \leq 1) + |r|^{-1}F(|r| \geq 1)$. Finally, *(C)* is obvious. Hence all of the results proved here under the assumption *(Q2)* on the charges are valid for Coulomb potentials; for general 3-particle Coulomb systems, the only assumption that remains to be verified is the implicit condition *(S)* below.

We also note that by Lemma A.1, $|r|^{-\alpha}(-\Delta_r + 1)^{-\frac{1}{2}}$ is bounded on $L^2(\mathbb{R}^2)$ if $\alpha < 1$, so that in the 2−dimensional case we may allow local singularities of the order $|r|^{-\alpha}$, $\alpha < 1$.

In the 3−dimensional dispersive case we will also need an implicit assumption on the spectrum of the *reduced Hamiltonians.* The notation below will be explained in Section 2.2; the condition *(S)* will be discussed in detail in Section 3.6.

*(S)* For all $a \in \mathcal{A}^{\mathrm{n}}$ and all $\eta_a^{\mathrm{n}} \in Y_a^{\mathrm{n}\prime}$, the point spectrum of $\tilde{H}^{a,\mathrm{n}}(\eta_a^{\mathrm{n}})$ is discrete.

**1.2.4. Self-adjointness.** Throughout this book, we will often have to prove that the operators under consideration are self-adjoint. We first verify in Proposition 1.2.1 that Hamiltonians of the form (1.2.1) are self-adjoint provided that *(V1)* holds: this turns out to be an easy consequence of Kato's inequality. For other (considerably more complicated) operators which will appear in the sequel, we will need to use Nelson's commutator theorem (Lemma 1.2.5). For future reference, we also state and prove a number of related technical lemmas. All results here are valid both for $d = 3$ and for $d = 2$. We state them only in the 3−dimensional case; the 2−dimensional case is identical except that $\mathrm{X}, \mathrm{x}$ are replaced by $\mathrm{Y}, \mathrm{y}$.

PROPOSITION 1.2.1. *Assume that (V1) holds. Then* $V$ *is* $\frac{1}{2}(D - \mathbf{A}\mathrm{x})^2$*-bounded with relative bound* 0. *Consequently,* $\mathrm{H} = \frac{1}{2}(D - \mathbf{A}\mathrm{x})^2 + V(\mathrm{x})$ *is self-adjoint and bounded from below with the magnetic Sobolev space*

$$H^2_{\mathbf{A}}(\mathrm{X}) := \{u \in L^2(\mathrm{X}) : \ (D - \mathbf{A}\mathrm{x})^2 u \in L^2(\mathrm{X})\}$$

*as its domain.*

PROOF. We recall the following version of Kato's inequality, whose proof may be found in Section 1.3 of [**CFKS**]:

$$|\mathrm{e}^{-\frac{1}{2}t(D_{\mathrm{x}} - \mathbf{A}\mathrm{x})^2} u|(\mathrm{x}) \leq \mathrm{e}^{-\frac{1}{2}tD_{\mathrm{x}}^2}|u|(\mathrm{x}). \tag{1.2.7}$$

Using that $(\mathrm{H} + \lambda)^{-1} = \int_0^\infty \mathrm{e}^{-t(\mathrm{H}+\lambda)}dt$, we obtain from (1.2.7) that

$$|(\frac{1}{2}(D_{\mathrm{x}} - \mathbf{A}\mathrm{x})^2 + \lambda)^{-1}u(\mathrm{x})| \leq (\frac{1}{2}D_{\mathrm{x}}^2 + \lambda)^{-1}|u(\mathrm{x})|, \ \lambda > 0.$$

Since $V_{ij}$ is a multiplication operator, this yields

$$
\begin{aligned}
&|V_{ij}(\mathrm{x}_i - \mathrm{x}_j)(\tfrac{1}{2}(D_\mathrm{x} - \mathbf{A}\mathrm{x})^2 + \lambda)^{-1}u(\mathrm{x})| \\
&\le |V_{ij}(\mathrm{x}_i - \mathrm{x}_j)|(\tfrac{1}{2}D_\mathrm{x}^2 + \lambda)^{-1}|u(\mathrm{x})|.
\end{aligned} \tag{1.2.8}
$$

By *(V1)*, this implies that $\lim_{\lambda\to+\infty} \|V(\frac{1}{2}(D_\mathrm{x} - \mathbf{A}\mathrm{x})^2 + \lambda)^{-1}\| = 0$, hence $V$ is $\frac{1}{2}(D_\mathrm{x} - \mathbf{A}\mathrm{x})^2$ - bounded with relative bound 0. □

We note the following lemma, which follows from the proof of Proposition 1.2.1 and an interpolation argument.

LEMMA 1.2.2. *Assume (V1). Then there is a constant $c_0$ such that for any function $F$ on* X, $0 \le s \le 1$*:*

$$
\|F(\mathrm{x})V(\mathrm{x})(\mathrm{H} + \mathrm{i})^{-s}\| \le c_0\|F(\mathrm{x})\,|V(\mathrm{x})|\,(\frac{1}{2}D_\mathrm{x}^2 + 1)^{-s}\|.
$$

We will need two more technical lemmas.

LEMMA 1.2.3. *Assume (V1). Then for $z \notin \sigma(\mathrm{H})$, the resolvent $(z-\mathrm{H})^{-1}$ preserves $\mathcal{D}(\langle \mathrm{x}\rangle)$, $\mathcal{D}(\langle D_\mathrm{x}\rangle)$, $\mathcal{D}(\langle \mathrm{x}\rangle^2)$ and $\mathcal{D}(D_\mathrm{x}^2 + \mathrm{x}^2)$. The operator norm of $(z - \mathrm{H})^{-1}$ on these spaces is bounded by $C_M|\mathrm{Im}\, z|^{-2}$ for $|\mathrm{Im}\, z| \le M$.*

PROOF. To prove that $(z - \mathrm{H})^{-1}$ preserves the domain of a self-adjoint operator $B$, we will use the identity

$$
\begin{aligned}
&s^{-1}(\mathrm{e}^{\mathrm{i}sB} - \mathbf{1})(z - \mathrm{H})^{-1} \\
&= s^{-1}(z - \mathrm{H}_s)^{-1}(\mathrm{H}_s - \mathrm{H})(z - \mathrm{H})^{-1}\mathrm{e}^{\mathrm{i}sB} \\
&+ s^{-1}(z - \mathrm{H})^{-1}(\mathrm{e}^{\mathrm{i}sB} - \mathbf{1}),
\end{aligned} \tag{1.2.9}
$$

where $\mathrm{H}_s = \mathrm{e}^{\mathrm{i}sB}\mathrm{H}\mathrm{e}^{-\mathrm{i}sB}$, and that $u \in \mathcal{D}(B)$ if and only if

$$
\sup_{|s|\le 1} s^{-1}\|(\mathrm{e}^{\mathrm{i}sB} - \mathbf{1})u\| < \infty, \tag{1.2.10}
$$

in which case $\|Bu\|$ is equal to the left-hand side of (1.2.10).

Let $B = \mathrm{x}_j$. Then $\mathrm{H}_s = \frac{1}{2}(D - \mathbf{A}\mathrm{x} + se_j)^2 + V(\mathrm{x})$, and

$$
\mathrm{H}_s - \mathrm{H} = \frac{1}{2}s^2 + s(D - \mathbf{A}\mathrm{x})_j. \tag{1.2.11}
$$

Hence by Proposition 1.2.1 $s^{-1}(z - \mathrm{H}_s)^{-1}(\mathrm{H}_s - \mathrm{H})(z - \mathrm{H})^{-1} = O(|\mathrm{Im}\, z|^{-2})$, uniformly for $|s| \le 1$. Plugging this into (1.2.9), we find that for $u \in \mathcal{D}(\mathrm{x}_j)$

$$
\|\mathrm{x}_j(z - \mathrm{H})^{-1}u\| \le C(|\mathrm{Im}\, z|^{-2}\|u\| + |\mathrm{Im}\, z|^{-1}\|\mathrm{x}_j u\|),
$$

hence $(z - \mathrm{H})^{-1}$ preserves $\mathcal{D}(\mathrm{x}_j)$ and

$$
\|(z - \mathrm{H})^{-1}u\|_{\mathcal{D}(\mathrm{x}_j)} \le C_M|\mathrm{Im}\, z|^{-2}\|u\|_{\mathcal{D}(\mathrm{x}_j)}
$$

for $|\mathrm{Im}\, z| \le M$. Since $j$ was arbitrary, the same is true with $\mathrm{x}_j$ replaced by $\langle \mathrm{x}\rangle$.

Let now $B = D_j$. Then $\mathrm{H}_s = \frac{1}{2}(D - \mathbf{A}\mathrm{x} + s\mathbf{A}e_j)^2 + V(\mathrm{x} + se_j)$, and:

$$\mathrm{H}_s - \mathrm{H} = \frac{1}{2}s^2(\mathbf{A}e_j)^2 + s\langle D - \mathbf{A}\mathrm{x}, \mathbf{A}e_j\rangle + V(\mathrm{x} + se_j) - V(\mathrm{x}).$$

By Proposition 1.2.1, $\langle D - \mathbf{A}\mathrm{x}, \mathbf{A}e_j\rangle(z - \mathrm{H})^{-1} = O(|\mathrm{Im}\, z|^{-1})$. Next, we write:

$$\begin{aligned} &V(\mathrm{x} + se_j) - V(\mathrm{x}) \\ &= (\mathbf{1} - \mathrm{e}^{-\mathrm{i}s(D - \mathbf{A}\mathrm{x})_j})V(\mathrm{x} + se_j) + V(\mathrm{x})(\mathrm{e}^{-\mathrm{i}s(D-\mathbf{A}\mathrm{x})_j} - \mathbf{1}). \end{aligned} \tag{1.2.12}$$

By Lemma 1.2.2, $V(\cdot + se_j)$ is $\mathrm{H}_0$-bounded with relative bound 0, hence $(z - \mathrm{H}_s)^{-1}(\mathrm{H}_0 + i) = O(|\mathrm{Im}\, z|^{-1})$, uniformly for $|s| \leq 1$. Since

$$\|s^{-1}(\mathrm{e}^{\mathrm{i}sC} - \mathbf{1})(C + i)^{-1}\| \leq 1$$

for any self-adjoint operator $C$, and since $(\mathrm{H}_0 + i)^{-1}(D - \mathbf{A}\mathrm{x})$ is bounded, we obtain from (1.2.12) that:

$$\|s^{-1}(z - \mathrm{H}_s)^{-1}\big(V(. + se_j) - V\big)(z - \mathrm{H})^{-1}\| \leq C|\mathrm{Im}\, z|^{-2}\|V(\mathrm{H}_0 + i)^{-1}\|.$$

Using (1.2.9) and (1.2.10), w conclude as above that $(z - \mathrm{H})^{-1}$ preserves $\mathcal{D}(\langle D\rangle)$ and has norm bounded by $C|\mathrm{Im}\, z|^{-2}$ for $|\mathrm{Im}\, z|$ bounded.

Let $B = \mathrm{x}_j$ again in (1.2.9), and let $s \to 0$. Then $\mathrm{e}^{\mathrm{i}sB} \to \mathbf{1}$ and $(z - \mathrm{H}_s)^{-1} \to (z - \mathrm{H})^{-1}$ strongly. Using also (1.2.11), we obtain the following identity on $\mathcal{D}(\mathrm{x}_j)$:

$$\mathrm{x}_j(z - \mathrm{H})^{-1} = (z - \mathrm{H})^{-1}\mathrm{x}_j + (z - \mathrm{H})^{-1}(D - \mathbf{A}\mathrm{x})_j(z - \mathrm{H})^{-1}. \tag{1.2.13}$$

Multiplying (1.2.13) from the left by $(D - \mathbf{A}\mathrm{x})_j$, we find that

$$(D - \mathbf{A}\mathrm{x})_j\, \mathrm{x}_j(z - \mathrm{H})^{-1}$$

is bounded on $\mathcal{D}(\mathrm{x}_j)$ with norm $O(|\mathrm{Im}\, z|^{-2})$ for $|\mathrm{Im}\, z|$ bounded. The same is true for $\mathrm{x}_j(D - \mathbf{A}\mathrm{x})(z - \mathrm{H})^{-1}$, since $[(D - \mathbf{A}\mathrm{x})_j, \mathrm{x}_j] = -\mathrm{i}\mathbf{1}$.

Using (1.2.13) again, we obtain that

$$\begin{aligned} s^{-1}(\mathrm{e}^{\mathrm{i}s\mathrm{x}_j} - \mathbf{1})\mathrm{x}_j(z - \mathrm{H})^{-1} &= s^{-1}(\mathrm{e}^{\mathrm{i}s\mathrm{x}_j} - \mathbf{1})(z - \mathrm{H})^{-1}\mathrm{x}_j \\ &+ s^{-1}(\mathrm{e}^{\mathrm{i}s\mathrm{x}_j} - \mathbf{1})(z - \mathrm{H})^{-1}(\mathrm{x}_j + i)^{-1}(\mathrm{x}_j + i)(D - \mathbf{A}\mathrm{x})_j(z - \mathrm{H})^{-1}. \end{aligned}$$

Therefore $(z - \mathrm{H})^{-1}$ preserves $\mathcal{D}(\mathrm{x}_j^2)$ for all $j$, and hence $\mathcal{D}(\mathrm{x}^2)$, with norm $O(|\mathrm{Im}\, z|^{-2})$.

Finally $(z - \mathrm{H})^{-1}$ maps $L^2(\mathrm{X})$ to $H^2_{\mathbf{A}}(\mathrm{X})$, and hence $\mathcal{D}(\mathrm{x}^2)$ to $\mathcal{D}(\mathrm{x}^2) \cap H^2_{\mathbf{A}}(\mathrm{X}) = \mathcal{D}(D_\mathrm{x}^2 + \mathrm{x}^2)$. This completes the proof of the lemma. □

The next lemma follows directly from Lemma 1.2.3 and the functional calculus formula in Subsection 1.2.6.

LEMMA 1.2.4. *Assume (V1), and let* $f \in S^{-\epsilon}_{\mathrm{cl}}(\mathbb{R})$, $\epsilon > 0$. *Then* $f(\mathrm{H})$ *preserves* $\mathcal{D}(\langle \mathrm{x}\rangle)$, $\mathcal{D}(\langle D_\mathrm{x}\rangle)$, $\mathcal{D}(\mathrm{x}^2)$, *and* $\mathcal{D}(D_\mathrm{x}^2 + \mathrm{x}^2)$.

In Chapter 3 we will use Nelson's commutator theorem, which we recall below together with an easy consequence.

LEMMA 1.2.5. *Let $\mathcal{H}$ be a Hilbert space, $N \geq c > 0$ – a self-adjoint operator on $\mathcal{H}$, $A$ – a symmetric operator on $\mathcal{H}$ such that $\mathcal{D}(N) \subset \mathcal{D}(A)$ and:*

$$(i) \qquad \|Au\| \leq C\|Nu\|,\ u \in \mathcal{D}(N),$$

$$(ii) \qquad |(Au, Nu) - (Nu, Au)| \leq C\|N^{1/2}u\|^2,\ u \in \mathcal{D}(N).$$

*Then $A$ is essentially self-adjoint on $\mathcal{D}(N)$. Moreover, if $u \in \mathcal{D}(\bar{A})$, where $\bar{A}$ is the unique self-adjoint extension of $A$, then $(\mathbf{1} + \mathrm{i}\epsilon N)^{-1}u$ converges to $u$ in the graph topology of $\mathcal{D}(\bar{A})$ as $\epsilon \to 0$.*

The operator $N$ in Lemma 1.2.5 will be called a *comparison operator.*

PROOF. The fact that $A$ is essentially self-adjoint on $\mathcal{D}(N)$ is Nelson's commutator theorem (see [**RS**, Theorem X.37]). Let now $u \in \mathcal{D}(\bar{A})$. We have

$$\begin{aligned}
&\bar{A}(1+\mathrm{i}\epsilon N)^{-1}u - (1+\mathrm{i}\epsilon N)^{-1}\bar{A}u \\
&= (1+\mathrm{i}\epsilon N)^{-1}\mathrm{i}\epsilon(N\bar{A} - \bar{A}N)(1+\mathrm{i}\epsilon N)^{-1}u \\
&= \mathrm{i}\epsilon(1+\mathrm{i}\epsilon N)^{-1}(NA - AN)(1+\mathrm{i}\epsilon N)^{-1}u,
\end{aligned}$$

since $\bar{A}_{|\mathcal{D}(N)} = A$. Using that $N^{-1/2}(AN - NA)N^{-1/2}$ is bounded and that

$$\|(1+i\epsilon N)^{-1}N^{1/2}\| \leq C\epsilon^{-1/2}, \tag{1.2.14}$$

we obtain that the operator $R_\epsilon = \bar{A}(1+\mathrm{i}\epsilon N)^{-1} - (1+\mathrm{i}\epsilon N)^{-1}\bar{A}$ is uniformly bounded. We claim that

$$\mathrm{s-}\lim_{\epsilon\to 0} R_\epsilon = 0;$$

this will imply the second statement of the lemma. Since $R_\epsilon$ is uniformly bounded, it suffices to prove the claim on $\mathcal{D}(N)$, which is dense in $\mathcal{H}$. Indeed, we have

$$\begin{aligned}
&\mathrm{i}\epsilon(1+\mathrm{i}\epsilon N)^{-1}(NA - AN)(1+\mathrm{i}\epsilon N)^{-1}N^{-1} \\
&= \mathrm{i}\epsilon(1+\mathrm{i}\epsilon N)^{-1}N^{1/2}N^{-1/2}(NA - AN)N^{-1/2}(1+\mathrm{i}\epsilon N)^{-1}N^{-1/2} \\
&= O(\epsilon^{1/2}),
\end{aligned}$$

by (1.2.14). □

REMARK 1.2.6. In Chapter 3, we will apply Lemma 1.2.5 in the following setup: $\mathcal{H} = L^2(X)$, $N = D_x^2 + x^2$, and $\mathcal{D}(A) = \mathcal{D}(_A0)$, where $A_0 = a_0(x, D_x)$ is a pseudodifferential operator whose symbol $a_0$ is a second order polynomial in $(x, \xi)$ with real coefficients. It is then straighforward to check that $A_0$ satisfies the hypotheses of Lemma 1.2.5.

**1.2.5. Weyl quantization.** In the sequel we will make extensive use of the *Weyl quantization*. The following definitions and facts are well known, see e.g., [**Hö**, Chapter 18] or [**DG1**, Appendix D]; we collect them here for the reader's convenience.

Let $X = \mathbb{R}^n$ for some $n$. Given a tempered distribution $a(x,\xi) \in \mathcal{S}'(T^*X)$, the pseudodifferential operator $a^{\mathrm{w}}(x, D_x)$ defined by

$$a^{\mathrm{w}}(x, D_x)u(x) := (2\pi)^{-n} \int\int \mathrm{e}^{\mathrm{i}\langle x-y,\xi\rangle} a(\frac{x+y}{2}, \xi)u(y)dyd\xi \tag{1.2.15}$$

is a continuous map from $\mathcal{S}(X)$ into $\mathcal{S}'(X)$. If $A = a^{\mathrm{w}}(x, D_x)$, we say that the distribution $a$ is the *Weyl symbol* of $A$ and $A$ is the *Weyl quantization* of $a$. Under appropriate conditions on $a$, for example if

$$|\partial_x^\alpha \partial_\xi^\beta a(x,\xi)| \leq C_{\alpha,\beta},\ \alpha, \beta \in \mathbb{N}^n, \tag{1.2.16}$$

$A$ extends to a bounded operator on $L^2(X)$.

Instead of the equation (1.2.15), we will sometimes use the identity [**DiS**]:

$$a^{\mathrm{w}}(x, D_x) = (2\pi^{-2n}) \int \hat{a}(\xi, \Xi)\mathrm{e}^{\mathrm{i}(\langle x,\xi\rangle + \langle D_x, \Xi\rangle)} d\xi d\Xi, \tag{1.2.17}$$

based on the Fourier inversion formula and valid, for example, if $a \in \mathcal{S}(T^*X)$.

A unitary operator

$$U : L^2(X) \to L^2(X)$$

is said to *implement* a linear symplectic transformation $\chi : T^*X \to T^*X$, if for any linear form $a(x,\xi)$ on $T^*X$ one has

$$U^*a(x, D_x)U = a \circ \chi(x, D_x). \tag{1.2.18}$$

It is a well known theorem of Segal ([**Se**]) that any linear symplectic transformation can be implemented by a (unique up to a multiplicative factor of modulus one) unitary transformation. The covariance property (1.2.18) extends to a much wider class of symbols, provided that one uses the Weyl quantization (1.2.15). Namely, one has

$$U^*a^{\mathrm{w}}(x, D_x)U = (a \circ \chi)^{\mathrm{w}}(x, D_x), \tag{1.2.19}$$

for $a \in \mathcal{S}'(T^*X)$. Note that by (1.2.18) $U$ preserves $\mathcal{S}(X)$ and $\mathcal{S}'(X)$, so that (1.2.19) makes sense as an identity between operators $\mathcal{S}(X) \to \mathcal{S}'(X)$.

Finally, we will need the following version of the sharp Gärding inequality.

PROPOSITION 1.2.7. *Let $a_t$ be a symbol such that for some $\rho, \delta \geq 0$:*

$$a(t, x, \xi) \geq 0,\ |\partial_x^\alpha \partial_\xi^\beta a_t(x,\xi)| \leq C_{\alpha,\beta} t^{-\rho|\beta| - \delta|\alpha|},\ \alpha, \beta \in \mathbb{N}^n.$$

*Then*

$$a_t^{\mathrm{w}}(x, D_x) \geq -Ct^{-\rho-\delta}.$$

We refer to [**DG1**, Proposition D.5.4] for a proof.

**1.2.6. Almost analytic extensions.** We will often express $C_0^\infty$ functions of operators in terms of their resolvent, using *almost analytic extensions*. The usefulness of the formula (1.2.20) below in a similar context was first noticed in [**HeSj**]. Here we only review briefly the facts that we will use in the sequel; for a more complete discussion of the subject, including an explicit construction of an almost analytic extension of any $f \in C_0^\infty(\mathbb{R})$ and a proof of the identity (1.2.20), the reader may consult e.g., Appendix C.2 of [**DG1**].

We embed $\mathbb{R}$ in $\mathbb{C}$ and denote $z = x + \mathrm{i}y$, $\partial_{\bar{z}} = \partial_x + \mathrm{i}\partial_y$, $dz \wedge d\bar{z} = 2\mathrm{i}dx\, dy$.

DEFINITION 1.2.8. Let $f \in C_0^\infty(\mathbb{R})$. An *almost analytic extension* of $f$ is a (non-unique) function $\tilde{f} \in C_0^\infty(\mathbb{C})$ such that:

(i) $\tilde{f}|_{\mathbb{R}} = f$;

(ii) $|\partial_{\bar{z}}\tilde{f}(z)| \leq C_n |\mathrm{Im}\, z|^n$ for any $n \in \mathbb{N}$.

Note that for an analytic function $F$ one has $\partial_{\bar{z}}F = 0$ by the Cauchy-Riemann equations. The function $\tilde{f}$ is "almost analytic" in the sense that $\partial_{\bar{z}}\tilde{f}$ vanishes of infinite order on the real axis.

Any function $f \in C_0^\infty(\mathbb{R})$ has an almost analytic extension $\tilde{f}$. Moreover, the support of $\tilde{f}$ can be taken arbitrarily close to $\mathrm{supp}\, f$. If $\tilde{f}$ is an almost analytic extension of $f$, then:

$$f(x) = \frac{\mathrm{i}}{2\pi} \int_{\mathbb{C}} \partial_{\bar{z}}\tilde{f}(z)(z - x)^{-1} dz \wedge d\bar{z}. \tag{1.2.20}$$

Consequently, for a self-adjoint operator $A$ we may represent $f(A)$ as

$$f(A) = \frac{\mathrm{i}}{2\pi} \int_{\mathbb{C}} \partial_{\bar{z}}\tilde{f}(z)(z - A)^{-1} dz \wedge d\bar{z}. \tag{1.2.21}$$

Indeed, using that

$$\|(z - A)^{-1}\| \leq |\mathrm{Im}\, z|^{-1}, z \in \mathbb{C}\backslash\mathbb{R}, \tag{1.2.22}$$

we see that the integral in (1.2.21) is norm convergent.

We will also need a similar functional calculus formula for functions $f \in S_{\mathrm{cl}}^\rho(\mathbb{R})$, $\rho < 0$. The following result, which we borrow from [**DG1**, Proposition C.2.2], guarantees the existence of almost analytic extensions of such functions.

PROPOSITION 1.2.9. *Let $f \in S_{\mathrm{cl}}^\rho(\mathbb{R})$, $\rho < 0$. Then there is a function $\tilde{f} \in C^\infty(\mathbb{C})$ such that*

$$\begin{aligned} &\tilde{f}|_{\mathbb{R}} = f, \\ &|\tfrac{\partial \tilde{f}}{\partial \bar{z}}(z)| \leq C_N |z|^{\rho - 1 - N} |\mathrm{Im}\, z|^N, \qquad N \in \mathbb{N}, \\ &\mathrm{supp}\, \tilde{f} \subset \{x + \mathrm{i}y \mid |y| \leq C\langle x \rangle\}. \end{aligned} \tag{1.2.23}$$

By (1.2.22), the integral in (1.2.21) is again convergent in norm, hence (1.2.21) holds also in this case.

**1.2.7. Direct integrals.** In this subsection we recall a few basic facts concerning direct integrals (see e.g., [**Di**, A69-A80]). Let $\mathcal{H}'$ be a separable Hilbert space and $(M, \mu)$ a measure space. Let

$$\mathcal{H} = L^2(M, \mu; \mathcal{H}') =: \int_M^{\oplus} \mathcal{H}' d\mu.$$

A function $M \ni k \mapsto H(k)$ with values in self-adjoint operators on $\mathcal{H}'$ is called *measurable* if for all $\phi, \psi \in \mathcal{H}'$, the function $M \ni k \mapsto (\psi, (H(k) + i)^{-1}\phi)$ is measurable. The operator $H$ on $\mathcal{H}$ defined by

$$\mathcal{D}(H) := \{\psi \in \mathcal{H} \mid \psi(k) \in \mathcal{D}(H(k)) \text{ a.e., } \textstyle\int_M \|H(k)\psi(k)\|^2 d\mu < \infty\},$$

$$(H\psi)(k) := H(k)\psi(k),$$

is self-adjoint; we call it the *direct integral* of the family $H(k)$ and write $H = \int_M^{\oplus} H(k) d\mu$. We associate to $H$ the set

$$\Sigma := \{(\lambda, k) \in \mathbb{R} \times M \mid \lambda \in \sigma(H(k))\},$$

which plays the role of an energy-momentum spectrum.

The spectrum of $H$ may be characterized in terms of the spectrum of $H(k)$: $\lambda \in \sigma(H)$ if and only if

$$\mu(\{k \in M: \ \sigma(H(k)) \cap (\lambda - \epsilon, \lambda + \epsilon)\}) > 0$$

for any $\epsilon > 0$, and $\lambda \in \sigma_{\text{pp}}(H)$ if and only if

$$\mu(\{k \in M: \ \lambda \in \sigma_{\text{pp}}(H(k))\}) > 0.$$

Of particular interest to us will be the following special case. Let $X = \mathbb{R}^{m+n}$ be an Euclidean space equipped with coordinates $x = (x', x'')$, $x' = (x_1, \dots, x_m)$, $x'' = (x_{m+1}, \dots, x_{m+n})$. Suppose that $H$ is a pseudodifferential operator on $L^2(X)$ of the form

$$H = h(D_{x'}; x'', D_{x''})$$

(i.e., $H$ does not depend explicitly on $x'$). Then a direct integral representation of $H$ may be written out explicitly in terms of the partial Fourier transform. Namely, let

$$\widehat{u}(k; x'') = (2\pi)^{-m/2} \int_{\mathbb{R}^m} \mathrm{e}^{-\mathrm{i}\langle k, x'\rangle} u(x', x'') dx'$$

be the partial Fourier transform in $x'$ of a function $u \in L^2(X)$. Then $\widehat{u} \in L^2(M; L^2(\mathbb{R}^n))$ for $M = (\mathbb{R}^m_{x'})'$, and

$$\widehat{Hu} = \int_M h(k; x'', D_{x''}) \widehat{u}(k; x'') dk,$$

where $h(k; x'', D_{x''})$ for a fixed value of $k$ is understood as a pseudodifferential operator on $\mathcal{H}' = L^2(\mathbb{R}^n_{x''})$.

## 1.3. Reducing transformations

A multiparticle system may either be in a bound state (i.e., the particles remain close together in a stable configuration), or break up, temporarily or permanently, into two or more smaller subsystems. Moreover, whenever a cluster of particles remains stable for long enough, one can distinguish between its "external" and "internal" dynamics. "External" motion is performed by the cluster as a whole. It is, essentially, what we observe when we look at it for a long time and from a large distance, so that individual particles can no longer be distinguished and the system (the atom, molecule, etc.) is perceived as one somewhat blurry moving object. "Internal" dynamics is observed from up close. Our camera travels together with the blurry object and transmits a magnified image of its internal structure. We are being informed of the relative position and velocities of all particles, or, if the number of the constituents is large, we may prefer the relevant statistical data.

Consider first an isolated (i.e., subject to no external forces) system of $N$ particles interacting via pair potentials $V_{ij}$ in $\mathbb{R}^3$. The Hamiltonian $H$ of such a system has the form

$$H = -\frac{1}{2}\Delta + V \text{ on } L^2(\mathbb{R}^{3N}),$$

where $V = \sum V_{ij}$. In this case, the simplest way to distinguish between internal and external dynamics is to separate the center of mass motion. One defines the center of mass of the system:

$$x_{\text{cm}} = \frac{\sum m_i x_i}{\sum m_i};$$

then $H$ may be rewritten as

$$H = \frac{1}{2}D^2_{x_{\text{cm}}} + H^{\text{cm}}, \tag{1.3.1}$$

where $D_{x_{\text{cm}}} = \sum D_{x_i}$ is the center of mass momentum, and the "reduced Hamiltonian" $H^{\text{cm}}$ commutes with $x_{\text{cm}}$ and $D_{x_{\text{cm}}}$. It describes the relative motion of particles, i.e., it acts on $L^2(X^{\text{cm}})$, where $X^{\text{cm}} = \{x \in X |\, x_{\text{cm}} = 0\}$ is the $3(N-1)$-dimensional configuration space in which all relative motion takes place. The external dynamics of the system is identified with that of the center of mass, and is governed by the free Hamiltonian $\frac{1}{2}D^2_{x_{\text{cm}}}$. Since $\frac{1}{2}D^2_{x_{\text{cm}}}$ and $H^{\text{cm}}$ act on different variables, they may be (and usually are) analyzed independently.

A complete description of the external dynamics follows readily from (1.3.1). The center of mass of the system will simply perform free motion in $\mathbb{R}^3$. All other information about the behavior of the physical system, or about the spectral properties of the Hamiltonian $H$, is contained in $H^{\text{cm}}$. It is therefore customary to restrict one's attention to $H^{\text{cm}}$ only (i.e., work in the "center of mass frame"), while the original Hamiltonian $H$ is all but forgotten.

If an external magnetic field is switched on, the situation becomes different in that magnetic Hamiltonians do not admit a center of mass separation. More precisely, it is not possible to write a magnetic Hamiltonian as a sum of two commuting operators governing the internal and external motion respectively. However, the general considerations which motivated the introduction of the center of mass frame in the first place are no less valid.

The first mathematical study of the separation of the center of mass in magnetic fields was done in [**AHS2**]. It was based on the fact that magnetic Hamiltonians have a constant of motion $K_{\text{cm}}$ called *total pseudomomentum* (see Subsections 1.3.2 and 1.3.3 for the definition). One may use this to define *reduced Hamiltonians*, which will act on fewer variables.

Since the magnetic field "does not act" in the z direction, we may simply separate the motion of the center of mass along the z axis using the conventional procedure described above; we will use the symbol $H$ to denote the obtained Hamiltonian. The reduction procedure in the transversal directions will be more complicated, and the details of it will depend on the *total electric charge* $Q = \sum q_i$ of the system. If $Q = 0$, all components of $K_{\text{cm}}$ commute and may be assigned exact values simultaneously. A *reduced Hamiltonian* is then simply $H$ with a value of the vector $K_{\text{cm}}$ fixed. On the other hand, if the system is charged ($Q \neq 0$), the commutator $i[K_{\text{cm}}^{(1)}, K_{\text{cm}}^{(2)}]$ is a non-zero multiple of the identity operator. One then finds a unitary transformation which maps $(K_{\text{cm}}^{(1)}, K_{\text{cm}}^{(2)})$ to a multiple of $(D_{y_0}, y_0)$ for a selected variable $y_0$. (A convenient choice is $y_0 = y_{\text{cm}}^{(1)}$). Such a transformation, applied to the Hamiltonian $H$, will reduce the number of the variables on which $H$ depends explicitly. The reduced Hamiltonian will be the "transformed" $H$, considered as an operator on a smaller space.

Contrary to what one might expect, the charged (non-commuting) case turns out to be much *easier* to study than the case $Q = 0$. This is because the dependence of the reduced Hamiltonians on the value of $K_{\text{cm}}$ for $Q = 0$ may be rather complicated. While partial results for $N = 2$ are available (see Section 3.5), there is no corresponding spectral theory for $N \geq 3$.

We may choose the reducing transformations so that the reduced Hamiltonians are operators on $L^2(X^{\text{cm}})$, where, as before, $X^{\text{cm}}$ is the configuration space of the relative motion of the particles. This reduction may then be used to analyze the internal dynamics of the system. However, to study the external dynamics one has to return to the Hamiltonian $H$. Unlike in cases when conventional center of mass separation is employed, this step is a rather non-trivial one; we discuss it in detail in Section 1.4. It turns out that if $Q \neq 0$, there is in fact *no* external motion across the field. The neutral case is much more complicated. Roughly speaking, for neutral systems one can *distinguish* between the "internal" and "external" dynamics, but one cannot *separate* them. This statement will be made precise in Subsection 1.4.2.

The reduction procedure described in this section applies to $N$-particle systems both in $\mathbb{R}^3$ and in $\mathbb{R}^2$. To avoid repetition, we will only present the case $d = 3$. The procedure for $d = 2$ is almost identical: one simply deletes the $z$ variables.

**1.3.1. Center of mass coordinates.** We will denote by $\mathrm{X}_{\mathrm{cm}}$ the 3-dimensional configuration space of the center of mass motion, and by $\mathrm{X}^{\mathrm{cm}}$ – the $(3N-3)$-dimensional configuration space of the relative motion of the particles:

$$\mathrm{X}_{\mathrm{cm}} := \{\mathrm{x} \in \mathrm{X} \mid \mathrm{x}_i = \mathrm{x}_j \text{ for all } i, j\},$$

$$\mathrm{X}^{\mathrm{cm}} := \{\mathrm{x} \in \mathrm{X} \mid \sum_{i=1}^{N} m_i \mathrm{x}_i = 0\}.$$

It is easy to check that $\mathrm{X}_{\mathrm{cm}}$ and $\mathrm{X}^{\mathrm{cm}}$ are orthogonal with respect to the scalar product $g(\cdot,\cdot)$ introduced in Section 1.2. We will write

$$\mathrm{Y}_{\mathrm{cm}} := \mathrm{X}_{\mathrm{cm}} \cap \mathrm{Y},\ \mathrm{Z}_{\mathrm{cm}} := \mathrm{X}_{\mathrm{cm}} \cap \mathrm{Z},$$

$$\mathrm{Y}^{\mathrm{cm}} := \mathrm{X}^{\mathrm{cm}} \cap \mathrm{Y},\ \mathrm{Z}^{\mathrm{cm}} := \mathrm{X}^{\mathrm{cm}} \cap \mathrm{Z}.$$

Note that the projections on $\mathrm{X}_{\mathrm{cm}}$, $\mathrm{X}^{\mathrm{cm}}$ commute with the projections on Y, Z, so that

$$\mathrm{X}_{\mathrm{cm}} = \mathrm{Y}_{\mathrm{cm}} \oplus \mathrm{Z}_{\mathrm{cm}},\ \mathrm{X}^{\mathrm{cm}} = \mathrm{Y}^{\mathrm{cm}} \oplus \mathrm{Z}^{\mathrm{cm}}.$$

The symbols

$$\mathrm{x}_{\mathrm{cm}},\ \mathrm{y}_{\mathrm{cm}},\ \mathrm{z}_{\mathrm{cm}},\ \mathrm{x}^{\mathrm{cm}},\ \mathrm{y}^{\mathrm{cm}},\ \mathrm{z}^{\mathrm{cm}},$$

will stand for the orthogonal projections of $\mathrm{x} \in \mathrm{X}$ on the above spaces. The spaces dual to $\mathrm{X}_{\mathrm{cm}}$ and $\mathrm{X}^{\mathrm{cm}}$ are:

$$\mathrm{X}'_{\mathrm{cm}} = \{\xi \in \mathrm{X}' \mid \frac{\xi_i}{m_i} = \frac{\xi_j}{m_j} \text{ for all } i, j\},$$

$$\mathrm{X}^{\mathrm{cm}\prime} = \{\xi \in \mathrm{X}' \mid \sum_{i=1}^{N} \xi_i = 0\}.$$

We have

$$|\mathrm{x}|^2 = |\mathrm{x}_{\mathrm{cm}}|^2 + |\mathrm{x}^{\mathrm{cm}}|^2,\ |\xi|^2 = |\xi_{\mathrm{cm}}|^2 + |\xi^{\mathrm{cm}}|^2, \tag{1.3.2}$$

and

$$\langle \xi, \mathrm{x} \rangle = \langle \xi_{\mathrm{cm}}, \mathrm{x}_{\mathrm{cm}} \rangle + \langle \xi^{\mathrm{cm}}, \mathrm{x}^{\mathrm{cm}} \rangle.$$

The spaces dual to $\mathrm{Y}_{\mathrm{cm}}$, $\mathrm{Z}_{\mathrm{cm}}$, ..., will be denoted by $\mathrm{Y}'_{\mathrm{cm}}$, $\mathrm{Z}'_{\mathrm{cm}}$, etc.

We now introduce the spaces that we will work with most of the time. We define:

$$Y = \mathrm{Y},\ Z = \mathrm{Z}^{\mathrm{cm}},\ X = Y \oplus Z,$$

$$y = \mathrm{y},\ z = \mathrm{z}^{\mathrm{cm}},\ x = y + z.$$

Hence

$$X = \{\mathrm{x} \in \mathrm{X} : \ \mathrm{z}_{\mathrm{cm}} = 0\}$$

is the configuration space of the system in a coordinate frame in which the z coordinate of the center of mass has been fixed at 0. We will also write:

$$Y_{\mathrm{cm}} = \mathrm{Y}_{\mathrm{cm}},\ Y^{\mathrm{cm}} = \mathrm{Y}^{\mathrm{cm}},\ Z_{\mathrm{cm}} = \{0\},\ Z^{\mathrm{cm}} = Z,$$

$$X_{\mathrm{cm}} = Y_{\mathrm{cm}},\ X^{\mathrm{cm}} = Y^{\mathrm{cm}} \oplus Z,$$

and similarly for the coordinates and adjoint spaces.

The purpose of this definition is to simplify the study of the Hamiltonian H by removing its trivial part corresponding to the free motion of the center of mass of the system in the z direction. Namely, we observe the invariance of both the vector potential and the scalar potential under translations in the z direction:

$$\mathbf{A}\mathrm{y} = \mathbf{A}y,\ V(\mathrm{x}) = V(x) = V(x^{\mathrm{cm}}).$$

Therefore the Hamiltonian H may be written as

$$\mathrm{H} = \frac{1}{2}D^2_{\mathrm{z}_{\mathrm{cm}}} + H,$$

where

$$H = \frac{1}{2}D_z^2 + \frac{1}{2}(D_y - Ay)^2 + V(x) \tag{1.3.3}$$

is an operator on $\mathcal{H} = L^2(X)$. The "trivial" part $\frac{1}{2}D^2_{\mathrm{z}_{\mathrm{cm}}}$ is of no interest to us; what we really wish to study is the Hamiltonian $H$ given by (1.3.3). Throughout most of this book we will consider only this operator.

We also change the notation accordingly in the 2–dimensional case and replace the symbols $\mathrm{Y}, \mathrm{y}, \mathrm{H}$ by $Y, y, H$. Since there are no z coordinates in the case $d = 2$, we simply have $\mathrm{Y} = Y$, $\mathrm{y} = y$, $\mathrm{z} = z$. The only reason why we use both the Roman and the italic symbols in this case is to keep the notation consistent.

Finally, we will use

$$\mathbf{A}^{\mathrm{cm}}{}_{\mathrm{cm}},\ \mathbf{A}_{\mathrm{cm}}{}^{\mathrm{cm}},\ \mathbf{A}_{\mathrm{cm}\,\mathrm{cm}},\ \mathbf{A}^{\mathrm{cm}\,\mathrm{cm}}$$

to denote the restrictions of the bilinear form $\mathbf{A}$ to $X^{\mathrm{cm}} \times X_{\mathrm{cm}}$, $X_{\mathrm{cm}} \times X^{\mathrm{cm}}$, $X_{\mathrm{cm}} \times X_{\mathrm{cm}}$, $X^{\mathrm{cm}} \times X^{\mathrm{cm}}$ respectively (for instance, $\mathbf{A}^{\mathrm{cm}}{}_{\mathrm{cm}}\, x_{\mathrm{cm}} = (\mathbf{A}x_{\mathrm{cm}})^{\mathrm{cm}}$, where the first projection is taken in $X$ and the second one in $X'$).

**1.3.2. Magnetic translations.** Consider first a system of $N$ non-interacting particles in a uniform magnetic field. The properties of such a system should be invariant under translations by vectors in $X$. Translation by a vector $x' \in X$ should be implemented by a unitary operator $U_{x'}$ such that:

$$U^*_{x'} x U_{x'} = x + x',\ U^*_{x'} v U_{x'} = v,$$

where $x$ is the *position observable* and $v = [\frac{1}{2}(D_x - Ax)^2, \mathrm{i}x]$ is the *velocity observable*. It is easy to check that the unitary operator

$$U_{x'}u(x) := \mathrm{e}^{-\mathrm{i}\langle x', \mathbf{A}x\rangle} u(x - x'),$$

called a *magnetic translation*, has these properties. We may write:

$$U_{x'} = \mathrm{e}^{-\mathrm{i}\langle x', K\rangle}, \; x' \in X,$$

where

$$K := D_x + \mathbf{A}x$$

is an observable called the *pseudomomentum* ([**AHS2**]). The invariance under magnetic translations follows from the identity

$$[K, \frac{1}{2}(D_x - \mathbf{A}x)^2] = 0. \tag{1.3.4}$$

A basic property of the pseudomomentum is that, in general, not all of its components commute. Namely, we have

$$[\langle x', K\rangle, \mathrm{i}\langle x'', K\rangle] = -2\langle x', \mathbf{A}x''\rangle. \tag{1.3.5}$$

Note also that the projection of $K$ on $Z'$ is equal to $D_z$. The part of $K$ which is perpendicular to the magnetic field $\vec{B}$, i.e., the projection $(K^{(1)}, K^{(2)}, 0)$ of $K$ on $Y'$, will be sometimes denoted by $K^{\perp}$.

**1.3.3. Pseudomomentum of the center of mass.** In the case of interacting particles, only those magnetic translations which correspond to translations of the system as a whole will commute with the Hamiltonian $H$. Accordingly, we define the *pseudomomentum of the center of mass*:

$$K_{\mathrm{cm}} := D_{x_{\mathrm{cm}}} + (\mathbf{A}x)_{\mathrm{cm}}.$$

$K_{\mathrm{cm}}$ is an observable in $X'_{\mathrm{cm}}$ and since $V(x + x') = V(x)$ for $x' \in X_{\mathrm{cm}}$, we have

$$[K_{\mathrm{cm}}, H] = 0. \tag{1.3.6}$$

Since we have separated the center of mass motion in the $z$ direction we have $K_{\mathrm{cm}} = K^{\perp}_{\mathrm{cm}} = (K^{(1)}_{\mathrm{cm}}, K^{(2)}_{\mathrm{cm}}, 0)$. From (1.3.5) we obtain the following commutation relation for the non-zero components of $K_{\mathrm{cm}}$:

$$[\langle y', K_{\mathrm{cm}}\rangle, \mathrm{i}\langle y'', K_{\mathrm{cm}}\rangle] = -2\langle y', \mathbf{A}_{\mathrm{cm\,cm}}\, y''\rangle, \; y', y'' \in Y_{\mathrm{cm}}. \tag{1.3.7}$$

For computational purposes, it is sometimes convenient to identify $Y'_{\mathrm{cm}}$ with $\mathbb{R}^2$ as follows:

$$Y'_{\mathrm{cm}} \ni \frac{1}{M}(m_1\nu, \ldots, m_N\nu) \to \nu \in \mathbb{R}^2, \tag{1.3.8}$$

where $M = \sum m_i$. (Note that the mapping (1.3.8) is *not* an isometry: we have $|\nu^*|_{Y'_{\mathrm{cm}}} = M^{-1/2}|\nu|_{\mathbb{R}^2}$, where $\nu^*$ is the vector on the left-hand side of

(1.3.8).) Under the identification (1.3.8), the projection $Y' \ni \eta \to \eta_{\rm cm} \in Y'_{\rm cm}$ becomes

$$Y' \ni (\eta_1, \ldots, \eta_N) \to \sum_1^N \eta_i \in \mathbb{R}^2,$$

hence $K_{\rm cm}$ corresponds to the observable

$$K_{\rm cm} = \sum_1^N (D_{y_i} + q_i J y_i), \tag{1.3.9}$$

with values in $\mathbb{R}^2$, and the commutation relation (1.3.7) becomes

$$[K^{(1)}_{\rm cm}, \mathrm{i}K^{(2)}_{\rm cm}] = 2Qb, \tag{1.3.10}$$

where $Q = \sum_{i=1}^N q_i$ is the *total charge* of the system.

**1.3.4. Neutral systems.** If the system is neutral, i.e., $Q = \sum q_i = 0$, we have $\mathbf{A}_{\rm cm\, cm} = 0$. Hence

$$K_{\rm cm} = \mathrm{e}^{-\mathrm{i}\langle y_{\rm cm}, \mathbf{A} y_{\rm cm}\rangle} D_{y_{\rm cm}} \mathrm{e}^{\mathrm{i}\langle y_{\rm cm}, \mathbf{A} y_{\rm cm}\rangle},$$

so that $K_{\rm cm}$ is unitarily equivalent to $D_{y_{\rm cm}}$. This implies ([**AHS2**, Thm. 4.1]) that there exists a (non-unique) pair $(U_{\rm cm}, \mathcal{H}^{\rm cm})$, where $\mathcal{H}^{\rm cm}$ is a Hilbert space and $U_{\rm cm}$ is a unitary transformation, such that:

$$\begin{aligned} U_{\rm cm} &: L^2(X) \mapsto \textstyle\int^{\oplus}_{Y'_{\rm cm}} \mathcal{H}^{\rm cm} d\eta_{\rm cm}, \\ U_{\rm cm} K_{\rm cm} U^{-1}_{\rm cm} &= \textstyle\int^{\oplus}_{Y'_{\rm cm}} \eta_{\rm cm} \mathbf{1} d\eta_{\rm cm}. \end{aligned} \tag{1.3.11}$$

Since $H$ commutes with $K_{\rm cm}$, we have

$$U_{\rm cm} H U^{-1}_{\rm cm} = \int^{\oplus}_{Y'_{\rm cm}} \tilde{H}(\eta_{\rm cm}) d\eta_{\rm cm},$$

where $Y'_{\rm cm} \ni \eta_{\rm cm} \mapsto \tilde{H}(\eta_{\rm cm})$ is a measurable family of self-adjoint operators (cf. [**Di**, A80]).

Any unitary transformation $U_{\rm cm}$ satisfying (1.3.11) is called a *reducing unitary transformation.* We will use reducing unitary transformations to define basic concepts such as bound states, scattering states, and wave operators. Although $U_{\rm cm}$ does appear in these definitions, we will show that the objects defined are independent of the choice of $U_{\rm cm}$.

To carry out the calculations in the sequel, we will need an explicitly defined pair $(U_{\rm cm}, \mathcal{H}^{\rm cm})$, which we now proceed to construct. In addition to (1.3.11), it has the property:

$$\mathcal{H}^{\rm cm} = L^2(\tilde{Y}) \otimes L^2(X^{\rm cm}),\ U_{\rm cm}\, x^{\rm cm}\, U^{-1}_{\rm cm} = x^{\rm cm}, \tag{1.3.12}$$

where $\tilde{Y}$ is an Euclidean space. This is particularly useful in applications.

We begin by constructing the corresponding classical linear symplectic transformation.

LEMMA 1.3.1. *If $Q = 0$, the following linear transformation is symplectic:*

$$\chi_{\mathrm{cm}} : T^*X \to T^*X$$
(1.3.13)
$$(x, \xi) \mapsto (\tilde{x}, \tilde{\xi}),$$

*where*

$$x = y_{\mathrm{cm}} + x^{\mathrm{cm}}, \ \xi = \eta_{\mathrm{cm}} + \xi^{\mathrm{cm}},$$

*and*

$$\left\{ \begin{array}{l} (\tilde{y}_{\mathrm{cm}}, \tilde{\eta}_{\mathrm{cm}}) = (y_{\mathrm{cm}}, \eta_{\mathrm{cm}} + (\mathbf{A}x)_{\mathrm{cm}}), \\ (\tilde{x}^{\mathrm{cm}}, \tilde{\xi}^{\mathrm{cm}}) = (x^{\mathrm{cm}}, \xi^{\mathrm{cm}} - \mathbf{A}^{\mathrm{cm}}{}_{\mathrm{cm}}\, x_{\mathrm{cm}}). \end{array} \right. \tag{1.3.14}$$

A proof of Lemma 1.3.1 is given in Subsection 1.3.7.

By Segal's theorem (see Subsection 1.2.4) there exists a unitary operator

$$U_{\mathrm{cm}} : L^2(X) \to L^2(X),$$

implementing this transformation. Moreover, it satisfies

$$U_{\mathrm{cm}}\, K_{\mathrm{cm}}\, U_{\mathrm{cm}}^* = D_{y_{\mathrm{cm}}}. \tag{1.3.15}$$

Using Lemma 1.3.1 and that $\mathbf{A}_{\mathrm{cm\,cm}} = 0$, we verify that the transformed Hamiltonian

$$\tilde{H} := U_{\mathrm{cm}}\, H U_{\mathrm{cm}}^* \tag{1.3.16}$$

has the form

$$\begin{aligned} \tilde{H} &= \tfrac{1}{2}(D_{y_{\mathrm{cm}}} - 2\mathbf{A}_{\mathrm{cm}}{}^{\mathrm{cm}}\, x^{\mathrm{cm}})^2 + \tfrac{1}{2}(D_{x^{\mathrm{cm}}} - \mathbf{A}^{\mathrm{cm\,cm}}\, x^{\mathrm{cm}})^2 \\ &\quad + V(x^{\mathrm{cm}}), \end{aligned} \tag{1.3.17}$$

acting on $L^2(X) = L^2(Y_{\mathrm{cm}} \times X^{\mathrm{cm}})$. Since $\tilde{H}$ does not depend explicitly on $y_{\mathrm{cm}}$, it commutes with $D_{y_{\mathrm{cm}}}$ and therefore may be written as a direct integral:

$$\tilde{H} = \int_{Y'_{\mathrm{cm}}}^{\oplus} \tilde{H}(\eta_{\mathrm{cm}}) d\eta_{\mathrm{cm}},$$

where

$$\tilde{H}(\eta_{\mathrm{cm}}) = \frac{1}{2}(\eta_{\mathrm{cm}} - 2\mathbf{A}_{\mathrm{cm}}{}^{\mathrm{cm}}\, x^{\mathrm{cm}})^2 + \frac{1}{2}(D_{x^{\mathrm{cm}}} - \mathbf{A}^{\mathrm{cm\,cm}}\, x^{\mathrm{cm}})^2 + V(x^{\mathrm{cm}})$$

acts on $L^2(X^{\mathrm{cm}})$. If *(V1)* is assumed, by Proposition 1.2.1 $V(x^{\mathrm{cm}})$ is $(D_{x^{\mathrm{cm}}} - \mathbf{A}^{\mathrm{cm\,cm}}\, x^{\mathrm{cm}})^2$- bounded with relative bound 0. Hence $\tilde{H}$ and $\tilde{H}(\eta_{\mathrm{cm}})$ are self-adjoint on the domains

$$\mathcal{D}(\tilde{H}) = \mathcal{D}(\frac{1}{2}(D_{y_{\mathrm{cm}}} - 2\mathbf{A}_{\mathrm{cm}}{}^{\mathrm{cm}}\, x^{\mathrm{cm}})^2 + \frac{1}{2}(D_{x^{\mathrm{cm}}} - \mathbf{A}^{\mathrm{cm\,cm}}\, x^{\mathrm{cm}})^2),$$

$$\mathcal{D}(\tilde{H}(\eta_{\mathrm{cm}})) = \mathcal{D}(2(\mathbf{A}_{\mathrm{cm}}{}^{\mathrm{cm}}\, x^{\mathrm{cm}})^2 + \frac{1}{2}(D_{x^{\mathrm{cm}}} - \mathbf{A}^{\mathrm{cm\,cm}}\, x^{\mathrm{cm}})^2),$$

respectively.

An operator $\tilde{H}(\eta_{\mathrm{cm}})$ for a fixed value of $\eta_{\mathrm{cm}} \in Y_{\mathrm{cm}}$, acting on $L^2(X^{\mathrm{cm}})$, is called a *reduced Hamiltonian.* The following proposition follows easily from the above characterization of the domain of $\tilde{H}(\eta_{\mathrm{cm}})$.

PROPOSITION 1.3.2. *Assume that (V1) holds. Then the mapping*

$$Y'_{\mathrm{cm}} \ni \eta_{\mathrm{cm}} \mapsto \tilde{H}(\eta_{\mathrm{cm}})$$

*is continuous in the norm resolvent sense.*

**1.3.5. Charged systems.** Assume now that the total charge $Q$ of the system is non-zero. Without loss of generality, by (1.2.2) we may assume that $Q > 0$. As in Subsection 1.3.3, we identify $L^2(X)$ with $L^2(X^{\mathrm{cm}} \times Y_{\mathrm{cm}})$. By Murray-von Neumann's theorem, there is ([**AHS2**, Theorem 5.3]) a (non-unique) pair $(U_{\mathrm{cm}}, \mathcal{H}^{\mathrm{cm}})$, where $\mathcal{H}^{\mathrm{cm}}$ is a Hilbert space and $U_{\mathrm{cm}}$ is a unitary transformation, such that:

$$\begin{aligned} &U_{\mathrm{cm}} : L^2(X^{\mathrm{cm}} \times Y_{\mathrm{cm}}) \to L^2(\mathbb{R}, dy^1) \otimes \mathcal{H}^{\mathrm{cm}}, \\ &U_{\mathrm{cm}}\, K_{\mathrm{cm}}^{(1)}\, U_{\mathrm{cm}}^{-1} = (2Qb)^{\frac{1}{2}} D_{y^1},\ U_{\mathrm{cm}}\, K_{\mathrm{cm}}^{(2)}\, U_{\mathrm{cm}}^{-1} = (2Qb)^{\frac{1}{2}} y^1, \\ &U_{\mathrm{cm}} H U_{\mathrm{cm}}^{-1} = \tilde{H}. \end{aligned} \tag{1.3.18}$$

Since $H$ commutes with $K_{\mathrm{cm}}^{(1)}, K_{\mathrm{cm}}^{(2)}$, and the von Neumann algebra generated by $y^1, D_{y^1}$ on $L^2(\mathbb{R}, dy^1)$ consists of all bounded operators, we see that

$$U_{\mathrm{cm}} H U_{\mathrm{cm}}^{-1} = \mathbf{1} \otimes \tilde{H},$$

where $\tilde{H}$ is a self-adjoint operator on $\mathcal{H}^{\mathrm{cm}}$. Any unitary transformation $U_{\mathrm{cm}}$ satisfying (1.3.18) will be called a *reducing unitary transformation.*

We now define explicitly a pair $(U_{\mathrm{cm}}, \mathcal{H}^{\mathrm{cm}})$ satisfying (1.3.12), which will be used in the computations in the sequel. Once more we begin by constructing the corresponding linear symplectic transformation.

LEMMA 1.3.3. *Suppose that $Q > 0$, and let $\kappa = (2Qb)^{-\frac{1}{2}}$. Then the following transformation is symplectic:*

$$\begin{aligned} &\chi_{\mathrm{cm}} : T^*X \to T^*\mathbb{R} \times T^*\mathbb{R} \times T^*X^{\mathrm{cm}} \\ &(x, \xi) \mapsto (\tilde{y}_{\mathrm{cm}}^{(1)}, \tilde{\eta}_{\mathrm{cm}}^{(1)}, \tilde{y}_{\mathrm{cm}}^{(2)}, \tilde{\eta}_{\mathrm{cm}}^{(2)}, \tilde{x}^{\mathrm{cm}}, \tilde{\xi}^{\mathrm{cm}}), \end{aligned} \tag{1.3.19}$$

*where:*

$$\begin{cases} (\tilde{\eta}_{\mathrm{cm}}^{(1)}, \tilde{y}_{\mathrm{cm}}^{(1)}) = \kappa(\eta_{\mathrm{cm}} + (\mathbf{A}x)_{\mathrm{cm}}), \\ (\tilde{y}_{\mathrm{cm}}^{(2)}, \tilde{\eta}_{\mathrm{cm}}^{(2)}) = \kappa(\eta_{\mathrm{cm}} - \mathbf{A}_{\mathrm{cm\,cm}}\, x_{\mathrm{cm}} + \mathbf{A}_{\mathrm{cm}}{}^{\mathrm{cm}}\, x^{\mathrm{cm}}), \\ (\tilde{x}^{\mathrm{cm}}, \tilde{\xi}^{\mathrm{cm}}) = (x^{\mathrm{cm}}, \xi^{\mathrm{cm}} - \mathbf{A}^{\mathrm{cm}}{}_{\mathrm{cm}}\, x_{\mathrm{cm}}) \end{cases} \tag{1.3.20}$$

*If $Q < 0$, the above transformation with $(\tilde{y}_{\mathrm{cm}}^{(1)}, \tilde{y}_{\mathrm{cm}}^{(2)})$ and $(\tilde{\eta}_{\mathrm{cm}}^{(1)}, \tilde{\eta}_{\mathrm{cm}}^{(2)})$ interchanged is symplectic.*

For a proof, see Subsection 1.3.7.

We denote by $Y_{\rm cm}^{(1)}$, $Y_{\rm cm}^{(2)}$ the two copies of $\mathbb{R}$ in (1.3.19). By Segal's theorem, the above transformation may be implemented by a unitary operator

$$U_{\rm cm} : L^2(X) \to L^2(Y_{\rm cm}^{(1)}) \otimes L^2(Y_{\rm cm}^{(2)}) \otimes L^2(X^{\rm cm}),$$

such that

$$U_{\rm cm}\, K_{\rm cm}\, U_{\rm cm}^* = (\kappa^{-1} D_{y_{\rm cm}^{(1)}}, \kappa^{-1} y_{\rm cm}^{(1)}). \tag{1.3.21}$$

The transformed Hamiltonian

$$\tilde{H} := U_{\rm cm} H U_{\rm cm}^* \tag{1.3.22}$$

has the form

$$\tilde{H} = R(y_{\rm cm}^{(2)}, D_{y_{\rm cm}^{(2)}}, x^{\rm cm}) + \frac{1}{2}(D_{x^{\rm cm}} - \mathbf{A}^{\rm cm\, cm}\, x^{\rm cm})^2 + V(x^{\rm cm}), \tag{1.3.23}$$

where

$$R(y_{\rm cm}^{(2)}, D_{y_{\rm cm}^{(2)}}, x^{\rm cm}) = \frac{1}{2M}\left(\kappa^{-1}(D_{y_{\rm cm}^{(2)}}, y_{\rm cm}^{(2)}) - 2\mathbf{A}_{\rm cm}{}^{\rm cm}\, x^{\rm cm}\right)^2.$$

$\tilde{H}$ is defined as an operator on $L^2(Y_{\rm cm}^{(1)}) \otimes L^2(Y_{\rm cm}^{(2)}) \otimes L^2(X^{\rm cm})$; however it does not contain either $y_{\rm cm}^{(1)}$ or $D_{y_{\rm cm}^{(1)}}$ explicitly, hence it commutes with all operators in $B(L^2(Y_{\rm cm}^{(1)}))$ and may be considered as an operator on $L^2(Y_{\rm cm}^{(2)}) \otimes L^2(X^{\rm cm})$. This operator, which we continue to denote by $\tilde{H}$, is called the *reduced Hamiltonian*. As in Subsection 1.3.4, $\tilde{H}$ is self-adjoint on the domain of $R(y_{\rm cm}^{(2)}, D_{y_{\rm cm}^{(2)}}, x^{\rm cm}) + \frac{1}{2}(D_{x^{\rm cm}} - \mathbf{A}^{\rm cm\, cm}\, x^{\rm cm})^2$.

**1.3.6. Cutoffs in the pseudomomentum.** We will often need to define functions of the pseudomomentum $K_{\rm cm}$. This is easy in the neutral case ($Q = 0$), since then all components of $K_{\rm cm}$ commute. However, for $Q \neq 0$ the non-zero components $K_{\rm cm}^{(1)}, K_{\rm cm}^{(2)}$ of $K_{\rm cm}$ do not commute, hence a suitable functional calculus must be used.

DEFINITION 1.3.4. For $F \in S^0(\mathbb{R}^2)$, we define

$$F(K_{\rm cm}) := \tilde{F}^{\rm w}(y, D_y),$$

where $\tilde{F}(y, \eta) = F(\eta_{\rm cm} + (\mathbf{A}y)_{\rm cm})$.

Note that if $F \in S^0(\mathbb{R}^2)$, then $\tilde{F} \in S^0(T^*Y)$, so that $F(K_{\rm cm})$ is a bounded operator on $L^2(X)$. It is also worthwhile to note the following alternative definition of $F(K_{\rm cm})$.

LEMMA 1.3.5. *For $F \in \mathcal{S}(\mathbb{R}^2)$, we have*

$$F(K_{\rm cm}) = (2\pi)^{-2} \int \hat{F}(y) {\rm e}^{{\rm i}\langle y, K_{\rm cm}\rangle} dy. \tag{1.3.24}$$

PROOF. We may assume that $Q > 0$. By (1.3.21), we have

$$U_{\mathrm{cm}}\,\mathrm{e}^{\mathrm{i}\langle y,K_{\mathrm{cm}}\rangle}U_{\mathrm{cm}}^{*}=\mathrm{e}^{\mathrm{i}\langle y,\kappa^{-1}(D_{y^{(1)}_{\mathrm{cm}}},y^{(1)}_{\mathrm{cm}})\rangle}.$$

Since $U_{\mathrm{cm}}$ implements $\chi_{\mathrm{cm}}$, we obtain that

$$U_{\mathrm{cm}}\,F(K_{\mathrm{cm}})U_{\mathrm{cm}}^{*}=(\tilde{F}\circ\chi_{\mathrm{cm}}^{-1})^{\mathrm{w}}(y^{(1)}_{\mathrm{cm}},D_{y^{(1)}_{\mathrm{cm}}}).$$

But by (1.3.20), $\tilde{F}\circ\chi_{\mathrm{cm}}^{-1}(y^{(1)}_{\mathrm{cm}},\eta^{(1)}_{\mathrm{cm}})=F(\kappa^{-1}y^{(1)}_{\mathrm{cm}},\kappa^{-1}\eta^{(1)}_{\mathrm{cm}})$. Hence the lemma follows from the identity (1.2.16). □

**1.3.7. Details of calculations.** Here we show the details of the calculations in Lemmas 1.3.1 and 1.3.3. We first introduce some auxiliary notation. We will write:

$$x=\begin{pmatrix}y^{(1)}\\ y^{(2)}\\ z\end{pmatrix},\quad y^{(i)}=\begin{pmatrix}y^{(i)}_1\\ y^{(i)}_2\\ \vdots\\ y^{(i)}_N\end{pmatrix},\quad z=\begin{pmatrix}z_1\\ z_2\\ \vdots\\ z_N\end{pmatrix},$$

$$\xi=\begin{pmatrix}\eta^{(1)}\\ \eta^{(2)}\\ \zeta\end{pmatrix},\quad \eta^{(i)}=\begin{pmatrix}\eta^{(i)}_1\\ \eta^{(i)}_2\\ \vdots\\ \eta^{(i)}_N\end{pmatrix},\quad \zeta=\begin{pmatrix}\zeta_1\\ \zeta_2\\ \vdots\\ \zeta_N\end{pmatrix}$$

(note the order of coordinates). Then:

$$x_{\mathrm{cm}}=\begin{pmatrix}T&0&0\\0&T&0\\0&0&T\end{pmatrix}x,\quad \xi_{\mathrm{cm}}=\begin{pmatrix}S&0&0\\0&S&0\\0&0&S\end{pmatrix}\xi, \tag{1.3.25}$$

$$x^{\mathrm{cm}}=x-x_{\mathrm{cm}},\quad \xi^{\mathrm{cm}}=\xi-\xi_{\mathrm{cm}}, \tag{1.3.26}$$

where:

$$T=\frac{1}{M}\begin{pmatrix}m_1&m_2&\dots&m_N\\ m_1&m_2&\dots&m_N\\ \vdots&\vdots&\ddots&\vdots\\ m_1&m_2&\dots&m_N\end{pmatrix}=\Omega\mathcal{M},$$

$$S=T^{\mathrm{t}}=\frac{1}{M}\begin{pmatrix}m_1&m_1&\dots&m_1\\ m_2&m_2&\dots&m_2\\ \vdots&\vdots&\ddots&\vdots\\ m_N&m_N&\dots&m_N\end{pmatrix}=\mathcal{M}\Omega,$$

$$\mathcal{M} = \frac{1}{M}\begin{pmatrix} m_1 & 0 & \dots & 0 \\ 0 & m_2 & \dots & 0 \\ \vdots & \vdots & \ddots & \vdots \\ 0 & 0 & \dots & m_N \end{pmatrix}, \quad \Omega = \begin{pmatrix} 1 & 1 & \dots & 1 \\ 1 & 1 & \dots & 1 \\ \vdots & \vdots & \ddots & \vdots \\ 1 & 1 & \dots & 1 \end{pmatrix},$$

and $M = \sum_{i=1}^N m_i$. Indeed, one can easily check that $x_{\rm cm}$, $x^{\rm cm}$, $\xi_{\rm cm}$, and $\xi^{\rm cm}$, given by (1.3.25) and (1.3.26), belong to $X_{\rm cm}$, $X^{\rm cm}$, $X'_{\rm cm}$, and $X^{\rm cm\prime}$ respectively. Recall also that the space $X$ is defined as

$$X = \{x : \ z_{\rm cm} = 0\} = \{x : \ Tz = 0\},$$

and consequently,

$$X' = \{\xi : \ \zeta_{\rm cm} = 0\} = \{\xi : \ S\zeta = 0\}.$$

The matrix of the mapping $\mathbf{A} : X \to X'$ in the above bases is:

$$\text{(1.3.27)} \qquad \mathbf{A} = \begin{pmatrix} 0 & -\mathcal{Q} & 0 \\ \mathcal{Q} & 0 & 0 \\ 0 & 0 & 0 \end{pmatrix}, \text{ where } \mathcal{Q} = b\begin{pmatrix} q_1 & 0 & \dots & 0 \\ 0 & q_2 & \dots & 0 \\ \vdots & \vdots & \ddots & \vdots \\ 0 & 0 & \dots & q_N \end{pmatrix}.$$

We will consider the linear transformations

$$\text{(1.3.28)} \qquad \chi_{\rm cm} : \ T^\star X \to T^\star X,$$

defined by

$$\text{(1.3.29)} \qquad \begin{cases} (\tilde{y}_{\rm cm}, \tilde{\eta}_{\rm cm}) = (y_{\rm cm}, \eta_{\rm cm} + (\mathbf{A}x)_{\rm cm}), \\ (\tilde{x}^{\rm cm}, \tilde{\xi}^{\rm cm}) = (x^{\rm cm}, \xi^{\rm cm} - \mathbf{A}^{\rm cm}{}_{\rm cm}\, x_{\rm cm}), \end{cases}$$

if $Q = 0$, and by

$$\text{(1.3.30)} \qquad \begin{cases} (\tilde{\eta}^{(1)}_{\rm cm}, \tilde{y}^{(1)}_{\rm cm}) = \kappa(\eta_{\rm cm} + (\mathbf{A}x)_{\rm cm}), \\ (\tilde{\eta}^{(2)}_{\rm cm}, \tilde{y}^{(2)}_{\rm cm}) = \kappa(\eta_{\rm cm} - \mathbf{A}_{\rm cm\,cm}\, x_{\rm cm} + \mathbf{A}_{\rm cm}{}^{\rm cm}\, x^{\rm cm}), \\ (\tilde{x}^{\rm cm}, \tilde{\xi}^{\rm cm}) = (x^{\rm cm}, \xi^{\rm cm} - \mathbf{A}^{\rm cm}{}_{\rm cm}\, x_{\rm cm}), \end{cases}$$

where $\kappa = (2\nu b|Q|)^{-1/2}$ and $\nu = \sum_{i=1}^N m_i^2/M^2$, if $Q > 0$. (If $Q < 0$, the transformation given by (1.3.30) with $\tilde{y}_{\rm cm}$ and $\tilde{\eta}_{\rm cm}$ interchanged should be considered instead.)

Note that the constant $\kappa$ in (1.3.30) differs from the corresponding constant in (1.3.20) by the factor $\nu^{-1/2}$. This is simply due to the fact that here $\chi_{\rm cm}$ is considered as a mapping $T^\star X \to T^\star X$, whereas in (1.3.20) it was a mapping with values in $T^*\mathbb{R} \times T^*\mathbb{R} \times T^* X^{\rm cm}$. Both points of view are obviously equivalent.

We need to prove that the transformations defined above are symplectic. To this end, let us first rewrite (1.3.29) and (1.3.30) using the notation introduced above. Then (1.3.29) becomes:

$$\begin{cases} S\tilde{\eta}^{(1)} = S\eta^{(1)} - S\mathcal{Q}y^{(2)}, \\ S\tilde{\eta}^{(2)} = S\eta^{(2)} + S\mathcal{Q}y^{(1)}, \\ T\tilde{y}^{(i)} = Ty^{(i)}, \\ (I-S)\tilde{\eta}^{(1)} = (I-S)\eta^{(1)} + (I-S)\mathcal{Q}Ty^{(2)}, \\ (I-S)\tilde{\eta}^{(2)} = (I-S)\eta^{(2)} - (I-S)\mathcal{Q}Ty^{(1)}, \\ (I-T)\tilde{x} = (I-T)x, \\ \tilde{\zeta} = \zeta, \tilde{z} = z, \end{cases}$$

and (1.3.30) takes the form:

$$\begin{cases} S\tilde{\eta}^{(1)} = \kappa(S\eta^{(1)} - S\mathcal{Q}y^{(2)}), \\ T\tilde{y}^{(1)} = \kappa(S\eta^{(2)} + S\mathcal{Q}y^{(1)}), \\ S\tilde{\eta}^{(2)} = \kappa(S\eta^{(2)} - S\mathcal{Q}Ty^{(1)} + S\mathcal{Q}(I-T)y^{(1)}), \\ T\tilde{y}^{(2)} = \kappa(S\eta^{(1)} + S\mathcal{Q}Ty^{(2)} - S\mathcal{Q}(I-T)y^{(2)}), \\ (I-S)\tilde{\eta}^{(1)} = (I-S)\eta^{(1)} + (I-S)\mathcal{Q}Ty^{(2)}, \\ (I-S)\tilde{\eta}^{(2)} = (I-S)\eta^{(2)} - (I-S)\mathcal{Q}Ty^{(1)}, \\ (I-T)\tilde{x} = (I-T)x, \\ \tilde{\zeta} = \zeta, \tilde{z} = z. \end{cases}$$

To prove that these transformations are symplectic, one has to show that they preserve the Poisson brackets involving $y^{(i)}$ and $\eta^{(i)}$. We first note that if $A = (a_{ij})$ and $B = (b_{ij})$ are two $N \times N$ matrices,

$$\{A\eta^{(i)}, By^{(i)}\} = \sum_{i,j} a_{ij}b_{ij}, \ \{A\eta^{(i)}, By^{(j)}\} = 0 \text{ if } i \neq j.$$

Using this formula, we verify the identities:

$$\begin{aligned} &\{S\eta^{(i)}, Ty^{(i)}\} = 1, \\ &\{S\eta^{(i)}, (I-T)y^{(i)}\} = \{(I-S)\eta^{(i)}, Ty^{(i)}\} = 0, \\ &\{S\eta^{(i)}, \mathcal{Q}(I-T)y^{(i)}\} = \{(I-S)\eta^{(i)}, \mathcal{Q}Ty^{(i)}\} = 0, \\ &\{S\eta^{(i)}, S\mathcal{Q}y^{(i)}\} = \{S\eta^{(i)}, S\mathcal{Q}Ty^{(i)}\} = \{\eta^{(i)}, S\mathcal{Q}Ty^{(i)}\} = \nu bQ, \\ &\{S\eta^{(i)}, \mathcal{Q}Ty^{(i)}\} = \{\eta^{(i)}, S\mathcal{Q}y^{(i)}\} = \tfrac{b}{M}\textstyle\sum m_i q_i. \end{aligned}$$

The invariance of the Poisson bracket under the above transformations follows now by direct computation.

We will now compute the symbol $h(\tilde{x}, \tilde{\xi}) = (h \circ \chi_{\mathrm{cm}})(x, \xi)$ of the transformed Hamiltonian. Suppose first that $Q = 0$. Then

$$(\mathbf{A}x_{\mathrm{cm}})_{\mathrm{cm}} = 0.$$

Indeed, this follows from (1.3.25), (1.3.27), and the from the fact that $SQT = 0$ if $Q = 0$. Hence $(\mathbf{A}x)_{\mathrm{cm}} = (\mathbf{A}x^{\mathrm{cm}})_{\mathrm{cm}}$. Using this and (1.3.2), we compute:

$$|\xi - \mathbf{A}x|^2 = |\zeta|^2 + |\eta_{\mathrm{cm}} - (\mathbf{A}x)_{\mathrm{cm}}|^2 + |\eta^{\mathrm{cm}} - (\mathbf{A}x)^{\mathrm{cm}}|^2$$

$$= |\zeta|^2 + |\eta_{\mathrm{cm}} + (\mathbf{A}x)_{\mathrm{cm}} - 2(\mathbf{A}x^{\mathrm{cm}})_{\mathrm{cm}}|^2 + |\eta^{\mathrm{cm}} - (\mathbf{A}x_{\mathrm{cm}})^{\mathrm{cm}} - (\mathbf{A}x^{\mathrm{cm}})^{\mathrm{cm}}|^2$$

$$= |\tilde{\zeta}|^2 + |\tilde{\eta}_{\mathrm{cm}} - 2(\mathbf{A}\tilde{x}^{\mathrm{cm}})_{\mathrm{cm}}|^2 + |\tilde{\eta}^{\mathrm{cm}} - (\mathbf{A}\tilde{x}^{\mathrm{cm}})^{\mathrm{cm}}|^2.$$

Hence

$$h(\tilde{\xi}, \tilde{x}) = \frac{1}{2}|\tilde{\zeta}|^2 + \frac{1}{2}|\tilde{\eta}_{\mathrm{cm}} - 2(\mathbf{A}\tilde{x}^{\mathrm{cm}})_{\mathrm{cm}}|^2 + \frac{1}{2}|\tilde{\eta}^{\mathrm{cm}} - (\mathbf{A}\tilde{x}^{\mathrm{cm}})^{\mathrm{cm}}|^2 + V(\tilde{x}^{\mathrm{cm}}).$$

In the case $Q > 0$, we have:

$$|\xi - \mathbf{A}x|^2 = |\zeta|^2 + |\eta_{\mathrm{cm}} - (\mathbf{A}x)_{\mathrm{cm}}|^2 + |\eta^{\mathrm{cm}} - (\mathbf{A}x)^{\mathrm{cm}}|^2$$

$$= |\zeta|^2 + |\eta_{\mathrm{cm}} - (\mathbf{A}x_{\mathrm{cm}})_{\mathrm{cm}} + (\mathbf{A}x^{\mathrm{cm}})_{\mathrm{cm}} - 2(\mathbf{A}x^{\mathrm{cm}})_{\mathrm{cm}}|^2$$

$$+ |\eta^{\mathrm{cm}} - (\mathbf{A}x_{\mathrm{cm}})^{\mathrm{cm}} - (\mathbf{A}x^{\mathrm{cm}})^{\mathrm{cm}}|^2$$

$$= |\tilde{\zeta}|^2 + |\kappa^{-1}(\tilde{\eta}^{(2)}_{\mathrm{cm}}, \tilde{y}^{(2)}_{\mathrm{cm}}) - 2(\mathbf{A}\tilde{x}^{\mathrm{cm}})_{\mathrm{cm}}|^2 + |\tilde{\eta}^{\mathrm{cm}} - (\mathbf{A}\tilde{x}^{\mathrm{cm}})^{\mathrm{cm}}|^2,$$

so that

$$h(\tilde{\xi}, \tilde{x}) = \frac{1}{2}|\tilde{\zeta}|^2 + r_{\mathrm{cm}}(\tilde{\eta}^{(2)}_{\mathrm{cm}}, \tilde{y}^{(2)}_{\mathrm{cm}}) + \frac{1}{2}|\tilde{\eta}^{\mathrm{cm}} - (\mathbf{A}\tilde{x}^{\mathrm{cm}})^{\mathrm{cm}}|^2,$$

where

$$r_{\mathrm{cm}}(\tilde{\eta}^{(2)}_{\mathrm{cm}}, \tilde{y}^{(2)}_{\mathrm{cm}}) = \frac{1}{2}|\kappa^{-1}(\tilde{\eta}^{(2)}_{\mathrm{cm}}, \tilde{y}^{(2)}_{\mathrm{cm}}) - 2(\mathbf{A}\tilde{x}^{\mathrm{cm}})_{\mathrm{cm}}|^2.$$

**1.3.8. One-particle magnetic Hamiltonians.** We conclude this section by discussing the reduction procedure in two simple cases which will be important in the sequel: those of a single charged particle and of two interacting particles of opposite charges.

We begin with the magnetic Hamiltonian of one particle in $\mathbb{R}^2$:

$$H = \frac{1}{2}(D_y - \mathbf{A}y)^2 \text{ on } L^2(\mathbb{R}^2) \tag{1.3.31}$$

(we omit the subscript 1 and assume for simplicity that $m = 1$, $q > 0$). The 3–dimensional case is identical, since for $N = 1$ we have $\mathrm{z} = \mathrm{z}_{\mathrm{cm}}$, $Z = \{0\}$, and $H$ also has the form (1.3.31).

It is well known that a 2–dimensional magnetic Hamiltonian $H$ is unitarily equivalent to a 1–dimensional harmonic oscillator. In fact, the transformation $U_{\mathrm{cm}}$ introduced in Subsection 1.3.5 implements this equivalence.

Indeed, for $N = 1$ we have $X_{\text{cm}} = X$ and $X^{\text{cm}} = \{0\}$, so that the symplectic transformation $\chi_{\text{cm}}$ defined in Lemma 1.3.3 is

$$\begin{cases} (\tilde{\eta}^{(1)}, \tilde{y}^{(1)}) = \kappa(\eta + \mathbf{A}x), \\ (\tilde{\eta}^{(2)}, \tilde{y}^{(2)}) = \kappa(\eta - \mathbf{A}x), \end{cases}$$

where $\kappa = (2bq)^{-1/2}$. Under this transformation, the symbol $h$ of $H$ becomes

$$\begin{aligned} h &= \tfrac{1}{2}(\eta - \mathbf{A}x)^2 = \tfrac{1}{2}\kappa^{-2}(|\tilde{\eta}^{(2)}|^2 + |\tilde{y}^{(2)}|^2) \\ &= bq(|\tilde{\eta}^{(2)}|^2 + |\tilde{y}^{(2)}|^2). \end{aligned}$$

The transformed quantum Hamiltonian $\tilde{H}$ is

$$\tilde{H} = bq(|D_{y^{(2)}}|^2 + |y^{(2)}|^2) \tag{1.3.32}$$

(a harmonic oscillator in $y^{(2)}$). Its spectrum consists of the *Landau levels*:

$$\Lambda_n = \frac{1}{2}|\vec{B}|q(2n+1), \ n = 0, 1, 2, \ldots.$$

Since $\tilde{H}$ does not act on the $y^{(1)}$ variable, each Landau level $\Lambda_n$ is an eigenvalue of infinite multiplicity. This degeneracy is removed if $\tilde{H}$ is considered as an operator acting only on $L^2(Y^{(2)})$.

**1.3.9. A neutral pair.** We consider now a pair of interacting particles whose electric charges are $q_1$ and $q_2 = -q_1$ respectively. Since the notation is simpler in the 2−dimensional case, we will treat this case first. The reduced Hamiltonian $\tilde{H}(\eta_{\text{cm}})$ has the form

$$\tilde{H}(\eta_{\text{cm}}) = \frac{1}{2}(\eta_{\text{cm}} - 2\mathbf{A}_{\text{cm}}{}^{\text{cm}}\, y^{\text{cm}})^2 + \frac{1}{2}(D_{y^{\text{cm}}} - \mathbf{A}^{\text{cm}\,\text{cm}}\, y^{\text{cm}})^2 + V(y^{\text{cm}}).$$

To shorten the notation, we will write $y^{(1)} = \bar{y}$, $y^{(2)} = \hat{y}$, $\eta^{(1)} = \bar{\eta}$, $\eta^{(2)} = \hat{\eta}$, so that

$$y = \begin{pmatrix} \bar{y}_1 \\ \bar{y}_2 \\ \hat{y}_1 \\ \hat{y}_2 \end{pmatrix}, \quad \eta = \begin{pmatrix} \bar{\eta}_1 \\ \bar{\eta}_2 \\ \hat{\eta}_1 \\ \hat{\eta}_2 \end{pmatrix}.$$

We define $e = q_1 b = -q_2 b$, $M = m_1 + m_2$, $k = \eta_1 + \eta_2$, $r = y_1 - y_2$, $\rho = M^{-1}(m_2\eta_1 - m_1\eta_2)$ (hence $k, r, \rho \in \mathbb{R}^2$). Using (1.3.25)–(1.3.27), we obtain that:

$$\eta^{\text{cm}} = \begin{pmatrix} \bar{\rho} \\ -\bar{\rho} \\ \hat{\rho} \\ -\hat{\rho} \end{pmatrix}, \quad \eta_{\text{cm}} = \frac{1}{M}\begin{pmatrix} m_1\bar{k} \\ m_2\bar{k} \\ m_1\hat{k} \\ m_2\hat{k} \end{pmatrix},$$

$$y^{\mathrm{cm}} = \frac{1}{M}\begin{pmatrix} m_2\bar{r} \\ -m_1\bar{r} \\ m_2\hat{r} \\ -m_1\hat{r} \end{pmatrix}, \quad \mathbf{A}y^{\mathrm{cm}} = \frac{e}{M}\begin{pmatrix} -m_2\hat{r} \\ -m_1\hat{r} \\ m_2\bar{r} \\ m_1\bar{r} \end{pmatrix},$$

$$\mathbf{A}_{\mathrm{cm}}{}^{\mathrm{cm}}\, y^{\mathrm{cm}} = \frac{e}{M}\begin{pmatrix} -m_1\hat{r} \\ -m_2\hat{r} \\ m_1\bar{r} \\ m_2\bar{r} \end{pmatrix}, \quad \mathbf{A}^{\mathrm{cm}\,\mathrm{cm}}\, y^{\mathrm{cm}} = \frac{e(m_1-m_2)}{M}\begin{pmatrix} \hat{r} \\ -\hat{r} \\ -\bar{r} \\ \bar{r} \end{pmatrix}$$

(in the last line, we used that $\mathbf{A}_{\mathrm{cm}}{}^{\mathrm{cm}}\, y = (\mathbf{A}y^{\mathrm{cm}})_{\mathrm{cm}}$ and $\mathbf{A}^{\mathrm{cm}\,\mathrm{cm}}\, x^{\mathrm{cm}} = (\mathbf{A}y^{\mathrm{cm}})^{\mathrm{cm}}$, where the first projection is taken in $Y$ and the second one in $Y'$). For future reference, we note that

$$(1.3.33) \qquad |\mathbf{A}_{\mathrm{cm}}{}^{\mathrm{cm}}\, y^{\mathrm{cm}}|^2 = \frac{e^2}{M}|r|^2 = \frac{e^2}{m_1m_2}|y^{\mathrm{cm}}|^2,$$

$$(1.3.34) \qquad |\mathbf{A}^{\mathrm{cm}\,\mathrm{cm}}\, y^{\mathrm{cm}}|^2 = \frac{e^2(m_1-m_2)^2}{Mm_1m_2}|r|^2 = \frac{e^2(m_1-m_2)^2}{m_1^2m_2^2}|y^{\mathrm{cm}}|^2,$$

where in each equality the first square is taken in $Y'$, the second – in $\mathbb{R}^2$, and the third – in $Y$.

A short computation shows now that

$$\begin{aligned} \tfrac{1}{2}(\eta_{\mathrm{cm}} - 2\mathbf{A}_{\mathrm{cm}}{}^{\mathrm{cm}}\, y^{\mathrm{cm}})^2 &= \tfrac{1}{2M}((\bar{k}+2e\hat{r})^2 + (\hat{k}-2e\bar{r})^2) \\ &= \tfrac{1}{2M}(k - 2q_1Jr)^2, \end{aligned}$$

and

$$\begin{aligned} &\tfrac{1}{2}(\eta^{\mathrm{cm}} - 2\mathbf{A}_{\mathrm{cm}}{}^{\mathrm{cm}}\, y^{\mathrm{cm}})^2 \\ &= \tfrac{M}{2m_1m_2}((\bar{\rho} + \tfrac{e(m_2-m_1)}{M}\hat{r})^2 + (\hat{\rho} - \tfrac{e(m_2-m_1)}{M}\bar{r})^2) \\ &= \tfrac{M}{2m_1m_2}(\rho - \tfrac{m_2-m_1}{M}q_1Jr)^2 \end{aligned}$$

(the squares on the left-hand side are taken in $Y'$, and on the right-hand side – in $\mathbb{R}^2$). Hence we may identify $\tilde{H}(\eta_{\mathrm{cm}})$ with an operator on $L^2(\mathbb{R}^2)$, equal to:

$$(1.3.35) \qquad \frac{1}{2M}(k-2q_1Jr)^2 + \frac{M}{2m_1m_2}(D_r - \frac{m_2-m_1}{M}q_1Jr)^2 + V(r).$$

In the 3–dimensional case, the same calculation shows that the reduced Hamiltonian can be understood as an operator on $L^2(\mathbb{R}^3)$ which also has the form (1.3.35), except that $r = x_1 - x_2$ is now a variable in $\mathbb{R}^3$ and $k$ is a fixed vector in $\mathbb{R}^3$ whose third component is zero.

## 1.4. Bound and scattering states

We are now ready to discuss the notion of bound and scattering states. Essentially, bound states are those which describe the particles remaining close together in a stable configuration (i.e., all relative coordinates $x_i - x_j$ are bounded uniformly in time), while scattering states correspond to a visible and permanent break-up of the system into several parts. Our goal here is to cast this intuitive understanding into precise mathematical terms.

In the case which is best understood from the point of view of scattering theory – that of a system of interacting particles subject to no external forces (in particular, no magnetic field is present) – this is routinely done via the center of mass separation. As explained in the introduction to Section 1.3, one may write the Hamiltonian $H$ of such a system as

$$H = \frac{1}{2} D_{x_{\mathrm{cm}}}^2 + H^{\mathrm{cm}},$$

where $H^{\mathrm{cm}}$, acting on $L^2(X^{\mathrm{cm}})$, describes only the relative motion of the particles. One then defines the bound and scattering states as eigenfunctions of $H^{\mathrm{cm}}$ and states from its continuous spectral subspace respectively:

$$\mathcal{H}_{\mathrm{bound}} := \mathcal{H}_{\mathrm{pp}}(H^{\mathrm{cm}}) \otimes L^2(X_{\mathrm{cm}}),$$

$$\mathcal{H}_{\mathrm{scatt}} := \mathcal{H}_{\mathrm{cont}}(H^{\mathrm{cm}}) \otimes L^2(X_{\mathrm{cm}}).$$

Many authors prefer to work in the "center of mass frame" and define the $N$-particle Hamiltonian as the operator denoted here by $H^{\mathrm{cm}}$. If this convention is followed, one simply has $\mathcal{H}_{\mathrm{bound}} = \mathcal{H}_{\mathrm{pp}}(H^{\mathrm{cm}})$ and $\mathcal{H}_{\mathrm{scatt}} = \mathcal{H}_{\mathrm{cont}}(H^{\mathrm{cm}})$.

The mathematical definition of the bound and scattering states of magnetic $N$-particle Hamiltonians is more complicated and will rely on the reduction procedure introduced in the previous section.

Since the form of the reduced Hamiltonians was different for neutral and charged systems, these two cases will need to be considered separately. One might be tempted to think that this distinction is merely an artifact of the mathematical formalism we have chosen. This, however, is not the case. We will see that the differences between the behavior of neutral and charged systems concern features meaningful from the point of view of physics, and clearly independent of the details of the mathematical description, such as the general shape of possible trajectories of bound states of the system and their long time average velocity.

In Subsection 1.4.1 we define the bound and scattering states of charged systems and investigate the basic properties of bound states. In particular, we will prove that bound states of charged systems can move freely in the direction of the magnetic field, but may perform only bounded motion in the transversal directions. This has been long known in the special case when the system under consideration is a single charged particle.

On the other hand, in Subsection 1.4.2 we prove that a bound state of a neutral system may travel to infinity across the field. However, the characteristics of such motion and the mechanism behind it are very different from those of the motion of the center of mass of an isolated system. *Noninteracting* charged particles are confined by the Lorentz forces to a bounded region in the transversal to the field directions and cannot travel to infinity across the field; whether their electric charges add up to zero or not is irrelevant. It is only when the mutual interactions partially neutralize the Lorentz forces that unbounded motion across the field becomes possible. In this case the external and internal dynamics of the system are interdependent. The external dynamics is governed by a certain family of *dispersive Hamiltonians* – a phenomenon which will later play an important role in the Mourre theory.

The following examples from classical mechanics may be used to illustrate the mechanism at work. Consider two classical particles, with masses $m_1 = m_2 = 1$ and electric charges $q_1 = 1$ and $q_2 = -1$, moving in a plane $Y$ perpendicular to a constant magnetic field $\vec{B} = (0, 0, 2b)$.

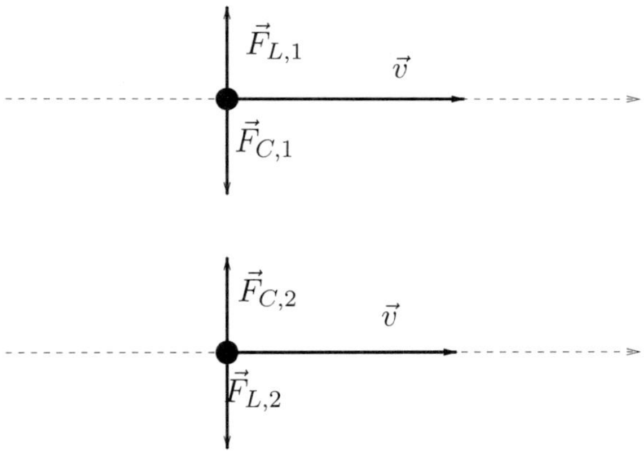


**Fig. 1.1**. Trajectories in Example 1

**Example 1.** For a generic pair potential $V = V_{12}$, we expect a family of solutions of the equations of motion in which the particles move across the field along parallel lines with the same velocity. Suppose for example that the particles interact via the Coulomb potential $V = -|y_1 - y_2|^{-1}$. Then we obtain the following family of solutions.

The trajectories are given by:

$$y_1(t) = \vec{v}t + \vec{r}_1,$$

$$y_2(t) = \vec{v}t + \vec{r}_2,$$

where the vector $\vec{r} = \vec{r}_1 - \vec{r}_2$ is perpendicular to the velocity $\vec{v}$. The Lorentz forces acting on the two particles are

$$\vec{F}_{L,1} = \vec{F}_L = e\vec{v} \times \vec{B},\ \vec{F}_{L,2} = -\vec{F}_L = -e\vec{v} \times \vec{B},$$

and the Coulomb forces are

$$\vec{F}_{C,1} = \vec{F}_C = \nabla V(\vec{r}),\ \vec{F}_{C,2} = -\vec{F}_C = -\nabla V(\vec{r}).$$

Since $\vec{F}_L$ and $\vec{F}_C$ act in the same direction, one can fix $\vec{v}$ and $\vec{r}$ so that $\vec{F}_L = -\vec{F}_C = -\nabla V(\vec{r})$. Then the Coulomb and Lorentz forces will cancel out.

It is clear that the Coulomb potential in the above example could be replaced by any potential $V$ with nonvanishing gradient. In particular, a similar 2−dimensional example could be constructed using a *repulsive* Coulomb potential.

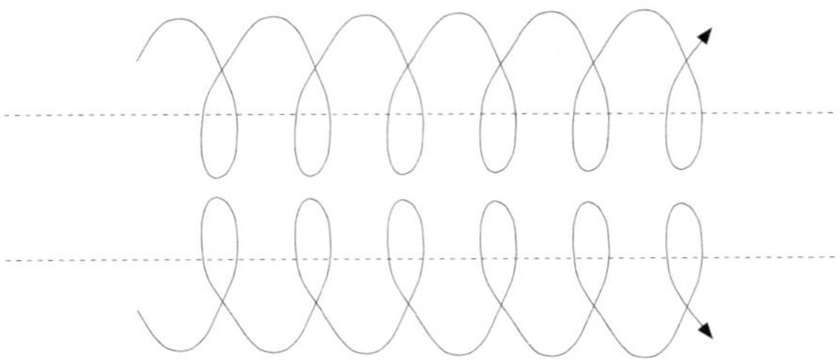

**Fig. 1.2.** Trajectories in Example 2

**Example 2.** We now consider the classical Hamiltonian

$$H = \sum_{i=1}^{2} (\eta_i - q_i J y_i)^2 + V(y_1 - y_2), \tag{1.4.1}$$

where $q_1 = 1$, $q_2 = -1$, and

$$y_i = \begin{pmatrix} \bar{y}_i \\ \hat{y}_i \end{pmatrix},\ q_i J y_i = \begin{pmatrix} -q_i b \hat{y}_i \\ q_i b \bar{y}_i \end{pmatrix},\ V(y_1 - y_2) = C|y_1 - y_2|^2. \tag{1.4.2}$$

The harmonic oscillator interaction may be somewhat unrealistic from the point of view of physics, but nevertheless it is often used as a good approximation to the true interactions, and it has the advantage of allowing us to compute the trajectories explicitly. Namely, we will seek solutions $(y_1, y_2)$ of the Hamilton equations

$$\frac{dy_i}{dt} = \frac{\partial H}{\partial \eta_i},\ \frac{d\eta_i}{dt} = -\frac{\partial H}{\partial y_i}, \tag{1.4.3}$$

of the form

$$y_i = \begin{pmatrix} vt + A\cos\omega t \\ q_i(B\sin\omega t + R) \end{pmatrix}, \tag{1.4.4}$$

where $v, A, B, \omega, R$ are constants. The first equation in (1.4.3) may be rewritten as

$$\frac{d}{dt}y_i = \eta_i - q_i J y_i,$$

from which $\eta_i$ may be computed if (1.4.4) is given. The second equation in (1.4.3) is then used to determine the constants $v, A, B, \omega, R$. It turns out that (1.4.4) solves (1.4.3) with $H$ given by (1.4.1)–(1.4.2) if

$$A = B = 0, \ v = 2CRb^{-1}, \tag{1.4.5}$$

or if

$$A \neq 0, \ \frac{B^2}{A^2} = 1 + \frac{C}{b^2}, \ \omega = -\frac{2B}{A}b, \ v = \frac{2CR}{b}. \tag{1.4.6}$$

The solutions corresponding to (1.4.5) are those found in Example 1 (with the Coulomb potential replaced by the harmonic oscillator interaction). More interesting to us at this time is the family of solutions with the constants as in (1.4.6). These solutions are expected to resemble the quantum trajectories more closely than those obtained in Example 1. In particular, we note the following two facts.

- The average velocity of the center of mass of the system, defined as

$$\lim_{t\to\infty} \frac{y_1(t) + y_2(t)}{2t}, \tag{1.4.7}$$

  is equal to $v$. Observe that by (1.4.6) $v$ is proportional to the average distance $R$ between the particles, and that by (1.4.2) $\nabla V$ is proportional to $|y_1 - y_2|$, so that the average value of $\nabla V$ is also proportional to $R$. This is consistent with the heuristic that the transversal motion depends on partial cancellations between the mutual interactions and the Lorentz forces.
- The actual center of mass velocity, $\frac{d}{dt}\frac{y_1(t)+y_2(t)}{2}$, is not constant along the given trajectories. Hence all statements such as "the system travels in the $y$ plane with nonzero average velocity" must be interpreted in terms of the limit (1.4.7).

After these preliminaries we begin a rigorous discussion of the quantum systems in question. Throughout this section, and especially in Definitions 1.4.1 and 1.4.4, we will use the reducing transformations $U_{\mathrm{cm}}$ defined in Section 1.3. While we will prove that the spaces $\mathcal{H}_{\mathrm{bound}}$ and $\mathcal{H}_{\mathrm{scatt}}$ defined below are independent of the choice of $U_{\mathrm{cm}}$, the exact form of the reduced Hamiltonians and some of their properties will depend on it. Whenever it is necessary to specify the reducing transformation being used, $U_{\mathrm{cm}}$ will be the transformation explicitly defined in Subsection 1.3.4 and 1.3.5 for neutral and charged systems respectively.

As in Section 1.3, we will only discuss the case $d = 3$. To obtain the corresponding results for $d = 2$, it suffices to replace $X$ and $x$ by $Y$ and $y$ respectively and to omit the $z$ variables.

**1.4.1. Charged systems.** If the system is charged, we have

$$U_{\rm cm} H U_{\rm cm}^{\star} = \tilde{H}, \tag{1.4.8}$$

where

$$\tilde{H} = R_{\rm cm}(D_{y_{\rm cm}^{(2)}}, y_{\rm cm}^{(2)}, x^{\rm cm}) + \frac{1}{2}(D^{\rm cm} - \mathbf{A}^{\rm cm\,cm}\, x^{\rm cm})^2 + V(x^{\rm cm}). \tag{1.4.9}$$

We will consider the reduced Hamiltonian $\tilde{H}$ as an operator acting only on $L^2(Y_{\rm cm}^{(2)} \times X^{\rm cm})$.

DEFINITION 1.4.1. Assume that $Q \neq 0$. We define:

$$\mathcal{H}_{\rm scatt} := U_{\rm cm}^{\star}\big(\mathcal{H}_{\rm cont}(\tilde{H}) \otimes L^2(Y_{\rm cm}^{(1)})\big),$$

$$\mathcal{H}_{\rm bound} := U_{\rm cm}^{\star}\big(\mathcal{H}_{\rm pp}(\tilde{H}) \otimes L^2(Y_{\rm cm}^{(1)})\big).$$

We have

$$L^2(X) = \mathcal{H}_{\rm scatt} \oplus \mathcal{H}_{\rm bound}.$$

Note that the factor $L^2(Y_{\rm cm}^{(1)})$ plays no significant role in the above definition and may be omitted if $\tilde{H}$ is understood as an operator on $L^2(Y_{\rm cm}^{(1)}) \otimes L^2(Y_{\rm cm}^{(2)}) \otimes L^2(X^{\rm cm})$.

LEMMA 1.4.2. *The spaces $\mathcal{H}_{\rm scatt}, \mathcal{H}_{\rm bound}$ are independent of the choice of the reducing unitary transformation $U_{\rm cm}$.*

PROOF. Consider two reducing unitary transformations $U_{\rm cm}, U'_{\rm cm}$. Since $\mathcal{H}^{\rm cm}$ and $\mathcal{H}^{\rm cm\prime}$ have the same dimension, we may assume that $\mathcal{H}^{\rm cm} = \mathcal{H}^{\rm cm\prime}$. The unitary transformation $U = U'_{\rm cm} U_{\rm cm}^{-1}$ commutes with $y^1, D_y^1$, so that it must act as the identity on $L^2(\mathbb{R}, dy^1)$. Hence $U\tilde{H} = \tilde{H}'U$ and $\mathcal{H}_{\rm pp}(\tilde{H}')$ is the image of $\mathcal{H}_{\rm pp}(\tilde{H})$ under $U$, which shows that $\mathcal{H}_{\rm bound}$ is independent of the choice of $U_{\rm cm}$. □

We will see that the above definition is consistent with the intuitive notion of a "bound state" as one for which all relative coordinates $x^{\rm cm}$ of the particles are bounded uniformly in time. Moreover, the magnetic field essentially confines the system to a bounded region of the transversal plane, i.e., all transversal coordinates $y$ remain bounded uniformly in $t$.

PROPOSITION 1.4.3. *Let $Q \neq 0$ and $u \in \mathcal{H}_{\rm bound}$. We have:*

$$(i)\ \lim_{R\to\infty} \sup_{t\in\mathbb{R}} \|F(\frac{|x^{\rm cm}|}{R} \geq 1)\mathrm{e}^{-\mathrm{i}tH}u\| = 0,$$

$$(ii)\ \lim_{R\to\infty}\sup_{t\in\mathbb{R}}\|F(\frac{|y_{\mathrm{cm}}|}{R}\geq 1)\mathrm{e}^{-\mathrm{i}tH}u\| = 0.$$

PROOF. By density, we need only consider vectors $u \in \mathcal{H}_{\mathrm{bound}}$ of the form $U_{\mathrm{cm}}^* v$, where $v = v_1 \otimes v_2$, $v_1 \in L^2(X^{\mathrm{cm}} \times Y_{\mathrm{cm}}^{(2)})$ is an eigenfunction of $\tilde{H}^{\mathrm{cm}}$ corresponding to an eigenvalue $E$, and $v_2 \in L^2(Y_{\mathrm{cm}}^{(1)})$. For such $u$, we have

$$\mathrm{e}^{-\mathrm{i}tH}u = U_{\mathrm{cm}}^*\mathrm{e}^{-\mathrm{i}t\tilde{H}}v = U_{\mathrm{cm}}^*(\mathrm{e}^{-\mathrm{i}tE}v_1 \otimes v_2).$$

Pick $F \in C_0^\infty(T^*Y_{\mathrm{cm}}^{(1)} \times T^*Y_{\mathrm{cm}}^{(2)} \times T^*X^{\mathrm{cm}})$ such that $F(0) = 1$, and let $F_R^{\mathrm{w}}$ be the Weyl quantization of $F_R(\cdot) = F(\frac{\cdot}{R})$. By standard pseudodifferential calculus,

$$\mathrm{s}-\lim_{R\to\infty} F_R^{\mathrm{w}} = \mathbf{1}.$$

Hence

$$\lim_{R\to\infty}\sup_{t\in\mathbb{R}}\|(\mathbf{1} - F_R^{\mathrm{w}})\mathrm{e}^{-\mathrm{i}t\tilde{H}}v\| = 0.$$

Inverting the transformation $U_{\mathrm{cm}}$ (see (1.3.20)) and using the covariance of the Weyl quantization under linear symplectic transformations, we obtain the proposition. □

**1.4.2. Neutral systems.** For a neutral system, we have

$$U_{\mathrm{cm}} H U_{\mathrm{cm}}^\star = \tilde{H}, \tag{1.4.10}$$

where

$$\begin{aligned}\tilde{H} &= \tfrac{1}{2}(D_{y_{\mathrm{cm}}} - 2\mathbf{A}_{\mathrm{cm}}{}^{\mathrm{cm}}\, x^{\mathrm{cm}})^2 + \tfrac{1}{2}(D_{x^{\mathrm{cm}}} - \mathbf{A}^{\mathrm{cm\,cm}}\, x^{\mathrm{cm}})^2 \\ &\quad + V(x^{\mathrm{cm}}).\end{aligned} \tag{1.4.11}$$

The transformed Hamiltonian $\tilde{H}$ may be written as a direct integral

$$\tilde{H} = \int_{Y'_{\mathrm{cm}}}^{\oplus} \tilde{H}(\eta_{\mathrm{cm}})d\eta_{\mathrm{cm}},$$

where

$$\tilde{H}(\eta_{\mathrm{cm}}) = \frac{1}{2}(\eta_{\mathrm{cm}} - 2\mathbf{A}_{\mathrm{cm}}{}^{\mathrm{cm}}\, x^{\mathrm{cm}})^2 + \frac{1}{2}(D_{x^{\mathrm{cm}}} - \mathbf{A}^{\mathrm{cm\,cm}}\, x^{\mathrm{cm}})^2 + V(x^{\mathrm{cm}})$$

acts on $L^2(X^{\mathrm{cm}})$. The bound and scattering states are then defined as direct integrals of eigenfunctions and of the states from the continuous spectral subspaces of the reduced Hamiltonians $\tilde{H}(\eta_{\mathrm{cm}})$ respectively.

DEFINITION 1.4.4. Assume that $Q = 0$. Then:

$$\mathcal{H}_{\mathrm{scatt}} := U_{\mathrm{cm}}^\star\left(\int_{Y'_{\mathrm{cm}}}^{\oplus} \mathcal{H}_{\mathrm{cont}}(\tilde{H}(\eta_{\mathrm{cm}}))d\eta_{\mathrm{cm}}\right),$$

$$\mathcal{H}_{\text{bound}} := U_{\text{cm}}^{\star}\left(\int_{Y'_{\text{cm}}}^{\oplus} \mathcal{H}_{\text{pp}}(\tilde{H}(\eta_{\text{cm}}))d\eta_{\text{cm}}\right).$$

We have

$$L^2(X) = \mathcal{H}_{\text{scatt}} \oplus \mathcal{H}_{\text{bound}}.$$

LEMMA 1.4.5. *The spaces $\mathcal{H}_{\text{scatt}}, \mathcal{H}_{\text{bound}}$ are independent of the choice of the reducing unitary transformation $U_{\text{cm}}$.*

PROOF. As in the proof of Lemma 1.4.2, we consider two reducing unitary transformations $U_{\text{cm}}, U'_{\text{cm}}$ and assume that $\mathcal{H}^{\text{cm}} = \mathcal{H}^{\text{cm}\prime}$. Then $U = U'_{\text{cm}}U_{\text{cm}}^{-1}$ is a unitary transformation on $\int_{X'_{\text{cm}}}^{\oplus} \mathcal{H}^{\text{cm}} d\eta_{\text{cm}}$ commuting with $\int_{X'_{\text{cm}}}^{\oplus} \eta_{\text{cm}} \cdot \mathbf{1} d\eta_{\text{cm}}$. By [**Di**, A80], this implies that $U = \int_{X'_{\text{cm}}}^{\oplus} U(\eta_{\text{cm}})d\eta_{\text{cm}}$, where $U(\eta_{\text{cm}})$ is a measurable family of unitary operators on $\mathcal{H}^{\text{cm}}$, with $U(\eta_{\text{cm}})\tilde{H}(\eta_{\text{cm}}) = \tilde{H}'(\eta_{\text{cm}})U(\eta_{\text{cm}})$.

Hence $\mathcal{H}_{\text{pp}}(\tilde{H}(\eta_{\text{cm}}))$ is the image of $\mathcal{H}_{\text{pp}}(\tilde{H}'(\eta_{\text{cm}}))$ under $U(\eta_{\text{cm}})$, which shows that $\mathcal{H}_{\text{bound}}$ is independent of the choice of $U_{\text{cm}}$. □

As expected, states in $\mathcal{H}_{\text{bound}}$ are indeed "bound states" in the sense that one has uniform in time bounds on all relative coordinates $x^{\text{cm}}$ of the particles. The motion of these bound states in the plane transversal to the field is governed by a family of so-called *effective Hamiltonians.* Namely, let $Y'_{\text{cm}} \ni \eta_{\text{cm}} \mapsto P(\eta_{\text{cm}})$ be a measurable family of eigenprojections of $\tilde{H}(\eta_{\text{cm}})$ corresponding to an eigenvalue $E(\eta_{\text{cm}})$, and let:

$$P = \int_{Y'_{\text{cm}}}^{\oplus} P(\eta_{\text{cm}})d\eta_{\text{cm}}.$$

Then the effective "external" kinetic energy of a bound state $u = U_{\text{cm}}^{*}v$, where $v = Pv$, is $E(\eta_{\text{cm}})$, i.e.,

$$\mathrm{e}^{-\mathrm{i}tH}u = U_{\text{cm}}^{\star}\int_{Y'_{\text{cm}}}^{\oplus} \mathrm{e}^{-\mathrm{i}tE(\eta_{\text{cm}})}v(\eta_{\text{cm}})d\eta_{\text{cm}}.$$

Proposition 1.4.6 *(ii)* below states that the average velocity of the motion of such a bound state across the field is $\nabla E(\eta_{\text{cm}})$, provided that the function $Y'_{\text{cm}} \ni \eta_{\text{cm}} \mapsto E(\eta_{\text{cm}})$ is $C^1$. We emphasize the word *average.* A simple calculation shows that the actual velocity of the center of mass of the system does not commute with the Hamiltonian and therefore is not constant; it is only the average velocity over long periods of time that becomes constant as $t \to \infty$. (See also Example 2 in the introduction to this section.)

PROPOSITION 1.4.6. *Assume that $Q = 0$. Then the following holds.*

*(i) for any $u \in \mathcal{H}_{\text{bound}}$ we have:*

$$\lim_{R\to\infty} \sup_{t\in\mathbb{R}} \|F(\frac{|x^{\text{cm}}|}{R} \geq 1)\mathrm{e}^{-\mathrm{i}tH}u\| = 0.$$

*(ii) let* $u \in \mathcal{H}_{\mathrm{bound}}$ *satisfy*

$$U_{\mathrm{cm}}u = PU_{\mathrm{cm}}u,$$

*and assume that:*

$$Y'_{\mathrm{cm}} \ni \eta_{\mathrm{cm}} \mapsto P(\eta_{\mathrm{cm}}) \in B(L^2(X^{\mathrm{cm}})) \text{ is measurable,}$$

$$Y'_{\mathrm{cm}} \ni \eta_{\mathrm{cm}} \mapsto E(\eta_{\mathrm{cm}}) \in \mathbb{R} \text{ is } C^1.$$

*Then for all* $g \in C_0^\infty(Y_{\mathrm{cm}})$ *we have*

$$\lim_{t\to+\infty} \big(g(\frac{y_{\mathrm{cm}}}{t}) - g(\nabla E(K_{\mathrm{cm}})\big)\mathrm{e}^{-\mathrm{i}tH}u = 0.$$

PROOF. *(i)* follows directly from the definition of $\mathcal{H}_{\mathrm{bound}}$ and from the fact that $U_{\mathrm{cm}}$ conserves $x^{\mathrm{cm}}$. It remains to prove *(ii)*. Let $u = U^\star_{\mathrm{cm}}v$, where $v = Pv \in L^2(X^{\mathrm{cm}} \times Y_{\mathrm{cm}})$. We have

$$\mathrm{e}^{-\mathrm{i}tH}u = U^\star_{\mathrm{cm}}\Big(\mathrm{e}^{-\mathrm{i}tE(D_{y_{\mathrm{cm}}})}v\Big).$$

Let $g \in C_0^\infty(Y_{\mathrm{cm}})$. We will prove that for any $\phi \in L^2(X^{\mathrm{cm}} \times Y_{\mathrm{cm}})$,

$$\lim_{t\to\infty} \|\mathrm{e}^{\mathrm{i}tE(D_{y_{\mathrm{cm}}})}g(\frac{y_{\mathrm{cm}}}{t})\mathrm{e}^{-\mathrm{i}tE(D_{y_{\mathrm{cm}}})}\phi - g(\nabla E(D_{y_{\mathrm{cm}}}))\phi\| = 0. \tag{1.4.12}$$

This implies *(ii)* by the covariance of the Weyl quantization under linear symplectic transformations.

It suffices to prove (1.4.12) for $\phi \in C_0^\infty(X^{\mathrm{cm}} \times Y_{\mathrm{cm}})$. In the computation below the variables $x^{\mathrm{cm}}$ will be only parameters, so that we will omit the subscript cm and write $D$, $y$ instead of $D_{y_{\mathrm{cm}}}$ and $y_{\mathrm{cm}}$. $\mathcal{F}$ and $\hat{}$ will denote the Fourier transform only in the $y_{\mathrm{cm}}$ variables. We have:

$$\begin{aligned}
&\mathcal{F}\big(\mathrm{e}^{\mathrm{i}tE(D)}g(\tfrac{y}{t})\mathrm{e}^{-\mathrm{i}tE(D)}\phi\big)\\
&= \mathrm{e}^{\mathrm{i}tE(\eta)}\mathcal{F}\big(g(\tfrac{y}{t})\mathrm{e}^{-\mathrm{i}tE(D)}\phi\big)\\
&= \mathrm{e}^{\mathrm{i}tE(\eta)}\big(\mathcal{F}(g(\tfrac{y}{t})) * \mathrm{e}^{-\mathrm{i}tE(\eta)}\hat{\phi}\big)\\
&= \mathrm{e}^{\mathrm{i}tE(\eta)}\textstyle\int t^2\hat{g}((\eta-\eta')t)\mathrm{e}^{-\mathrm{i}tE(\eta')}\hat{\phi}(\eta')d\eta'\\
&= \textstyle\int \mathrm{e}^{\mathrm{i}t(E(\eta)-E(\eta'))}t^2\hat{g}((\eta-\eta')t)\hat{\phi}(\eta')d\eta',
\end{aligned}$$

where we used that $\mathcal{F}(g(y/t)) = t^2\hat{g}(\eta t)$. Substituting $(\eta-\eta')t = \theta$, we see that the last integral is equal to

$$\int \mathrm{e}^{\mathrm{i}t(E(\eta)-E(\eta-\theta/t))}\hat{g}(\theta)\hat{\phi}(\eta-\theta/t)d\theta. \tag{1.4.13}$$

As $t \to \infty$, the integrand in (1.4.13) converges to

$$\mathrm{e}^{\mathrm{i}\theta\cdot\nabla E(\eta)}\hat{g}(\theta)\hat{\phi}(\eta).$$

Using the rapid decay of $\hat{g}$ and the dominated convergence theorem, we see that (1.4.13) converges in $L^2$ as $t \to \infty$ to:

$$\int e^{i\theta \cdot \nabla E(\eta)} \hat{g}(\theta)\hat{\phi}(\eta) d\theta = g(\nabla E(\eta))\hat{\phi}(\eta) = \mathcal{F}(g(\nabla E(D))\phi),$$

which proves (1.4.12). □

CHAPTER 2

# Geometrical Methods I

In this chapter we introduce and explain the $N$-body notation and geometry. We also prove a number of technical results which will be used throughout the rest of this book.

The main characteristic feature of $N$-body problems in general, with or without an external field, is that there are certain subspaces $X_a$ of the configuration space X along which the $N$-body potential $V = \sum V_{ij}(\mathrm{x}_i - \mathrm{x}_j)$ does not vanish at infinity, namely the hyperplanes $\{\mathrm{x}_i = \mathrm{x}_j\}$ and their intersections. This provides a mathematical basis for the phenomenon of *multichannel scattering*: there are many possible outcomes of the scattering process, called *scattering channels* and usually described in terms of *cluster decompositions* and *channel Hamiltonians.* A *cluster decomposition* $a$ is a break-up of $\{1, \ldots, N\}$ into disjoint subsystems called clusters. There is a correspondence between cluster decompositions $a$ and the subspaces $X_a$ just mentioned: the geometrical statement that the trajectory of the system is localized near $X_a$ means that the particles belonging to the same cluster in $a$ are close together. A *channel Hamiltonian* $H_a$ describes the dynamics of the system with the interactions between different clusters in $a$ deleted; the basic idea of the geometrical approach to $N$-body scattering is that the full Hamiltonian $H$ is similar to $H_a$ in the region of configuration space where particles belonging to different clusters of $a$ are far apart.

The use of cluster decompositions and channel Hamiltonians in scattering theory is traditional and goes back to the 1950s and 60s ([**Ja**], [**F**], [**Hu**]). The significance of geometrical methods was recognized considerably later ([**E1**], [**Sim1**]; this is also where $N$-body partitions of unity similar to those in Section 2.3 were probably first used). The geometrical approach was crucial in the subsequent advances in the field: see e.g., [**PSS**], [**SS1**], [**De1**], and especially [**Gr**], where the understanding of the $N$-body geometry was raised to a new level. (A geometrical construction similar in spirit to that of [**Gr**], but more complicated technically, will be presented in Chapter 5 of this book.)

Our goal in this chapter is to reconcile the basic definitions and results of the geometrical approach with the additional structure due to the magnetic field. In particular, we must distinguish between the directions parallel to the field ($z$) and perpendicular to it ($y$), classify the cluster decompositions according to the total electric charges of their clusters, and extend the reduction procedure of Section 1.3 to more general cluster decompositions.

Section 2.2 generally follows [**GŁ1**]. The construction of the partitions of unity in Section 2.3 is standard in scattering theory, as are the non-magnetic analogues of the arguments in Subsection 2.3.3 (cf. [**HuSi**], [**De1**], [**DG1**]). A much deeper geometrical analysis will be required later on in the proofs of the *propagation estimates* of Chapters 4 and 5.

The readers already familiar with $N$-body theory will note that we will *not* work with an abstract generalization of $N$-body Hamiltonians known as *Agmon Hamiltonians* ([**DG1**], Section 6.1). Although the abstract Agmon framework captures all the essential geometry of the $N$-body potential, it does not reflect any of the additional features of magnetic Hamiltonians; consequently, those of our arguments in which the magnetic field appears in a nontrivial way cannot be extended to the Agmon setting. Finding an Agmon-type abstract generalization of the magnetic Hamiltonians to which the results of this book would naturally extend remains an open problem.

## 2.1. N-body geometry

We begin with a (rather long!) list of notation. Much of it is standard in the $N$-body theory (see e.g., [**DG1**]); however, we also need to take into account the additional structures due to the presence of the magnetic field. We will only consider the case of an $N$-particle system in $\mathbb{R}^3$; in the 2−dimensional case the same analysis applies except that the z variables should be omitted.

A non-empty subset $C = \{i_1, \dots, i_k\}$ of $\{1, 2, \dots, N\}$ is called a *cluster*. We will denote by $\mathrm{X}[C]$ the configuration space for the cluster $C$:

$$\mathrm{X}[C] = \{\mathrm{x} \in \mathrm{X} : \ \mathrm{x}_j = 0 \text{ if } j \notin C\}.$$

Note that $\mathrm{X}[C]$ is isomorphic to $\mathbb{R}^{3k}$. The spaces $\mathrm{Y}[C]$ and $\mathrm{Z}[C]$ are defined in a similar way.

We may introduce the center of mass coordinates (i.e., repeat the procedure of Subsection 1.3.1) on the cluster configuration space $\mathrm{X}[C]$. Namely the configuration space of the motion of the center of mass of $C$,

$$\frac{\sum_{i\in C} m_i \mathrm{x}_i}{\sum_{i\in C} m_i},$$

is

$$\mathrm{X}_{\mathrm{cm}}[C] := \{\mathrm{x} \in \mathrm{X} : \ \mathrm{x}_i = \mathrm{x}_j \text{ if } i, j \in C, \ \mathrm{x}_l = 0 \text{ if } l \notin C\}, \tag{2.1.1}$$

and the configuration space of the relative motion within $C$ is

$$\mathrm{X}^{\mathrm{cm}}[C] := \{\mathrm{x} \in \mathrm{X} : \ \sum_{i\in C} m_i \mathrm{x}_i = 0, \ \mathrm{x}_l = 0 \text{ if } l \notin C\}. \tag{2.1.2}$$

We have $\dim(\mathrm{X}_{\mathrm{cm}}[C]) = 3$ and $\dim(\mathrm{X}^{\mathrm{cm}}[C]) = 3(k-1)$, where $k$ is the number of the particles in $C$.

Similarly we define $\mathrm{Y}[C]$, $\mathrm{Z}[C]$, etc. We then set:

$$Y[C] = \mathrm{Y}[C], \ Z[C] = \mathrm{Z}^{\mathrm{cm}}[C], \ X[C] = Y[C] \oplus Z[C].$$

The total electric charge of the cluster $C$ will be denoted by $Q[C]$:

$$Q[C] := \sum_{i \in C} q_i.$$

We will call the cluster $C$ *neutral* if $Q[C] = 0$ and *charged* if $Q[C] \neq 0$.

A *cluster decomposition* is a decomposition of the system into disjoint clusters. Mathematically this is described by a partition $a$ of the set $\{1, 2, \ldots, N\}$ into non-empty disjoint subsets:

$$a = \{C_1, C_2, \ldots, C_l\}, \ C_i \cap C_j = \emptyset \text{ if } i \neq j, \ \bigcup_{i=1}^{l} C_i = \{1, 2, \ldots, N\}.$$

Important special cases are the cluster decompositions

$$a_{\max} := \{\{1, \ldots, N\}\}, \ a_{\min} := \{\{1\}, \ldots, \{N\}\},$$

and, for any pair of indices $i, j \in \{1, \ldots, N\}$, $i \neq j$,

$$(ij) = \{\{i, j\}, \{1\}, \{2\}, \ldots, \{i-1\}, \{i+1\}, \ldots, \{j-1\}, \{j+1\}, \ldots, \{N\}\},$$

which consists of the two-particle cluster $\{i, j\}$ and $(N-2)$ remaining one-particle clusters.

The number of the clusters in the decomposition $a$ will be denoted by $\sharp a$. For example, $\sharp a_{\max} = 1$, $\sharp a_{\min} = N$, and $\sharp(ij) = N - 1$.

We will denote by $\mathcal{A}$ the set of all cluster decompositions of the $N$-particle system. There is a natural partial ordering on $\mathcal{A}$, defined as follows. We will say that $a$ is a *refinement* of $b$ and write

$$a \leq b,$$

if all clusters of $a$ are subsets of clusters of $b$. For instance, if $N = 5$ and

$$a = \{\{1, 2\}, \{3\}, \{4\}, \{5\}\}, \ b = \{\{1, 2, 3\}, \{4, 5\}\},$$

then $a \leq b$.

Note that $a \leq b$ implies that $\sharp a \geq \sharp b$, and that $(ij) \leq a$ means that the $i$-th and $j$-th particles belong to the same cluster of $a$. We will write $a < b$ if $a \leq b$ and $a \neq b$. The relation $\leq$ defines a lattice structure on $\mathcal{A}$ with the maximal and minimal elements $a_{\max}$ and $a_{\min}$.

Given two cluster decompositions $a, b \in \mathcal{A}$, $a \vee b$ will denote the unique cluster decomposition such that:

- $a \leq a \vee b$ and $b \leq a \vee b$,
- if $c$ is any other cluster decomposition such that $a \leq c$ and $b \leq c$, then $a \vee b \leq c$.

For example, if

$$N = 4, \ a = \{\{1, 2\}, \{3\}, \{4\}\}, \ b = \{\{1\}, \{2, 3\}, \{4\}\},$$

then

$$a \vee b = \{\{1, 2, 3\}, \{4\}\}.$$

We have $a \leq b$ if and only if $a \vee b = b$.

Let

$$a = \{C_1, \ldots, C_l\}$$

be a cluster decomposition. Then the cluster configuration spaces $\mathrm{X}[C_j]$ are mutually orthogonal and

$$\mathrm{X} = \bigoplus_{j=1}^{l} \mathrm{X}[C_j].$$

Writing each $\mathrm{X}[C_j]$ as $\mathrm{X}_{\mathrm{cm}}[C_j] \oplus \mathrm{X}^{\mathrm{cm}}[C_j]$, we obtain the decomposition

$$\mathrm{X} = \mathrm{X}_a \oplus \mathrm{X}^a,$$

where

$$\mathrm{X}_a := \bigoplus_{j=1}^{l} \mathrm{X}_{\mathrm{cm}}[C_j], \ \mathrm{X}^a := \bigoplus_{j=1}^{l} \mathrm{X}^{\mathrm{cm}}[C_j]. \tag{2.1.3}$$

$\mathrm{X}_a$ and $\mathrm{X}^a$ are the configuration spaces of the centers of mass of the clusters in $a$ and of the relative motion within clusters. They are orthogonal in the metric $g$ introduced in Section 1.2. They may be written more explicitly as:

$$\begin{aligned} \mathrm{X}_a &= \{\mathrm{x} \in \mathrm{X} |\ \mathrm{x}_i = \mathrm{x}_j \text{ for all } i, j \text{ such that } (ij) \leq a\}, \\ \mathrm{X}^a &= \{\mathrm{x} \in \mathrm{X} |\ \textstyle\sum_{i \in C_k} m_i \mathrm{x}_i = 0 \text{ for all } C_k \in a\}. \end{aligned} \tag{2.1.4}$$

It is straightforward to verify that:

$$\begin{aligned} &a \leq b \text{ iff } \mathrm{X}_b \subset \mathrm{X}_a, \\ &\mathrm{X}_{a_{\max}} = \mathrm{X}_{\mathrm{cm}}, \ \mathrm{X}_{a_{\min}} = \mathrm{X}, \\ &\mathrm{X}_{a \vee b} = \mathrm{X}_a \cap \mathrm{X}_b. \end{aligned} \tag{2.1.5}$$

The orthogonal projections onto $\mathrm{X}_a$ and $\mathrm{X}^a$ will be denoted by $\Pi_a$ and $\Pi^a$ respectively. We will also write $\mathrm{x}_a = \Pi_a \mathrm{x}$ and $\mathrm{x}^a = \Pi^a \mathrm{x}$.

If $a \leq b$, then by (2.1.5) all of the projections $\Pi_a$, $\Pi^a$, $\Pi_b$, $\Pi^b$ commute, and:

$$\Pi_a \Pi_b = \Pi_b \Pi_a = \Pi_b, \ \Pi^a \Pi^b = \Pi^b \Pi^a = \Pi^a. \tag{2.1.6}$$

We also define for $a \leq b$:

$$\Pi_a^b = \Pi^b \Pi_a = \Pi_a \Pi^b, \ \mathrm{X}_a^b = \Pi_a^b \mathrm{X} = \mathrm{X}^b \cap \mathrm{X}_a, \ \mathrm{x}_a^b = \Pi_a^b \mathrm{x} = \mathrm{x}^b - \mathrm{x}^a.$$

As in Chapter 1, we will now fix the z coordinate of the center of mass of the system at 0, and consider the spaces

$$Y = \mathrm{Y}, \ Z = \mathrm{Z}^{\mathrm{cm}}, \ X = Y \oplus Z.$$

By (2.1.5)–(2.1.6), we may define:

$$\begin{aligned} Y_a &= \mathrm{Y}_a, \ Z_a = \mathrm{Z}_a^{\mathrm{cm}}, \ X_a = Y_a \oplus Z_a, \\ Y^a &= \mathrm{Y}^a, \ Z^a = \mathrm{Z}^a, \ X^a = Y^a \oplus Z^a, \end{aligned}$$

and

$$\mathcal{Y}_a^b = \mathrm{Y}_a^b, \ \mathcal{Z}_a^b = \mathrm{Z}_a^b, \ X_a^b = Y_a^b \oplus Z_a^b \text{ for } a \le b.$$

A similar convention will be observed for coordinates: we will write $y_a = \mathrm{y}_a$, $z_a = \mathrm{z}_a^{\mathrm{cm}}$, $x_a = y_a + z_a$, etc. The projections on the spaces $X_a$, $X^a$, $X_a^b$ will be denoted by $\pi_a$, $\pi^a$, $\pi_a^b$ respectively. Note that these projections commute with the orthogonal projections onto the spaces $Y$ and $Z$, hence we may write $Y_a = \pi_a Y$, $Y^a = \pi^a Y$, etc. The symbols $\mathbf{A}^a{}_a$, $\mathbf{A}_a{}^a$, $\mathbf{A}_{a\,a}$, $\mathbf{A}^{a\,a}$ will be used to denote the restrictions of the bilinear form $\mathbf{A}$ to $X^a \times X_a$, $X_a \times X^a$, $X_a \times X_a$, $X^a \times X^a$ respectively.

It will sometimes be convenient to add to the family $\{X_a\}_{a \in \mathcal{A}}$ the subspace $\{0\} =: X_\emptyset$. Here $\emptyset$ is, formally, an "empty cluster decomposition" for which $\sharp\emptyset = 0$ and $\emptyset > a$ for all $a \in \mathcal{A}$. Then

$$\mathrm{X}^\emptyset = \mathrm{X}, \ \pi_\emptyset = \mathbf{1}, \ \pi^\emptyset = 0,$$

so that $\mathrm{x}_\emptyset = \mathrm{x}$, $\mathrm{x}^\emptyset = 0$. The augmented set of cluster decompositions $\mathcal{A} \cup \{\emptyset\}$ will be denoted by $\mathcal{A}_*$.

In scattering theory it will be important to distinguish between those cluster decompositions which contain neutral clusters and those which do not. This is because, given the results of Section 1.4, we expect charged and neutral clusters to display rather different large time asymptotic behaviour.

Consider a cluster decomposition $a = \{C_1, \cdots C_l\}$. We will say that:

$$\begin{aligned} &a \in \mathcal{A}^{\mathrm{c}} \text{ if } Q[C_i] \neq 0 \text{ for all } C_i \in a, \\ &a \in \mathcal{A}^{\mathrm{n}} \text{ otherwise.} \end{aligned} \tag{2.1.7}$$

The cluster decompositions in $\mathcal{A}^{\mathrm{c}}$ (those which consist of only charged clusters) will be sometimes referred to as *charged*. An important subset of $\mathcal{A}^{\mathrm{c}}$ is

$$\mathcal{A}^{\mathrm{cc}} = \{a \in \mathcal{A}^{\mathrm{c}} : \ b \in \mathcal{A}^{\mathrm{c}} \text{ for all } b \le a\}. \tag{2.1.8}$$

$\mathcal{A}^{\mathrm{cc}}$ consists of cluster decompositions $a$ such that any nonempty subset of any cluster of $a$ is charged, i.e., $\sum_{k=1}^{j} q_{i_k} \neq 0$ for any particles $i_1, \dots, i_k$ which belong to the same cluster of $a$. Such decompositions will be called *strongly charged.*

The "empty cluster decomposition" $\emptyset$ contains no clusters and, by convention, belongs to neither $\mathcal{A}^{\mathrm{n}}$ nor $\mathcal{A}^{\mathrm{c}}$. However, the condition in (2.1.8) makes sense for $a = \emptyset$, provided that we replace $b \le a$ by $b < a$. We will say that $\emptyset$ is strongly charged if $b \in \mathcal{A}^{\mathrm{c}}$ for all $b < \emptyset$, i.e., for all $b \in \mathcal{A}$. This condition is equivalent to

$$q_{i_1} + q_{i_2} + \cdots + q_{i_k} \neq 0 \text{ for any } \{i_1, i_2, \dots, i_k\} \subset \{1, \dots, N\}. \tag{2.1.9}$$

We will write:

$$\begin{aligned} &\mathcal{A}_*^{\mathrm{cc}} = \mathcal{A}^{\mathrm{c}} \cup \{\emptyset\} \quad &&\text{if } \emptyset \text{ is strongly charged (i.e., (2.1.9) holds),} \\ &\mathcal{A}_*^{\mathrm{cc}} = \mathcal{A}^{\mathrm{c}} &&\text{otherwise.} \end{aligned}$$

Let $a = \{C_1, \ldots, C_k\} \in \mathcal{A}$. We can decompose the configuration space X into the configuration spaces $\mathrm{X}^{\mathrm{n}}(a)$ and $\mathrm{X}^{\mathrm{c}}(a)$ of all neutral and all charged clusters of $a$:

$$\mathrm{X}^{\mathrm{n}}(a) := \bigoplus_{C_l \text{ neutral}} \mathrm{X}[C_l], \ \mathrm{X}^{\mathrm{c}}(a) := \bigoplus_{C_l \text{ charged}} \mathrm{X}[C_l]. \tag{2.1.10}$$

We have $\dim(\mathrm{X}^{\mathrm{n}}(a)) = 3\sharp^{\mathrm{n}}a$ and $\dim(\mathrm{X}^{\mathrm{c}}(a)) = 3\sharp^{\mathrm{c}}a$, where $\sharp^{\mathrm{n}}a$ and $\sharp^{\mathrm{c}}a$ stand for the total number of neutral and charged clusters of $a$ respectively. (If $a \in \mathcal{A}^{\mathrm{c}}$, $\sharp^{\mathrm{n}}a = 0$ and $\mathrm{X}^{\mathrm{n}}(a) = \{0\}$.) We also define:

$$\begin{aligned} &\mathrm{X}_a^{\mathrm{n}} := \oplus_{C_l \text{ neutral}} \mathrm{X}_{\mathrm{cm}}[C_l], \ \mathrm{X}_a^{\mathrm{c}} := \oplus_{C_l \text{ charged}} \mathrm{X}_{\mathrm{cm}}[C_l], \\ &\mathrm{X}^{a,\mathrm{n}} = \oplus_{C_l \text{ neutral}} \mathrm{X}^{\mathrm{cm}}[C_l], \ \mathrm{X}^{a,\mathrm{c}} = \oplus_{C_l \text{ charged}} \mathrm{X}^{\mathrm{cm}}[C_l], \end{aligned} \tag{2.1.11}$$

where $\mathrm{X}^{\mathrm{cm}}[C_l]$ and $\mathrm{X}_{\mathrm{cm}}[C^l]$ are defined in (2.1.1)–(2.1.2). Then

$$\mathrm{X}_a = \mathrm{X}_a^{\mathrm{n}} \oplus \mathrm{X}_a^{\mathrm{c}}, \ \mathrm{X}^a = \mathrm{X}^{a,\mathrm{n}} \oplus \mathrm{X}^{a,\mathrm{c}},$$

and

$$\mathrm{X}^{\mathrm{n}}(a) = \mathrm{X}_a^{\mathrm{n}} \oplus \mathrm{X}^{a,\mathrm{n}}, \ \mathrm{X}^{\mathrm{c}}(a) = \mathrm{X}_a^{\mathrm{c}} \oplus \mathrm{X}^{a,\mathrm{c}}.$$

We denote by $\mathrm{x}^{\mathrm{n}}(a)$, $\mathrm{x}_a^{\mathrm{n}}$, $\mathrm{x}^{a,\mathrm{n}}$, etc., the orthogonal projections of $\mathrm{x} \in \mathrm{X}$ on the corresponding spaces. In particular, we have $\mathrm{x}_{\mathrm{cm}} = \mathrm{x}_{\mathrm{cm}}^{\mathrm{n}}$ or $\mathrm{x}_{\mathrm{cm}} = \mathrm{x}_{\mathrm{cm}}^{\mathrm{c}}$, depending on whether the system is neutral or charged.

Observe that the orthogonal projections on $\mathrm{Y} = Y$ and on $\mathrm{X}[C_l]$ commute, so that we may repeat the above procedure and define the spaces

$$Y^{\mathrm{n}}(a), \ Y^{\mathrm{c}}(a), \ Y_a^{\mathrm{n}}, \ Y_a^{\mathrm{c}}, \ Y^{a,\mathrm{n}}, \ Y^{a,\mathrm{c}};$$

the projections of $y \in Y$ on the above spaces will be denoted by $y^{\mathrm{n}}(a)$, $y^{\mathrm{c}}(a)$, $y_a^{\mathrm{n}}$, etc. We also note that $\mathrm{X}^{a,\mathrm{n}}$ and $\mathrm{X}^{a,\mathrm{c}}$ are subspaces of $\mathrm{X}^a = X^a$, so that the notation

$$X^{a,\mathrm{n}} = \mathrm{X}^{a,\mathrm{n}}, \ X^{a,\mathrm{c}} = \mathrm{X}^{a,\mathrm{c}},$$

is non-ambiguous. (Note, however, that the projections on $\mathrm{X}^{\mathrm{n}}(a)$, $\mathrm{X}^{\mathrm{c}}(a)$, $\mathrm{X}_a^{\mathrm{n}}$, $\mathrm{X}_a^{\mathrm{c}}$ do *not* commute with the projection on $X$.)

Finally, to avoid triple superscripts (as in $y^{(1)a,\mathrm{c}}$) we will write $y^{(1)} = \bar{y}$, $y^{(2)} = \hat{y}$, $\eta^{(1)} = \bar{\eta}$, $\eta^{(2)} = \hat{\eta}$.

## 2.2. Channel Hamiltonians

**2.2.1. Definition.** Consider a cluster decomposition $a = (C_1, \cdots C_{\sharp a})$. We define the *channel Hamiltonian* $\mathrm{H}_a$ as

$$\mathrm{H}_a := \sum_{l=1}^{\sharp a} \mathrm{H}[C_l], \tag{2.2.1}$$

where $\mathrm{H}[C_l]$ is the Hamiltonian of the cluster $C_l$:

$$\mathrm{H}[C_l] = \sum_{i \in C_l} \frac{1}{2m_i}(D_i - q_i J \mathrm{x}_i)^2 + \sum_{i,j \in C_l} V_{ij}(\mathrm{x}_i - \mathrm{x}_j), \tag{2.2.2}$$

acting on $L^2(\mathrm{X}[C_l])$. In other words, $\mathrm{H}_a$ is the Hamiltonian of the *non-interacting* clusters of $a$. We also define

$$H_a := \mathrm{H}_a - \frac{1}{2}D^2_{z_{\mathrm{cm}}}.$$

Note that H and $H$ can be obtained from H and $H$, respectively, by removing the interactions between particles belonging to different clusters. More precisely, if we define the *intercluster interaction* $I_a$ as

$$I_a(x) := \sum_{(ij)\not\leq a} V_{ij}(x_i - x_j), \tag{2.2.3}$$

we have:

$$\mathrm{H}_a = \mathrm{H} - I_a, \; H_a = H - I_a. \tag{2.2.4}$$

$H_a$ may also be written as

$$H_a = \frac{1}{2}(D - \mathbf{A}x)^2 + V^a(x^a), \tag{2.2.5}$$

where

$$V^a(x^a) := \sum_{(ij)\leq a} V_{ij}(x_i - x_j), \tag{2.2.6}$$

Observe that $V^a$ commutes with $D_{z_a}$. Hence

$$\mathrm{H}_a = \frac{1}{2}D^2_{\mathrm{z}_a} + H^a, \; H_a = \frac{1}{2}D^2_{z_a} + H^a, \tag{2.2.7}$$

where

$$H^a := \frac{1}{2}D^2_{z^a} + \frac{1}{2}(D_y - Ay)^2 + V^a(x^a) = \sum_{l=1}^{\#a} H[C_l], \tag{2.2.8}$$

and

$$H[C_l] = \mathrm{H}[C_l] - \frac{1}{2}D^2_{\mathrm{z}_{\mathrm{cm}}[C_l]}, \; l = 1, \dots, \#a,$$

acting on $L^2(X[C_l])$, are the Hamiltonians of the clusters $C_l$ with the center of mass motion in the z direction separated.

We will often have to split up the "internal" Hamiltonian $H^a$ into parts corresponding to the neutral and charged clusters. For this purpose we introduce the decomposition

$$H^a = H^{a,\mathrm{n}} + H^{a,\mathrm{c}}, \tag{2.2.9}$$

where

$$H^{a,\mathrm{n}} := \sum_{C_l \text{ neutral}} H[C_l], \; H^{a,\mathrm{c}} := \sum_{C_l \text{ charged}} H[C_l]. \tag{2.2.10}$$

$H^{a,\mathrm{n}}$ and $H^{a,\mathrm{c}}$ act on $L^2(X^{a,\mathrm{n}})$ and $L^2(X^{a,\mathrm{c}})$ respectively.

**2.2.2. Channel reducing transformations.** As we saw in Subsection 2.2.1, the channel Hamiltonian $H_a$ may be written as

$$H_a = \frac{1}{2} D_{z_a}^2 + \sum_{i=1}^{\sharp a} H[C_i].$$

Each $H[C_i]$ can be understood simply as an $n_i$-particle magnetic Hamiltonian with the center of mass motion in the $z$ variables separated out, acting on $L^2(X[C_i])$. (We denote by $n_i$ the number of particles in the cluster $C_i$.)

All considerations of Sections 1.2–1.4 apply to the $H[C_i]$'s. In particular, for each $C_i$ there is a reducing transformation $U[C_i]$, satisfying (1.3.11) if $C_i$ is neutral and (1.3.18) if it is charged. A unitary transformation $U_a : L^2(X) \to L^2(X)$ of the form

$$U_a = \bigotimes_{i=1}^{\sharp a} U[C_i], \tag{2.2.11}$$

where $U[C_i]$ is a reducing transformation for the cluster Hamiltonian $H[C_i]$, will be called a *channel reducing transformation* for the cluster decomposition $a$.

As in Subsections 1.3.4 and 1.3.5, we will construct explicitly a channel reducing transformation which will be used in the sequel. Once more, we start with the corresponding classical symplectic transformation. We write

$$T^*X = T^*Z_a \oplus \bigoplus T^*(X[C_i]),$$

and on each $T^*(X[C_i])$ we define a symplectic transformation $\chi[C_i]$ as in (1.3.13–1.3.14) or (1.3.19–1.3.20), depending on whether the cluster $C_i$ is neutral or charged. We then put

$$\chi_a = \mathbf{1} \otimes \bigotimes_{i=1}^{\sharp a} \chi[C_i].$$

The symplectic transformation obtained in this way has the form

$$\chi_a : (x, \xi) \mapsto (x', \xi'), \tag{2.2.12}$$

where:

$$\begin{cases} x' = (x'_a, x^{a\prime}), \ x'_a = (\bar{y}_a^{\mathrm{c}\prime}, \hat{y}_a^{\mathrm{c}\prime}, y_a^{\mathrm{n}\prime}, z'_a), \\ \xi' = (\xi'_a, \xi^{a\prime}), \ \xi'_a = (\bar{\eta}_a^{\mathrm{c}\prime}, \hat{\eta}_a^{\mathrm{c}\prime}, \eta_a^{\mathrm{n}\prime}, \zeta'_a), \end{cases} \tag{2.2.13}$$

and

$$(2.2.14)\qquad \begin{cases} (\bar\eta_a^{c\prime}, \bar y_a^{c\prime}) = \bar M_a^{c}(\eta_a^{c} + (\mathbf{A}x)_a^{c}), \\ (\hat\eta_a^{c\prime}, \hat y_a^{c\prime}) = \hat M_a^{c}(\eta_a^{c} + (\mathbf{A}_a{}^a\, x^a)^{c} - (\mathbf{A}_{a\,a}\, x_a)^{c}), \\ (\eta_a^{n\prime}, y_a^{n\prime}) = (\eta_a^{n} + (\mathbf{A}x)_a^{n}, y_a^{n}), \\ (\zeta_a', z_a') = (\zeta_a, z_a), \\ (\xi^{a\prime}, x^{a\prime}) = (\xi^a - \mathbf{A}^a{}_a\, x_a, x^a). \end{cases}$$

We consider $\chi_a$ as a symplectic map

$$\chi_a : T^*X \mapsto T^*Z_a \times T^*Y_a^{n} \times T^*\bar Y_a^{c} \times T^*\hat Y_a^{c} \times T^*X^a,$$

where $\bar Y_a^{c} = \hat Y_a^{c} = \mathbb{R}^{\sharp_{c}a}$. Here $\bar M_a^{c} : Y_a^{c\prime} \to T^\star\mathbb{R}^{\sharp_{c}a}$ and $\hat M_a^{c} : Y_a^{c\prime} \to T^\star\mathbb{R}^{\sharp_{c}a}$ are linear isomorphisms whose exact form may be easily seen from (1.3.20).

Let

$$U_a : L^2(X) \mapsto L^2(Z_a) \otimes L^2(Y_a^{n}) \otimes L^2(\bar Y_a^{c}) \otimes L^2(\hat Y_a^{c}) \otimes L^2(X^a)$$

be a unitary transformation implementing $\chi_a$. Then

$$(2.2.15)\qquad U_a H_a U_a^\star =: \tilde H_a = \frac{1}{2} D_{z_a}^2 + \tilde H^a,$$

where

$$(2.2.16)\qquad \begin{aligned} \tilde H^a &= U_a H^a U_a^* \\ &= R_a^{w}(D_{\hat y_a^{c}}, \hat y_a^{c}, x^a) + \tfrac12 (D_{y_a^{n}} - 2(\mathbf{A}_a{}^a\, x^a)^{n})^2 \\ &\quad + \tfrac12 (D_{x^a} - \mathbf{A}^{a\,a}\, x^a)^2 + V^a(x^a), \end{aligned}$$

and $R_a(\hat\eta_a^{c}, \hat y_a^{c}, x^a)$ is a second order polynomial equal to

$$(2.2.17)\qquad R_a(\hat\eta_a^{c}, \hat y_a^{c}, x^a) = \frac12 \left( (\hat M_a^{c})^{-1}(\hat\eta_a^{c}, \hat y_a^{c}) - 2(\mathbf{A}_a{}^a\, x^a)^{c} \right)^2.$$

The transformed Hamiltonian $\tilde H^a$ may be written as a direct integral:

$$\tilde H^a = \int_{Y_a^{n\prime}}^{\oplus} \tilde H^a(\eta_a^{n}) d\eta_a^{n},$$

where

$$(2.2.18)\qquad \begin{aligned} \tilde H^a(\eta_a^{n}) = \ & R_a^{w}(D_{\hat y_a^{c}}, \hat y_a^{c}, x^a) + \tfrac12 (\eta_a^{n} - 2(\mathbf{A}_a{}^a\, x^a)^{n})^2 \\ & + \tfrac12 (D_{x^a} - \mathbf{A}^{a\,a}\, x^a)^2 + V^a(x^a) \end{aligned}$$

acts on $L^2(\hat Y_a^c \times X^a)$. As in Subsections 1.3.4, 1.3.5, $\tilde H^a(\eta_a^{n})$ is self-adjoint on the domain of $R_a^{w}(D_{\hat y_a^{c}}, \hat y_a^{c}, x^a) + \frac12(\eta_a^{n} - 2(\mathbf{A}_a{}^a\, x^a)^{n})^2 + \frac12(D_{x^a} - \mathbf{A}^{a\,a}\, x^a)^2$.

REMARK 2.2.1. The transformations $U_{cm}$ constructed in Section 1.3 are examples of channel reducing transformations for the cluster decomposition $a = a_{max}$. If $a_{max}$ is charged, the spaces $\bar Y_{cm}, \hat Y_{cm}$ are identical to the $Y_{cm}^{(1)}$ and $Y_{cm}^{(2)}$ of Subsections 1.3.5.

## 2.3. Partitions of unity

Fix a cluster decomposition $c \in \mathcal{A}$. An *$N$-body partition of unity on $X^c$* is, roughly speaking, a partition

$$1_{X^c} = \sum_{a \leq c} (j_a^c)^2(x^c),$$

such that each $j_a^c$ is a $C^\infty(X^c)$ function supported in the region where:

- for each cluster $C_i$ of $a$, all particles in $C_i$ are close together;
- particles belonging to different clusters of $a$ are far apart.

The details of the construction and additional requirements imposed on $j_a^c$ may vary, and so may the precise mathematical meaning of the terms "close together" and "far apart". The partition we construct below in Proposition 2.3.1 is "cylindrical" in the sense that both the upper bounds on the size of clusters of $a$ and the lower bounds on the distance between them are fixed constants. The partition in Proposition 2.3.4 is "conical", i.e., the bounds are proportional to $|x^c|$.

### 2.3.1. Cylindrical partition of unity.

PROPOSITION 2.3.1. *Let $c \in \mathcal{A}$. There are positive constants $\{\delta_a, \epsilon_a\}_{a \leq c}$, which can be taken smaller than any $\epsilon > 0$, and a partition of unity $1 = \sum_{a \leq c} (q_a^c)^2(x^c)$ on $X^c$, such that:*

*(i) $q_a^c \in S^0(X^c)$, $0 \leq q_a^c \leq 1$ for all $a \leq c$;*

*(ii) $q_c^c \in C_0^\infty(\{|x^c| \leq \delta_c\})$;*

*(iii)* $\operatorname{supp} q_a^c \subset \{x^c \in X^c : |x^a| \leq \delta_a,\ |x^b| \geq \epsilon_a$ *if $b \leq c, b \not\leq a\}$ for all $a < c$;*

*(iv) if $a < c$, $b < c$, $a \not\leq b$ and $b \not\leq a$, then $q_a^c q_b^c = 0$.*

*The functions $q_a^{a_{\max}}$ are denoted simply by $q_a$. We may also extend $q_a^c$ to functions on $X$ constant in the $x_c$ variables, i.e., set $q_a^c(x) = q_a^c(x^c)$ for $x \in X$.*

The condition *(iii)* implies that on $\operatorname{supp} q_a^c$ one has

$$|x|_a^c \geq \epsilon_a' > 0,$$

where

$$|x|_a^c := \min_{(ij) \leq c, (ij) \not\leq a} |x_i - x_j| \tag{2.3.1}$$

is the shortest distance between particles belonging to the same clusters of $c$ and different clusters of $a$. Indeed, this follows immediately from the fact that for any pair of indices $i \neq j$ we have

$$|x^{(ij)}| = c_{ij} |x_i - x_j|$$

(the constant $c_{ij}$ can be computed explicitly: $c_{ij}^2 = (m_i^2 + m_j^2)/(m_i + m_j)$).

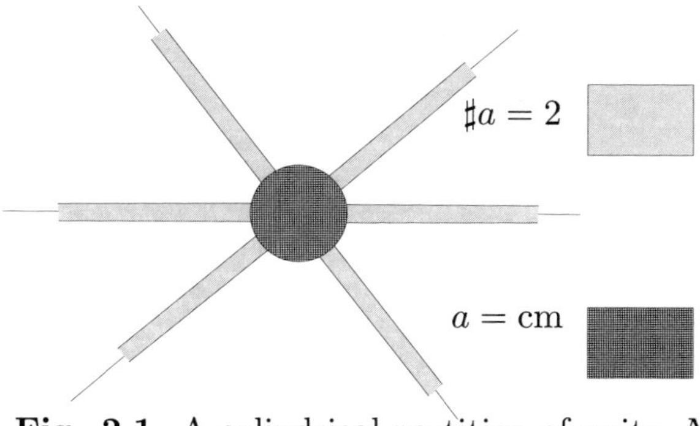


**Fig. 2.1.** A cylindrical partition of unity, $N = 3$

The proof of Proposition 2.3.1 is based on the following two lemmas. The cluster decomposition $c \in \mathcal{A}$ is fixed and remains the same throughout the proof.

LEMMA 2.3.2. *Let*

$$F_a = \{x^c \in X^c : \ |x^a| \leq \delta_a\}, \ a \leq c.$$

*We may choose the constants* $\delta_a > 0$, $a \leq c$, *so that*

$$F_a \cap F_b \subset \frac{1}{3} F_{a \vee b} \textit{ if } a, b \leq c, a \not\leq b, b \not\leq a,$$

*where* $\alpha F_a$ *denotes the set* $\{x^c \in X^c : \ |x^a| \leq \alpha \delta_a\}$

PROOF. Since $X^{a \vee b} = \text{Span}(X^a, X^b)$, there is a constant $c_0$ (depending only on the masses $m_i$ of the particles) such that

$$|x^{a \vee b}| \leq c_0(|x^a| + |x^b|) \tag{2.3.2}$$

for all $a, b \in \mathcal{A}_*$. We may assume that $c_0 > 1$. For $a \leq c$ and $0 < \epsilon < 1$, we define

$$\delta_a = \epsilon(6c_0)^{-\sharp a}. \tag{2.3.3}$$

Note that $\delta_a \leq \epsilon < 1$ for all $a \leq c$.

Suppose that $a \not\leq b$ and $b \not\leq a$; then $a \vee b \neq a$ and $a \vee b \neq b$, so that

$$\sharp(a \vee b) \leq \min(\sharp a, \sharp b) - 1. \tag{2.3.4}$$

We have

$$\begin{aligned} F_a \cap F_b &= \{|x^a| \leq \delta_a, \ |x^b| \leq \delta_b\} \\ &\subset \{|x^{a \vee b}| \leq c_0(\delta_a + \delta_b)\}, \end{aligned}$$

and, by (2.3.4),

$$\begin{aligned} c_0(\delta_a+\delta_b) &\le 2\epsilon c_0\cdot(6c_0)^{-\min(\sharp a,\sharp b)} \\ &\le \tfrac{\epsilon}{3}\cdot(4c_0)^{-\sharp(a\vee b)}=\tfrac{1}{3}\delta_{a\vee b}. \end{aligned}$$

Hence

$$F_a\cap F_b\subset\{|x^{a\vee b}|\le\frac{1}{3}\delta_{a\vee b}\}=\frac{1}{3}F_{a\vee b}.$$

□

LEMMA 2.3.3. *Let*

$$T_a=\frac{1}{2}F_a\setminus\bigcup_{a<b\le c}\frac{1}{2}F_b. \tag{2.3.5}$$

*Then:*

*(i) the sets* $\{T_a:\ a\le c\}$ *are disjoint and* $\bigcup_{a\le c}T_a=X^c$*;*

*(ii)* $T_a\subset\{|x^a|\le\frac{1}{2}\delta_a,\ |x_b|\ge\frac{1}{2}\delta_b$ *if* $b\not\le a\}$.

PROOF. *(i)* Let $x\in X^c$. Since $x^{a_{\min}}=0$, we have $x\in\frac{1}{2}F_{a_{\min}}$. Hence the set of cluster decompositions $a\le c$ such that $x\in\frac{1}{2}F_c$ is nonempty. Let $a$ be a cluster decomposition with the smallest possible value of $\sharp a$ such that $x\in\frac{1}{2}F_a$. Then $x\notin\frac{1}{2}F_b$ for any $b>a$, since $b>a$ implies that $\sharp b<\sharp a$. Hence $x\in T_a$.

Suppose now that $x\in T_a\cap T_b$, $a,b<c$. If $a<b$ or $b<a$, this contradicts (2.3.5). If $a\not\le b$ and $b\not\le a$, then $a\vee b>a$ and, by Lemma 2.3.2,

$$x\in T_a\cap T_b\subset F_a\cap F_b\subset\frac{1}{2}F_{a\vee b},$$

which again contradicts (2.3.5). Hence $a=b$.

*(ii)* If $x\in T_a$, by definition $|x^a|\le\frac{1}{2}\delta_a$ and $|x^b|\ge\frac{1}{2}\delta_b$ if $a<b\le c$. Suppose that $a,b\le c$, $a\not\le b$, $b\not\le a$, $x\in T_a$ and $|x^b|\le\frac{1}{2}\delta_b$. Then $a\vee b>a$ and

$$x\in F_a\cap F_b\subset\frac{1}{2}F_{a\vee b}$$

by Lemma 2.3.2, which contradicts (2.3.5). □

PROOF OF PROPOSITION 2.3.1. Let $\chi_a$ be the characteristic function of $T_a$, and let $\omega$ be a function in $C_0^\infty(X^c)$ such that:

$$\operatorname{supp}\omega\subset\{|x^c|<\epsilon^*\},\quad\int_{X^c}\omega(x^c)dx^c=1, \tag{2.3.6}$$

where $\epsilon^*=\frac{1}{100}\min_{a\le c}\delta_a$. Since $\sum_{a\le c}\chi_a\equiv 1$ by Lemma 2.3.3*(i)*, we have $\sum_{a\le c}\chi_a*\omega\equiv 1$. We define

$$q_a^c=\frac{\chi_a*\omega}{(\sum(\chi_a*\omega)^2)^{\frac{1}{2}}}, \tag{2.3.7}$$

so that $\sum(q_a^c)^2=1$.

Then *(i)* follows immediately from the definition of $q_a^c$. To prove *(ii)* and *(iii)*, we note that by (2.3.7) and (2.3.6), and by Lemma 2.3.3*(ii)*,

$$\operatorname{supp} q_c^c \subset \{|x^c| \le \tfrac{1}{2}\delta_c + \epsilon^*\},$$

$$\operatorname{supp} q_a^c \subset \{|x^a| \le \tfrac{1}{2}\delta_a + \epsilon^*,\ |x^b| \ge \tfrac{1}{2}\delta_b - \epsilon^*\}.$$

By the choice of $\epsilon^*$ in (2.3.6), we have for all $a \le c$:

$$\frac{1}{2}\delta_a + \epsilon^* < \delta_a,\ \frac{1}{2}\delta_a - \epsilon^* > \frac{1}{3}\delta_a.$$

Hence *(ii)* and *(iii)* are satisfied with $\delta_a$ defined by (2.3.3) and with

$$\epsilon_a = \min_{b \le c, b \not\le a} \frac{1}{3}\delta_b.$$

Finally, suppose that $a, b \le c$, $a \not\le b$ and $b \not\le a$. Since $\operatorname{supp} q_a^c \subset F_a$ by *(iii)*, we have

$$\operatorname{supp} q_a^c q_b^c \subset F_a \cap F_b \subset \frac{1}{3}F_{a\vee b} = \{|x^{a\vee b}| \le \frac{1}{3}\delta_{a\vee b}\},$$

by Lemma 2.3.2. But, since $a \vee b > a$, we have

$$|x_{a\vee b}| \ge \frac{1}{2}\delta_{a\vee b} - \epsilon^* > \frac{1}{3}\delta_{a\vee b}$$

on $\operatorname{supp} q_a^c$. Hence $q_a^c q_b^c \equiv 0$. □

### 2.3.2. Conical partition of unity.

PROPOSITION 2.3.4. *Let $c \in \mathcal{A}$. There are positive constants $\{\delta_a, \epsilon_a\}_{a \le c}$, which may be taken smaller than any $\epsilon > 0$, and a partition of unity $1 = \sum_{a \le c}(j_a^c)^2(x^c)$ on $X^c$, such that:*

*(i) $j_a^c \in C^\infty(X^c)$ and $0 \le j_a^c \le 1$ for all $a \le c$;*

*(ii) $|x^c| \le \delta_c$ on $\operatorname{supp} j_c^c$;*

*(iii) $\operatorname{supp} j_a^c \subset \{x^c \in X^c : |x^a| \le \delta_a |x^c|,\ |x^c| \ge \epsilon_c,\ |x^b| \ge \epsilon_a |x^c|$ if $b \le c, b \not\le a\}$ for all $a < c$;*

*(iv) if $a < c$, $b < c$, $a \not\le b$ and $b \not\le a$, then $j_a^c j_b^c = 0$;*

*(v) for $a < c$, $j_a^c$ is homogeneous degree zero for $|x^c| \ge R$, provided that $R > 0$ is large enough. In particular,*

$$|\partial^\alpha j_a^c| \le C_\alpha \langle x^c \rangle^{-\alpha},\ |\alpha| \ge 0.$$

*We will write $j_a^{a_{\max}} = j_a$. The functions $j_a^c$ may be extended to functions on $X$ equal to $j_a^c(x) = j_a^c(x^c)$ for $x \in X$.*

Note that *(iii)* implies that on $\operatorname{supp} j_a^c$ one has

$$|x|_a^c \ge \epsilon_a' |x^c|,$$

where $|x|_a^c$ is defined in (2.3.1).

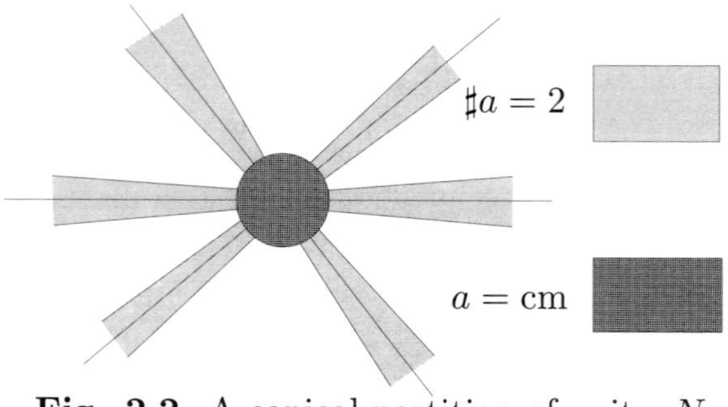


**Fig. 2.2.** A conical partition of unity, $N = 3$.

The proof of Proposition 2.3.4 will use Lemma 2.3.2 from the previous subsection and Lemma 2.3.5 below.

LEMMA 2.3.5. *Let $F_a$, $a \le c$, be the sets defined in Lemma 2.3.2, and let*

$$G_a = Int(F_a \setminus \bigcup_{a<b\le c} \frac{1}{2} F_b). \tag{2.3.8}$$

*Then the open sets $\{G_a : \ a \le c\}$ have the following properties:*
*(i)* $\bigcup_{a\le c} G_a = X^c$;
*(ii)* $G_c = \{|x^c| < \delta_c\}$;
*(iii)* $G_a \subset \{|x^a| \le \delta_a,\ |x^b| \ge \frac{1}{2}\delta_b$ *for all* $b \le c,\ b \not\le a\}$ *if* $a < c$;
*(iv)* $G_a \cap G_b = \emptyset$ *if* $a < c,\ b < c,\ a \not\le b$ *and* $b \not\le a$.

PROOF. *(i)* Since $G_a \subset T_a$, this follows from Lemma 2.3.3*(i)*.

*(ii)* is obvious.

*(iii)* Let $a < c$ and $x \in G_a$. Suppose that

$$|x^b| \le \frac{1}{2}\delta_b,$$

i.e., $x \in \frac{1}{2}F_b$, for some $b \le c$, $b \not\le a$. If $a < b$, this is not possible by (2.3.8). If $b \not\le a$ and $a \not\le b$, we have $a \vee b > a$ and

$$x \in F_a \cap F_b \subset \frac{1}{2}F_{a\vee b},$$

which again contradicts (2.3.8).

*(iv)* Suppose that $x \in G_a \cap G_b$, $a, b < c$, $a \not\leq b$, $b \not\leq a$. Then $a \vee b > a$ and

$$x \in G_a \cap G_b \subset F_a \cap F_b \subset \frac{1}{2} F_{a \vee b},$$

which again is impossible by (2.3.8). □

PROOF OF PROPOSITION 2.3.4. Let $S^c = \{x^c \in X^c : \ |x^c| = 1\}$ be the unit sphere in $X^c$. Since $\{G_a \cap S^c\}_{a<c}$ is an open covering of $S^c$, there is a partition of unity $\{\phi_a\}_{a<c}$ such that $\sum \phi_a^2 \equiv 1$, $\operatorname{supp} \phi_a \subset G_a \cap S^c$, $\phi_a \in C_0^\infty(S^c)$, $0 \leq \phi_a \leq 1$. Let $j_c^c$ be a function such that $j_c^c \in C_0^\infty(G^c)$, $j_c^c \equiv 1$ for $|x^c| \leq \frac{1}{2}\delta_c$, $0 \leq j_c^c \leq 1$, and let

$$j_a^c(x^c) = \phi_a(\frac{x^c}{|x^c|}) \cdot (1 - j_c^c(x^c)).$$

Then *(i)*, *(ii)*, and *(v)* follow immediately from the definition of $j_a$, and *(iii)*, *(iv)* are consequences of Lemma 2.3.5*(iii)*, *(iv)*. □

**2.3.3. A few technical results.** We will often rely on the following easy consequences of the support properties of $q_a$ and $j_a$.

LEMMA 2.3.6. *Assume that (V1) holds, and let $q_a$ and $j_a$ be as in Propositions 2.3.1 and 2.3.4 respectively. Let $-c \notin \sigma(H)$ (e.g., $\operatorname{Im} c \neq 0$).*

*(i) Assume also (V2), then $\|(H+c)^{-1} q_a(\frac{x}{R}) I_a\| = o(1)$ as $R \to \infty$.*

*(ii) Assume also (V3), then $\|\langle x^{\mathrm{cm}} \rangle^\mu j_a I_a (H+c)^{-1}\| < \infty$.*

The next lemma, which is easily proved by direct computation, will be needed in Chapter 3.

LEMMA 2.3.7. *Let $\{q_a\}_{a \in \mathcal{A}}$ be $C^\infty$ functions such that $\sum q_a^2 \equiv 1$, and let:*

$$A = \frac{1}{2}(\langle z, D_z \rangle + \langle D_z, z \rangle).$$

*Then $A = \sum q_a A q_a$.*

LEMMA 2.3.8. *Assume that (V1), (V2) hold, and let $f \in C_0^\infty(\mathbb{R})$.*

*(i) Let $q_a$, $a \in \mathcal{A}$, be a cylindrical partition of unity as in Proposition 2.3.1. Then:*

$$q_a(\frac{x}{R})(f(H) - f(H_a))(H+c) = o(1) \ \text{as } R \to \infty.$$

*(ii) Let $q \in S^0(X^{\mathrm{cm}})$, then:*

$$[q(\frac{x}{R}), f(H)](H+c) = O(R^{-1}) \ \text{as } R \to \infty.$$

Heuristically, the reason why *(i)* holds is that on the support of $q_a(\cdot/R)$ the distances $x_i - x_j$ between particles belonging to different clusters of $a$ are large, so that the intercluster interactions $I_a$ should be small and $H$ and $H_a$ are almost the same. Part *(ii)* is an additional technical result which will be needed later.

PROOF. By (1.2.21), we have:

$$\begin{aligned} & q_a(\tfrac{x}{R})(f(H) - f(H_a))(H+c) \\ = \; & q_a(\tfrac{x}{R})\tfrac{i}{2\pi}\int_{\mathbb{C}} \partial_{\bar z}\tilde f(z)((z-H)^{-1} - (z-H_a)^{-1})(H+c)dz\wedge d\bar z \\ = \; & \tfrac{i}{2\pi}\int_{\mathbb{C}} \partial_{\bar z}\tilde f(z) q_a(\tfrac{x}{R})(z-H)^{-1}I_a(z-H_a)^{-1}(H+c)dz\wedge d\bar z. \end{aligned}$$

We write

$$q_a(\frac{x}{R})(z-H)^{-1}I_a(z-H_a)^{-1}(H+c) = I_1 + I_2,$$

where:

$$\begin{aligned} I_1 &= (z-H)^{-1}q_a(\tfrac{x}{R})I_a(z-H_a)^{-1}(H+c), \\ I_2 &= [q_a(\tfrac{x}{R}), (z-H)^{-1}]I_a(z-H_a)^{-1}(H+c). \end{aligned}$$

By Lemma 2.3.6*(i)* and Proposition 1.2.1,

$$\begin{aligned} \|I_1\| \le \; & \|(z-H)^{-1}(H+c)\| \cdot \|(H+c)^{-1}q_a(\tfrac{x}{R})I_a\| \cdot \|(z-H_a)^{-1}(H+c)\| \\ = \; & O(|\mathrm{Im}\, z|^{-2}) \cdot o(R^0). \end{aligned}$$

Let $q \in S^0(X^{\mathrm{cm}})$. Then:

$$\begin{aligned} & [q(\tfrac{x}{R}), (z-H)^{-1}](H+c) \\ = \; & (z-H)^{-1}[H, q(\tfrac{x}{R})](z-H)^{-1}(H+c) \\ = \; & (z-H)^{-1}\Big((D-\mathbf{A}x)[D-\mathbf{A}x, q(\tfrac{x}{R})] + \mathrm{h.c.}\Big)(z-H)^{-1}(H+c). \\ = \; & (z-H)^{-1}\Big(2(D-\mathbf{A}x)[D-\mathbf{A}x, q(\tfrac{x}{R})] \\ & +[[D-\mathbf{A}x, q(\tfrac{x}{R})], D-\mathbf{A}x]\Big)(z-H)^{-1}(H+c). \end{aligned}$$

By Proposition 1.2.1,

$$\|(z-H)^{-1}(D-\mathbf{A}x)\| = O(|\mathrm{Im}\, z|^{-1}),$$

and by Proposition 2.3.1*(i)*,

$$\begin{aligned} & \|[D-\mathbf{A}x, q_a(\tfrac{x}{R})]\| = \|\tfrac{1}{R}(\nabla q)(\tfrac{x}{R})\| = O(R^{-1}), \\ & \|[[D-\mathbf{A}x, q(\tfrac{x}{R})], D-\mathbf{A}x]\| = O(R^{-1}), \end{aligned} \tag{2.3.9}$$

Hence

$$[q(\frac{x}{R}), (z-H)^{-1}](H+c) = O(R^{-1})|\mathrm{Im}\, z|^{-2}. \tag{2.3.10}$$

Letting $q = q_a$ and using again *(V1)* and Proposition 1.2.1, we obtain that

$$\|I_2\| = O(R^{-1}|\operatorname{Im} z|^{-3}).$$

Combining (2.3.10) with the above estimate, and using (1.2.21) and Definition 1.2.8*(ii)*, we obtain part *(i)* of the lemma. Part *(ii)* follows from (2.3.10), (1.2.21), and Definition 1.2.8*(ii)*. □

We also need an analogue of Lemma 2.3.8 with a conical partition of unity and functions in $S^\delta_{\mathrm{cl}}(X^{\mathrm{cm}})$.

LEMMA 2.3.9. *Assume (V1), (V3), and let $f \in C_0^\infty(\mathbb{R})$. Then:*

*(i) Let $j_a$ be a conical partition of unity as in Proposition 2.3.4. Then:*

$$j_a(f(H) - f(H_a))(H + c)\langle x^{\mathrm{cm}}\rangle^\mu, \ (H + c)j_a(f(H) - f(H_a))\langle x^{\mathrm{cm}}\rangle^\mu$$

*are bounded.*

*(ii) Let $j \in S^\delta_{\mathrm{cl}}(X^{\mathrm{cm}})$, $\delta \le 0$, and $0 \le s \le 1 - \delta$. Then:*

$$\langle x^{\mathrm{cm}}\rangle^s [j, f(H)](H + c)\langle x^{\mathrm{cm}}\rangle^{1-\delta-s} \text{ is bounded.}$$

*(iii) Let $q_1, q_2 \in S^0_{\mathrm{cl}}(X^{\mathrm{cm}})$, $\operatorname{supp} q_1 \cap \operatorname{supp} q_2 = \emptyset$. Then:*

$$q_1 f(H) q_2 \langle x^{\mathrm{cm}}\rangle^2 (H + c) \text{ is bounded.}$$

PROOF. As in the proof of Lemma 2.3.8, it suffices to prove *(i)–(iii)* with $f(H)$ replaced by $(z - H)^{-1}$ for $|\operatorname{Im} z| \le M$. We will use that $(z - H)^{-1}$ preserves the domain of $\langle x^{\mathrm{cm}}\rangle^s$ for $0 \le s \le 2$; this follows from Lemma 1.2.3 by interpolation.

We begin with *(ii)*. Let $j \in S^\delta_{\mathrm{cl}}(X^{\mathrm{cm}})$. As in the proof of (2.3.10), we obtain:

$$\begin{aligned}
&\|\langle x^{\mathrm{cm}}\rangle^{1-\delta}(H + c)[(z - H)^{-1}, j]\| \\
&= \|\langle x^{\mathrm{cm}}\rangle^{1-\delta}(H + c)(z - H)^{-1}[H, j](z - H)^{-1}\| \\
&\le C_1|\operatorname{Im} z|^{-2}\|\langle x^{\mathrm{cm}}\rangle^{1-\delta}[H, j](z - H)^{-1}\| \\
&\le C_2|\operatorname{Im} z|^{-3},
\end{aligned} \tag{2.3.11}$$

where instead of (2.3.9) we used that $\nabla j = O(\langle x^{\mathrm{cm}}\rangle^{-1+\delta})$. Similarly,

$$\langle x^{\mathrm{cm}}\rangle^{1-\delta}[(z - H)^{-1}, j](H + c). \tag{2.3.12}$$

This proves *(ii)* for $s = 0$ and for $s = 1 - \delta$; for $0 < s < 1 - \delta$ the result follows by interpolation.

To prove the first statement in *(i)*, we write

$$\begin{aligned}
&j_a((z - H)^{-1} - (z - H_a)^{-1})(H + c)\langle x^{\mathrm{cm}}\rangle^\mu \\
&= -j_a(z - H_a)^{-1} I_a (z - H)^{-1}(H + c)\langle x^{\mathrm{cm}}\rangle^\mu \\
&= I_1 + I_2,
\end{aligned}$$

where

$$I_1 = -(z - H_a)^{-1} j_a I_a (z - H)^{-1}(H + c)\langle x^{\rm cm}\rangle^\mu,$$

$$I_2 = [(z - H_a)^{-1}, j_a] I_a (z - H)^{-1}(H + c)\langle x^{\rm cm}\rangle^\mu.$$

From Lemma 1.2.3 and Lemma 2.3.6*(ii)*, we have

$$\begin{aligned} \|I_1^*\| &\le \|\langle x^{\rm cm}\rangle^\mu \Big((c+z)(z-H)^{-1} - \mathbf{1}\Big) j_a I_a (z - H_a)^{-1}\| \\ &\le C_3 |\operatorname{Im} z|^{-2} \|\langle x^{\rm cm}\rangle^\mu j_a I_a (z - H_a)^{-1}\| \qquad (2.3.13) \\ &\le C_4 |\operatorname{Im} z|^{-3}, \end{aligned}$$

hence $I_1 = O(|\operatorname{Im} z|^{-3})$. Similarly, we have

$$\|I_2^*\| \le C_5 |\operatorname{Im} z|^{-1} \|\langle x^{\rm cm}\rangle I_a [(z - H_a)^{-1}, j_a]\|. \qquad (2.3.14)$$

By Proposition 2.3.4*(v)*, (2.3.11) holds for $j_a$ with $\delta = 0$ and with $H$ replaced by $H_a$. Combining this with (2.3.14) and Proposition 1.2.1, we obtain that

$$\|I_3^*\| \le C_6 |\operatorname{Im} z|^{-4},$$

which together with (2.3.13) implies the first part of *(i)*. The proof of the second part is similar. We write

$$\begin{aligned} & (H + c) j_a ((z - H)^{-1} - (z - H_a)^{-1})\langle x^{\rm cm}\rangle^\mu \\ = \; & -(H + c) j_a (z - H_a)^{-1} I_a (z - H)^{-1} \langle x^{\rm cm}\rangle^\mu \\ = \; & I_1' + I_2', \end{aligned}$$

where:

$$I_1' = -(H + c)(z - H_a)^{-1} j_a I_a (z - H)^{-1}\langle x^{\rm cm}\rangle^\mu,$$

$$I_2' = (H + c)[(z - H_a)^{-1}, j_a] I_a (z - H)^{-1}\langle x^{\rm cm}\rangle^\mu.$$

Using *(V3)* and (2.3.12) with $j = \langle x^{\rm cm}\rangle^{-\mu}$, we obtain that

$$\begin{aligned} \|I_1'\| \le \; & \|(H + c)(z - H_a)^{-1}\| \|j_a I_a \langle x^{\rm cm}\rangle^\mu (H + c)^{-1}\| \\ & \times \|(H + c)\langle x^{\rm cm}\rangle^{-\mu}(z - H)^{-1}\langle x^{\rm cm}\rangle^\mu\| \\ = \; & O(|\operatorname{Im} z|^{-4}), \end{aligned}$$

Similarly, applying (2.3.11) to $H_a$ and $j_a$ we obtain

$$\|(H + c)[(z - H_a)^{-1}, j_a]\langle x^{\rm cm}\rangle\| = O(|\operatorname{Im} z||^{-3}),$$

hence

$$\|I_2'\| \le C |\operatorname{Im} z|^{-3} \|\langle x^{\rm cm}\rangle^{-1} I_a (z - H)^{-1}\langle x^{\rm cm}\rangle^\mu\|.$$

By (2.3.11) with $j = \langle x^{\rm cm}\rangle^{-1}$, we have

$$\|(H + c)\langle x^{\rm cm}\rangle^{-1}(z - H)^{-1}\langle x^{\rm cm}\rangle^\mu\| = O(|\operatorname{Im} z|^{-3});$$

using also *(V1)* we find that $\|I_2'\| \le C|\operatorname{Im} z|^{-6}\|I_a (H + c)^{-1}\| = O(|\operatorname{Im} z|^{-6})$. This completes the proof of *(i)*.

Finally let $q_1, q_2$ be as in *(iii)*. Since $q_1 q_2 = 0$ and $[q_1, H]q_2 = 0$, we have the identity:

$$\begin{aligned} & q_1(z-H)^{-1}q_2\langle x^{\mathrm{cm}}\rangle^2(H+c) \\ = \ & (z-H)^{-1}[q_1,H](z-H)^{-1}q_2\langle x^{\mathrm{cm}}\rangle^2(H+c) \\ = \ & (z-H)^{-1}[q_1,H](z-H)^{-1}[H,q_2](z-H)^{-1}\langle x^{\mathrm{cm}}\rangle^2(H+c). \end{aligned}$$

Now *(iii)* follows from essentially the same argument as before: we use that $\nabla q_i = O(\langle x^{\mathrm{cm}}\rangle^{-1})$, $i = 1, 2$, and that by Lemma 1.2.3 we may commute $\langle x^{\mathrm{cm}}\rangle$ and $\langle x^{\mathrm{cm}}\rangle^2$ through $(z-H)^{-1}$. □

CHAPTER 3

# The Mourre Theory

This chapter is devoted to the Mourre theory for $N$-particle magnetic Hamiltonians.

The starting point for the Mourre theory is the *Mourre estimate*: given the Hamiltonian $H$, one finds a (usually self-adjoint) operator $A$, called a *conjugate operator*, such that the commutator $[H, iA]$ restricted to fixed energy intervals is positive modulo compact errors. Under additional technical assumptions (which will be discussed in detail later on), the Mourre estimate may be used to establish basic spectral properties of $H$, such as the absence of singular continuous spectrum, and to prove estimates known as *local decay* and *minimal velocity estimates*, which are essential in scattering theory.

The Mourre theory originated in [**M1**], [**M2**], where it was also applied to 2- and 3-body quantum Hamiltonians. Subsequently, several abstract frameworks for the Mourre theory were constructed, see e.g., [**PSS**], [**CFKS**] (with the corrections pointed out in [**GG**]), [**BG**], and [**ABG**]; the latter framework will be especially useful for our purposes.

Nowhere has the power of the Mourre theory been more manifest than in the $N$-body theory. The Mourre estimate was first extended to the $N$-particle case in [**PSS**]; simpler proofs were given later in [**FH1**], [**Sk1**]. It constitutes an essential part of all presently known proofs of $N$-body asymptotic completeness (see [**SS1**], [**Gr**], [**De1**], [**SS5**]), including $N$-body systems in external fields ([**AT**], [**Sk2**], [**HMS1**], [**HMS2**], [**Sk4**], and the present work) and hard-core interactions ([**BGS**], [**Gri**], [**If**]). It has also been used to prove resolvent estimates and the limiting absorption principle ([**JMP**], [**BP**], [**GIS**]). More recently, it has been applied to the problems of spectral theory and asymptotic completeness in certain models of field theory ([**HüSp**],[**Sk4**], [**BFSS**], [**DG2**], [**DG3**]); this required extending the Mourre theory to cases when the conjugate operator $A$ is not self-adjoint ([**HüSp**], [**Sk4**]) and the commutator $[H, iA]$ is not relatively bounded ([**Sk4**]). Another class of "$N$-body type" systems to which the Mourre theory has been succesfully applied is dispersive Hamiltonians, i.e., Hamiltonians in which the kinetic energy term $-\Delta$ is replaced by a more general pseudodifferential operator $\omega(D)$ ([**De2**], [**G1**], [**GN1**]). However, asymptotic completeness for such systems has been proved only for the 2-body case and for $N$-body systems at energy levels below the 3-cluster threshold ([**G1**]). One of the main results of this book is asymptotic completeness for a class of physically meaningful, and mathematically quite complicated, 3-body dispersive Hamiltonians.

Other applications of the Mourre estimate include the theory of resonances ([**GS**], [**SW1**], [**MeS**]), wave equations on manifolds ([**FHi**], [**FHP**], [**DHS**]), and certain nonlinear problems ([**Sig2**]). The literature related to the Mourre theory is rather extensive, and this short review should not be construed as complete.

In this chapter we will prove the Mourre estimate for the following classes of $N$-body magnetic Hamiltonians (cf. Subsection 1.2.3):

- *(Strongly charged case)* $N$ arbitrary, *(Q2)* holds (i.e., all proper subsystems are charged);
- *(Dispersive case)* $N = 3$, *(Q1)* holds, and the implicit assumption *(S)* is satisfied if $d = 3$. (We expect *(S)* to hold for generic 3-body systems.)

The latter case turns out to be much more difficult, even though we are only considering 3-particle systems; the general $N$-body case (without additional assumptions on the charges of subsystems) remains an open problem.

The reasons why neutral subsystems are so hard to handle become clear when one reviews the typical strategies used to prove the Mourre estimate and measures them against the problem at hand. Our goal is to find a conjugate operator $A$ such that $[H, \mathrm{i}A] > c > 0$, provided that the energy $H$ is appropriately localized. The usual choice of $A$ is a self-adjoint modification of $x \cdot v$, where $x$ and $v = [H, \mathrm{i}x]$ are the position and velocity of the system. Then

$$[H, \mathrm{i}x \cdot v] = [H, \mathrm{i}x] \cdot v + x \cdot [H, \mathrm{i}v] = v^2 + x \cdot [H, \mathrm{i}v] =: v^2 + K,$$

and the error term $K$ can be made small, provided that $v$ comes close enough to commuting with $H$. Consider for instance one-body Hamiltonians $H = \frac{1}{2}D^2 + V$, where $V$ and $x \cdot \nabla V$ are $D^2$-compact; then $v = D$, $A = \frac{1}{2}(\langle x, D\rangle + \langle D, x\rangle)$, and

$$[H, \mathrm{i}A] = D^2 + K,$$

where $K = -x \cdot \nabla V$ is relatively compact. Applying localization in energy, one may remove the error $K$ and prove that $D^2$ is not only nonnegative but in fact bounded from below by a positive constant. $N$-particle potentials present additional difficulties, since $x \cdot \nabla V$ does not vanish at infinity and cannot be dismissed as a small error. To circumvent this obstacle one may use cluster expansions ([**PSS**]), induction in subsystems ([**FH1**]), or replace the $x$ in $x \cdot v$ by a vector field constructed in [**Gr**] ([**Sk1**]).

Magnetic Hamiltonians do not lend themselves easily to this approach. The velocity of a system of charged particles in a magnetic field is nowhere close to constant; we must therefore find another observable which has similar physical meaning but is "better behaved". Moreover, many standard techniques (such as the center of mass separation) are not allowed and have to be replaced by other arguments.

For 3−dimensional $N$-particle systems whose all proper subsystems are charged, the situation is actually similar to the standard $N$-body case. One uses the reduction procedure of Sections 1.3, 2.2 instead of the usual $N$-body geometric arguments, and essentially proves that the continuous spectrum of the reduced Hamiltonian is due to the relative motion of the clusters in the $z$ direction. Thus the proof of the Mourre estimate is similar to that in [**FH1**], with the conjugate operator $A = \frac{1}{2}(\langle z, D_z \rangle + \langle D_z, z \rangle)$. The 2−dimensional case is even simpler: we will prove in Section 6.5 that $\mathcal{H}_{\text{scatt}} = \{0\}$, hence no Mourre estimate is either possible or needed.

This reasoning, though, does not apply to systems containing neutral clusters. We saw in Section 1.4 that a neutral cluster may travel transversally to the magnetic field; hence the "velocity operator" and the conjugate operator must contain a part acting in the transversal directions. The correct expression for the velocity of the neutral clusters turns out to be that given in Proposition 1.4.6. That, however, leads to several difficult issues which must be resolved. Different conjugate operators will be needed in different regions of the configuration space, and one must be careful to ensure their compatibility at the boundaries of these regions. Furthermore, it becomes highly non-trivial to prove the self-adjointness of the "composite" conjugate operator and to verify the technical conditions needed later on in the proof of asymptotic completeness. We are able to do this only for 3-body systems; even so, this part of the book is unfortunately very technical and requires rather long and tedious calculations.

We begin by discussing some of the well known definitions and arguments of the Mourre theory, due mostly to Mourre [**M1**], [**M2**] and Froese-Herbst [**FH1**], in Section 3.1. (The presentation follows [**GŁ3**].) In Section 3.2 we review the abstract framework of [**ABG**] which will be used in the sequel. Section 3.3 is devoted to the proof of the Mourre estimate in the "charged" case. The result was first obtained in [**GŁ1**]; the proof combines the arguments of [**FH1**] with the reduction procedure of [**GŁ1**] (explained here in Chapters 1 and 2). The rest of this chapter deals with the 3-body "dispersive" case. Our construction of the channel conjugate operators will rely on a result of [**GN1**], which we review in Section 3.4. In Section 3.5 we present a detailed analysis of the reduced Hamiltonians of a neutral pair; some of these results (notably Theorem 3.5.3) were first proved in [**AHS2**] under slightly different assumptions on the pair potentials, others first appeared in [**GŁ3**]. The channel conjugate operators are constructed in Section 3.6, and the "global" Mourre estimate is proved in Sections 3.7, 3.8. Our presentation of the 3−dimensional case differs substantially from that in [**GŁ3**], where the main results were first obtained. Using the results of [**GN1**], we were able to simplify the proof of the Mourre estimate for channel Hamiltonians (Section 3.6). We have also replaced an implicit assumption appearing in [**GŁ3**] by the (simpler and probably easier to verify) condition *(S)*, included the details missing from [**GŁ3**], and corrected several technical errors. The 2−dimensional results are published here for the first time.

## 3.1. Local inequalities for operators

DEFINITION 3.1.1. Let $H$ be a self-adjoint operator on a Hilbert space $\mathcal{H}$, and let $A, B$ be two symmetric quadratic forms on $\mathcal{H}$ such that $\mathcal{D}(H) \subset \mathcal{D}(A) \cap \mathcal{D}(B)$. Let $c \in \mathbb{R}$. We will say that:

(i) $B \geq c$ at $H = \lambda$ if there is a $\delta > 0$ such that for any function $f \in C_0^\infty(\mathbb{R})$ supported in $(\lambda - \delta, \lambda + \delta)$ we have

$$f(H)Bf(H) \geq cf^2(H); \tag{3.1.1}$$

*(ii)* $B$ is *almost positive* ($B \geq_* 0$) *at* $H = \lambda$ if $B \geq -\epsilon$ at $\lambda$ for any $\epsilon > 0$;

*(iii)* $B \geq A$ ($B \geq_* A$) *at* $H = \lambda$ if $B - A \geq 0$ ($B - A \geq_* 0$) at $H = \lambda$.

DEFINITION 3.1.2. Let $B(k, \lambda)$ be a measurable family of symmetric quadratic forms and $H(k)$ – a measurable family of self-adjoint operators on $\mathcal{H}$, with $\mathcal{D}(H(k)) \subset \mathcal{D}(B(k, \lambda))$ for $k \in \Omega_1 \subset \mathbb{R}^m$, $\lambda \in I \subset \mathbb{R}$. Let

$$H = \int_{\Omega_1}^{\oplus} H(k)dk, \; B = B(\lambda) = \int_{\Omega_1}^{\oplus} B(k, \lambda)dk.$$

We will say that:

(i) $B \geq c$ *locally in* $H$ *uniformly for* $(k, \lambda) \in \Omega \subset \Omega_1 \times I$ if there is a $\delta > 0$ such that for any $f \in C_0^\infty(\mathbb{R})$ supported in $(-\delta, \delta)$ we have

$$f(H(k) - \lambda)B(k, \lambda)f(H(k) - \lambda) \geq cf^2(H(k) - \lambda) \text{ for all } (k, \lambda) \in \Omega;$$

(ii) $B \geq_* 0$ *locally in* $H$ *uniformly for* $(k, \lambda) \in \Omega \subset \Omega_1 \times I$ if for any $\epsilon > 0$ we have $B \geq -\epsilon$ locally in $H$ uniformly for $(k, \lambda) \in \Omega$.

In the important special case when $H$ and $B$ are independent of parameters, but the energy level $\lambda$ is allowed to vary, Definition 3.1.2 says that $B \geq c$ at $H = \lambda$ uniformly for $\lambda \in I$ if there is a $\delta > 0$ such that for any $f \in C_0^\infty(\mathbb{R})$ supported in $(-\delta, \delta)$ we have

$$f(H - \lambda)Bf(H - \lambda) \geq cf^2(H - \lambda) \text{ for all } \lambda \in I.$$

LEMMA 3.1.3. *Let $B(k, \lambda)$ and $H(k)$ be as in Definition 3.1.2. Assume that the mappings:*

$$\Omega_1 \ni k \to f(H(k)),$$

$$\Omega_1 \times I \ni (k, \lambda) \to f(H(k))B(k, \lambda)f(H(k))$$

*are norm continuous for any $f \in C_0^\infty(\mathbb{R})$. Let $\Omega$ be an open subset of $\Omega_1 \times I$.*

*(i) If $B(k_0, \lambda_0) \geq_* 0$ at $H(k_0) = \lambda_0$ for some $(k_0, \lambda_0) \in \Omega$, then for any $\epsilon > 0$ there is an open neighbourhood $\mathcal{U}$ of $(k_0, \lambda_0)$ such that $B \geq -\epsilon$ locally in $H$ uniformly for $(k, \lambda) \in \mathcal{U}$.*

*(ii) If $B \geq_* 0$ locally in $H$ pointwise on $\Omega$ (i.e., $B(k, \lambda) \geq_* 0$ at $H(k) = \lambda$ for any $(k, \lambda) \in \Omega$), then $B \geq_* 0$ locally in $H$ uniformly on any compact subset of $\Omega$.*

PROOF. *(i)* Pick $\epsilon > 0$. Then there is a $\delta_0 > 0$ such that for any $C_0^\infty$ function $f_0$ supported in $(\lambda_0 - \delta_0, \lambda_0 + \delta_0)$ we have

$$f_0(H(k_0))B(k_0, \lambda_0)f_0(H(k_0)) \geq -\frac{\epsilon}{2}f_0^2(H(k_0)).$$

We choose $f_0$ so that $f_0 = 1$ on $(\lambda_0 - \delta_0/2, \lambda_0 + \delta_0/2)$. By the continuity assumptions, there is an open neighbourhood $\mathcal{U}_1$ of $k_0$ and $\delta_1 > 0$ such that for all $k \in \mathcal{U}_1$, $|\lambda - \lambda_0| < \delta_1$:

$$f_0(H(k))B(k, \lambda)f_0(H(k)) \geq -\frac{\epsilon}{2}f_0^2(H(k)) - \frac{\epsilon}{2}. \tag{3.1.2}$$

Let $\delta = \frac{1}{10}\min(\delta_0, \delta_1)$ and $\Delta = (\lambda_0 - \delta, \lambda_0 + \delta)$. Then for any $\lambda \in \Delta$ and for any function $f$ supported in $(\lambda - \delta, \lambda + \delta)$ we have $ff_0 = f$. Multiplying (3.1.2) from both sides by $f$, we obtain that for all $k \in \mathcal{U}_1$:

$$f(H(k))B(k, \lambda)f(H(k)) \geq -\epsilon f^2(H(k)).$$

Hence we can take $\mathcal{U} = \mathcal{U}_1 \times \Delta$.

*(ii)* This follows immediately from *(i)* by a covering argument. □

The main tool which we will use to establish local almost positivity of operators is the following lemma, which in fact is an abstract version of a well known argument in the Mourre theory (see [FH]).

LEMMA 3.1.4. *Let $H$ be a self-adjoint operator and $B$ a symmetric quadratic form on $\mathcal{H}$ such that $\mathcal{D}(H) \subset \mathcal{D}(B)$. Let $c$ be a positive constant, and fix $\lambda \in \mathbb{R}$. Suppose that for any $\epsilon > 0$ there is a compact operator $K_\epsilon$ such that*

$$B \geq c - \epsilon + K_\epsilon \text{ at } H = \lambda. \tag{3.1.3}$$

*(i) If $\lambda$ is not an eigenvalue of $H$, we have $B \geq_* c$ at $H = \lambda$.*

*(ii) Suppose that $\lambda$ is an eigenvalue of $H$. Let $P = \mathbf{1}_{\{\lambda\}}(H)$ be the corresponding spectral projection, and let $c_0 = \|PBP\|$. Then*

$$B \geq_* c(1 - P) - 2c_0P \text{ at } H = \lambda.$$

PROOF. We first note that, instead of arbitrary functions $f \in C_0^\infty(\mathbb{R})$ supported in $(\lambda - \delta, \lambda + \delta)$, it suffices to consider those satisfying $0 \leq f \leq 1$ and $f = 1$ on $(\lambda - \delta/2, \lambda + \delta/2)$. Throughout the proof, $f_\delta$ will denote a function for which these conditions hold.

We first prove *(i)*. Given $\epsilon > 0$, we may choose $\delta_0 > 0$ and a compact operator $K_\epsilon$ such that

$$f_{\delta_0}(H)Bf_{\delta_0}(H) \geq (c - \epsilon)f_{\delta_0}^2(H) + f_{\delta_0}(H)K_\epsilon f_{\delta_0}(H). \tag{3.1.4}$$

Multiplying then (3.1.4) from both sides by $f_\delta(H)$ for $\delta < \delta_0/2$, and using that $f_\delta(H) \to 0$ strongly if $\lambda$ is not an eigenvalue of $H$, we obtain that for sufficiently small $\delta$:

$$\begin{aligned} f_\delta(H)Bf_\delta(H) &\geq (c-\epsilon)f_\delta^2(H) + f_\delta(H)K_\epsilon f_\delta(H) \\ &\geq (c-\epsilon)f_\delta^2(H) - \epsilon f_\delta^2(H), \end{aligned}$$

which shows that $B \geq_* c$ at $H = \delta$.

Let us now prove *(ii)*. Let $P^n$ be a sequence of finite-dimensional projections such that $P = \mathrm{s-}\lim P^n$ and $P^n \leq P$. We write

$$B = C_1 + C_2 + C_2^* + C_3,$$

where:

$$\begin{aligned} C_1 &= (1-P^n)B(1-P^n), \\ C_2 &= (1-P)BP^n, \qquad (3.1.5) \\ C_3 &= PBP - (P-P^n)B(P-P^n) \geq -2c_0 P. \end{aligned}$$

Fix $\epsilon > 0$, and choose $\delta > 0$ and $K_\epsilon$ such that (3.1.3) holds for $f = f_\delta$. Then:

$$\begin{aligned} &f_\delta(H)C_1 f_\delta(H) \\ &\geq (c-\epsilon)f_\delta^2(H)(1-P^n) - f_\delta(H)(1-P^n)K_\epsilon(1-P^n)f_\delta(H) \qquad (3.1.6) \\ &\geq (c-\epsilon)f_\delta^2(H)(1-P) - f_\delta(H)(1-P^n)K_\epsilon(1-P^n)f_\delta(H), \end{aligned}$$

where we have used that $P^n \leq P$. Since $K_\epsilon$ is compact and $P^n \to P$ strongly, for sufficiently large $n$ we have

$$\|(1-P^n)K_\epsilon(1-P^n) - (1-P)K_\epsilon(1-P)\| < \epsilon. \qquad (3.1.7)$$

As in the proof of *(i)*, we now use that $f_\delta(H)(1-P) \to 0$ strongly as $\delta \to 0$ to obtain that for $\delta < \delta_0$,

$$\|f_\delta(H)(1-P)K_\epsilon(1-P)f_\delta(H)\| < \epsilon. \qquad (3.1.8)$$

Hence for large enough $n$ and for $\delta < \delta_0$,

$$f_\delta(H)C_1 f_\delta(H) \geq (c-\epsilon)f_\delta^2(H)(1-P) - 2\epsilon,$$

so that

$$C_1 \geq_* c(1-P) \text{ at } H = \lambda. \qquad (3.1.9)$$

Finally, consider the term $C_2 + C_2^*$. Using the inequality

$$(\epsilon^{-1/2}C_2 + \epsilon^{1/2}P^n)(\epsilon^{-1/2}C_2^* + \epsilon^{1/2}P^n) \geq 0,$$

we obtain that

$$C_2 + C_2^* \geq -\epsilon P^n - \epsilon^{-1}C_2 C_2^*.$$

Note that the operator

$$\begin{aligned} &f_\delta(H)C_2C_2^* f_\delta(H) \\ &= f_\delta(H)(1-P)BP^n f_\delta^2(H)B(1-P)f_\delta(H) \end{aligned}$$

is compact since $P^n$ is a finite rank projection. Once more, we use that $f_\delta(H)(1-P) \to 0$ strongly as $\delta \to 0$ and argue as in *(i)* to show that

$$C_2 + C_2^* \geq_* 0 \text{ at } H = \lambda. \tag{3.1.10}$$

This, together with (3.1.5) and (3.1.9), implies part *(ii)* of the lemma. □

## 3.2. Basics of the Mourre theory

The essence of the Mourre theory is the *Mourre estimate*: one finds a self-adjoint operator $A$, called the *conjugate operator*, such that the commutator $[H, iA]$ is locally positive in the sense of Definition 3.1.1. The exact statement of the Mourre estimate itself is given in Definition 3.2.5; however, there are additional technicalities that must be taken care of first. Since both $H$ and $A$ are unbounded, one has to pay attention to domain questions in the definition of $[H, iA]$ and in the computations that follow. Throughout this book we will rely on the abstract framework constructed in [**ABG**], which we now review briefly.

Let $H, A$ be two self-adjoint operators on a Hilbert space $\mathcal{H}$. For $k \in \mathbb{N}$, we say that $H \in C^k(A)$ if for some $z \notin \sigma(H)$ the map

$$\mathbb{R} \ni s \mapsto \mathrm{e}^{\mathrm{i}sA}(z-H)^{-1}\mathrm{e}^{-\mathrm{i}sA} =: T_{A,H}(s) \in B(\mathcal{H}) \tag{3.2.1}$$

is $C^k$ in the strong topology of $B(\mathcal{H})$. We equip $C^k(A)$ with the norm:

$$\|H\|_{C^k(A)} = \sum_{l \leq k} \|\partial_s^l T_{A,H}(0)\|_{B(\mathcal{H})}.$$

Let $H \in C^k(A)$ and $0 \leq l \leq k$. Then the quadratic forms $\mathrm{ad}_A^l(z-H)^{-1}$, defined inductively on $\mathcal{D}(A)$ by

$$\mathrm{ad}_A(z-H)^{-1} = [(z-H)^{-1}, A], \ \mathrm{ad}_A^l(z-H)^{-1} := [\mathrm{ad}_A^{l-1}(z-H)^{-1}, A],$$

are bounded in the topology of $\mathcal{H}$ and therefore extend to bounded operators on $\mathcal{H}$, which we will continue to denote by $\mathrm{ad}_A^l(z-H)^{-1}$. Moreover,

$$\partial_s^l T_{A,H}(0) = \mathrm{i}^{-k}\mathrm{ad}_A^l((z-H)^{-1}).$$

For $0 < \epsilon < 1$, we say that $H \in C^\epsilon(A)$ if for some $z \notin \sigma(H)$ the map (3.2.1) is Hölder continuous of order $\epsilon$ in the *operator* topology of $B(\mathcal{H})$. We equip $C^\epsilon(A)$ with the norm

$$\|H\|_{C^\epsilon(A)} = \sup_{s,s' \in \mathbb{R}} \frac{\|T_{A,H}(s) - T_{A,H}(s')\|}{|s-s'|^\epsilon} + \sup_{s \in \mathbb{R}} \|T_{A,H}(s)\|.$$

Finally, for $\sigma = k + \epsilon$, $k \in \mathbb{N}$, $0 < \epsilon < 1$, we say that $H \in C^\sigma(A)$ if $H \in C^k(A)$ and the map

$$\mathbb{R} \ni s \mapsto \mathrm{e}^{\mathrm{i}sA}\mathrm{ad}_A^k((z-H)^{-1})\mathrm{e}^{-\mathrm{i}sA} =: T_{A,H}^{(k)}(s) \in B(\mathcal{H}) \tag{3.2.2}$$

is Hölder continuous of order $\epsilon$ in the operator topology of $B(\mathcal{H})$. The norm in $C^\sigma(A)$ is given by

$$\|H\|_{C^\sigma(A)} = \sum_{l\leq k} \|\partial_s^l T(0)\|_{B(\mathcal{H})} + \sup_{s,s'\in\mathbb{R}} \frac{\|T^{(k)}_{A,H}(s) - T^{(k)}_{A,H}(s')\|}{|s-s'|^\epsilon}.$$

The condition that $H \in C^1(A)$ will be a part of our definition of the Mourre estimate; for the application to the proof of asymptotic completeness, we will also require that $H \in C^{1+\epsilon}(A)$ for some $\epsilon > 0$.

We will often use the following alternative characterizations of operators in $C^1(A)$ [**ABG**, Theorem 6.2.10].

PROPOSITION 3.2.1. *Let $H$, $A$ be self-adjoint operators on $\mathcal{H}$. Then the following are equivalent:*

- *$H \in C^1(A)$;*
- *for some $z \in \mathbb{C}\setminus\sigma(H)$:*

$$|((\overline{z}-H)^{-1}u, Au) - (Au, (z-H)^{-1}u)| \leq C\|u\|^2_{\mathcal{H}},\ u \in \mathcal{D}(A); \tag{3.2.3}$$

- *for all $z \in \mathbb{C}\setminus\sigma(H)$, (3.2.3) holds;*
- *the following two conditions hold:*
  *(i) there is a constant $C < \infty$ such that*
  $$|(Au, Hu) - (Hu, Au)| \leq C\|(H+\mathrm{i})u\|^2,\ u \in \mathcal{D}(H)\cap\mathcal{D}(A);$$
  *(ii) there is a $z \in \mathbb{C}\setminus\sigma(H)$ such that*
  $$\{u \in \mathcal{D}(A)|\ (z-H)^{-1}u \in \mathcal{D}(A), (\overline{z}-H)^{-1}u \in \mathcal{D}(A)\}$$
  *is a core for $A$.*

Unless the domains of both $H$ and $A$ are explicitly known, it may be difficult to verify the conditions of Proposition 3.2.1 directly and to compute the quadratic form $[H, iA]$. In our case, this problem will be solved using the following lemma.

LEMMA 3.2.2. *Let $H, H_0, N$ be self-adjoint operators on $\mathcal{H}$ such that $N \geq c_0 > 0$, $\mathcal{D}(H) = \mathcal{D}(H_0)$ as Banach spaces, and $(z-H)^{-1}$ maps $\mathcal{D}(N)$ into itself. Let $A$ be a symmetric operator on $\mathcal{D}(N)$. Assume that $H_0, A$ satisfy the hypotheses of Lemma 1.2.5 with comparison operator $N$ (hence $A$ has a unique self-adjoint extension, denoted also by $A$). Assume moreover that*

$$|(Au, Hu) - (Hu, Au)| \leq C(\|Hu\|^2 + \|u\|^2),\ u \in \mathcal{D}(N). \tag{3.2.4}$$

*Then:*

*(i) $\mathcal{D}(N)$ is dense in $\mathcal{D}(A)\cap\mathcal{D}(H)$ (with the norm $\|Hu\| + \|Au\| + \|u\|$);*

*(ii) the commutator $[H, \mathrm{i}A]$, defined as a quadratic form on $\mathcal{D}(A)\cap\mathcal{D}(H)$, is the unique extension of the quadratic form $[H, \mathrm{i}A]$ defined by (3.2.4) on $\mathcal{D}(N)$;*

*(iii) $H \in C^1(A)$.*

PROOF. By Lemma 1.2.5 applied to $A$ and $H_0$, for all $u \in \mathcal{D}(A) \cap \mathcal{D}(H_0)$ we have:

$$\lim_{\epsilon \to 0} \|(1 + \mathrm{i}\epsilon N)^{-1} u - u\|_{\mathcal{D}(A) \cap \mathcal{D}(H_0)} = 0.$$

This implies *(i)* and hence *(ii)*. To prove *(iii)*, we will verify that *(i)–(ii)* of Proposition 3.2.1 are satisfied. Indeed, from *(ii)* above and (3.2.4) we obtain *(i)*; *(ii)* follows since $(z - H)^{-1}$ preserves $\mathcal{D}(N)$, which is a core for $A$. □

The condition that $H \in C^1(A)$ will often provide rigorous justification for calculations which would be trivial, were it not for domain questions. In particular, we will need the following results of [**ABG**] (Theorem 6.2.10, Proposition 7.2.10), together with two simple consequences *(iv), (v)*.

PROPOSITION 3.2.3. *Let $H, A$ be self-adjoint operators such that $H \in C^1(A)$. Then:*

*(i) $\mathcal{D}(A) \cap \mathcal{D}(H)$ is a core for $H$, hence the quadratic form $[H, iA]$ on $\mathcal{D}(A) \cap \mathcal{D}(H)$ extends uniquely to a bounded operator $\mathcal{D}(H) \to \mathcal{D}(H)^*$, denoted again by $[H, iA]$;*

*(ii) the* virial relation *holds:*

$$\mathbf{1}_{\{\lambda\}}(H)[H, \mathrm{i}A]\mathbf{1}_{\{\lambda\}}(H) = 0; \tag{3.2.5}$$

*(iii) $[(z - H)^{-1}, \mathrm{i}A] = (z - H)^{-1}[H, \mathrm{i}A](z - H)^{-1}$ for $z \in \mathbb{C} \backslash \sigma(H)$;*

*(iv) $[f(H), A]$ is bounded for $f \in C_0^\infty(\mathbb{R})$;*

*(v) for $z \notin \sigma(H)$, $(z - H)^{-1}$ preserves $\mathcal{D}(A)$.*

REMARK 3.2.4. Generally, commutators $[A, B]$ of two symmetric operators $A, B$ are defined as quadratic forms on $\mathcal{D}(A) \cap \mathcal{D}(B)$. In this setting, the Leibniz rule does not always hold: it is not necessarily true that $[A, B_1 B_2] = [A, B_1]B_2 + B_1[A, B_2]$, since the right-hand side need not be defined. However, it is legitimate to use the Leibniz rule under certain additional conditions. Let $\mathcal{D}$ be a dense subspace of $\mathcal{H}$, $A, B$ – symmetric operators with $\mathcal{D} \subset \mathcal{D}(A) \cap \mathcal{D}(B)$, and let $C$ be a bounded operator such that $C : \mathcal{D} \mapsto \mathcal{D}$; then the expansion

$$[A, CBC] = [A, C]BC + C[A, B]C + CB[A, C],$$

is justified as an identity between quadratic forms on $\mathcal{D}$, since $[A, C]$ is well defined as a bounded operator on $\mathcal{D}$.

We may now give a rigorous definition of the Mourre estimate. Observe that by Proposition 3.2.3*(i)*, $f(H)[H, \mathrm{i}A]f(H) \in B(\mathcal{H})$ for $f \in C_0^\infty(\mathbb{R})$.

DEFINITION 3.2.5. Let $H$ and $A$ be self-adjoint operators on $\mathcal{H}$ such that $H \in C^1(A)$. We will say that:

(i) $H$ satisfies the *Mourre estimate* with the *conjugate operator* $A$ at energy $\lambda$, if:

$$[H, \mathrm{i}A] \geq c - K \text{ at } H = \lambda,$$

where $c > 0$ and $K$ is a compact operator on $\mathcal{H}$;

(ii) $H$ satisfies the *strict Mourre estimate* with the *conjugate operator* $A$ at energy $\lambda$, if:

$$[H, \mathrm{i}A] \geq c > 0 \text{ at } H = \lambda.$$

We end this section by stating two simple consequences of the Mourre estimate.

LEMMA 3.2.6. *(i) If $H$ satisfies the strict Mourre estimate at $H = \lambda$, then there is an open neighbourhood of $\lambda$ in which $H$ has no eigenvalues.*

*(ii) If $H$ satisfies the Mourre estimate at $H = \lambda$, then eigenvalues of $H$, counted with multiplicity, cannot accumulate at $\lambda$. (In particular, $\lambda$ is not an eigenvalue of $H$ of infinite multiplicity).*

PROOF. *(i)* Assume that $[H, \mathrm{i}A] > c > 0$ at $H = \lambda$. Then there is a function $f \in C_0^\infty(\mathbb{R})$, equal to 1 in an open neighbourhood $\mathcal{U}$ of $\lambda$, such that

$$f(H)[H, \mathrm{i}A]f(H) \geq cf^2(H). \tag{3.2.6}$$

Suppose that $E_0 \in \mathcal{U}$ is an eigenvalue of $H$. Multiplying both sides of (3.2.6) by $\mathbf{1}_{\{E_0\}}(H)$ we obtain that

$$\mathbf{1}_{\{E_0\}}(H)[H, \mathrm{i}A]\mathbf{1}_{\{E_0\}}(H) \geq c\mathbf{1}_{\{E_0\}}(H).$$

However, this contradicts the virial relation (3.2.5).

*(ii)* Suppose that $\psi_i$, $i = 1, 2, \ldots$, is an infinite sequence of eigenfunctions of $H$ such that $\|\psi_i\| = 1$, $\psi_i \to 0$ weakly as $i \to \infty$, and $H\psi_i = E_i\psi_i$, $E_i \to \lambda$. By the Mourre estimate,

$$\langle \psi_i, [H, \mathrm{i}A]\psi_i \rangle \geq c - \langle \psi_i, K\psi_i \rangle \tag{3.2.7}$$

for some $c > 0$ and $K$ compact. But the left-hand side of (3.2.7) is zero by the virial relation, and $|\langle \psi_i, K\psi_i \rangle| \leq \|K\psi_i\| \to 0$, since $K$ is compact and $\psi_i \to 0$ weakly. Hence a contradiction. □

## 3.3. The strongly charged case

In this section we prove the Mourre estimate for $H$ with the conjugate operator

$$A = \frac{1}{2}(\langle z, D_z \rangle + \langle D_z, z \rangle),$$

under the assumption that all clusters consisting of less than $N$ particles are charged, i.e.,

(Q2) $\quad q_{i_1} + q_{i_2} + \cdots + q_{i_n} \neq 0$ for all $\{i_1, \ldots, i_n\} \subset \{1, \ldots, N\}$, $n < N$.

The total charge $Q$ of the system may be arbitrary.

LEMMA 3.3.1. *Assume that (V1), (M1) hold. Then $H \in C^1(A)$.*

PROOF. It suffices to verify that $H$ and $A$ satisfy the assumptions of Lemma 3.2.2 with the comparison operator $N = D_x^2 + x^2$. We have

$$[H, \mathrm{i}A] = D_z^2 - z\nabla_z V(x^{\mathrm{cm}}),$$

where all terms are understood as quadratic forms on $\mathcal{D}(N)$; by *(M1)*, the last term is bounded as a quadratic form on $\mathcal{D}(H)$. Hence the lemma follows from Lemma 3.2.2. □

For a cluster decomposition $a$, we define its associated *threshold set*:

$$\tau_a = \bigcup_{b<a} \sigma_{\mathrm{pp}}(\tilde{H}^b),$$

and put

$$d_a(\lambda) = \begin{cases} \mathrm{dist}\,(\lambda, \tau_a), & \lambda \geq \inf \tau_a, \\ 0, & \lambda < \inf \tau_a. \end{cases}$$

We will write $\tau_{\mathrm{cm}} = \tau_{a_{\max}}$ and $d_{\mathrm{cm}}(\lambda) = d_{a_{\max}}(\lambda)$. Readers already familiar with the Mourre theory should note that we *do not* follow the convention adopted by many authors, according to which $d(\lambda)$ denotes the distance to the thresholds to the left of $\lambda$ (i.e., $\mathrm{dist}\,(\lambda, \tau_a \cup (-\infty, \lambda))$); this will allow us to avoid numerous technical complications due to the lack of continuity of the latter function. We feel that the benefits of this simplification more than outweigh the price that we will pay for it – that the constant in our Mourre estimate will not be optimal.

The main result of this section is the following theorem.

THEOREM 3.3.2. *Assume (V1), (V2), (M1), (Q2).*

- *If $Q \neq 0$:*

  *(i) the set $\sigma_{\mathrm{pp}}(H)$ may accumulate only at $\tau_{\mathrm{cm}}$. Consequently, $\tau_{\mathrm{cm}} \cup \sigma_{\mathrm{pp}}(H)$ is a closed countable set;*

  *(ii) for any $\lambda \notin \tau_{\mathrm{cm}} \cup \sigma_{\mathrm{pp}}(H)$, there is a $c(\lambda) > 0$ such that*

$$[H, \mathrm{i}A] \geq c(\lambda) \text{ at } H = \lambda.$$

- *If $Q = 0$:*

  *(iii) for $\eta_{\mathrm{cm}} \in Y'_{\mathrm{cm}}$, the set $\sigma_{\mathrm{pp}}(\tilde{H}(\eta_{\mathrm{cm}}))$ can accumulate only at $\tau_{\mathrm{cm}}$. Consequently, $\tau_{\mathrm{cm}} \cup \sigma_{\mathrm{pp}}(\tilde{H}(\eta_{\mathrm{cm}}))$ is a closed countable set;*

  *(iv) for any $\lambda \notin \tau_{\mathrm{cm}} \cup \sigma_{\mathrm{pp}}(\tilde{H}(\eta_{\mathrm{cm}}))$, there is a $c(\lambda) > 0$ such that*

$$[\tilde{H}(\eta_{\mathrm{cm}}), \mathrm{i}A] \geq c(\lambda) \text{ at } H = \lambda.$$

Theorem 3.3.2 will be a consequence of Theorem 3.3.4 below. To formulate the latter, we need the notion of *$a$−compact operators*; a similar definition was introduced in [**PSS**] for standard $N$−particle Hamiltonians.

DEFINITION 3.3.3. Let $a \in \mathcal{A}$. An operator $\tilde{K}^a$ acting on $L^2(Y_a^n \times \bar{Y}_a^c \times \hat{Y}_a^c \times X^a)$ is called *$\tilde{a}$-compact* if

$$\tilde{K}^a = 1_{\bar{Y}_a^c} \otimes \int_{Y_a^{n\prime}}^{\oplus} \tilde{K}^a(\eta_a^n) d\eta_a^n,$$

where each $\tilde{K}^a(\eta_a^n) \in B(L^2(\hat{Y}_a^c \times X^a))$ satisfies:

$$(i)\ \tilde{K}^a(\eta_a^n) \text{ is compact,}$$

$$(ii)\ \eta_a^n \to \tilde{K}^a(\eta_a^n) \text{ is norm continuous,}$$

$$(iii)\ \lim_{\eta_a^n \to \infty} \tilde{K}^a(\eta_a^n) = 0.$$

An operator $K^a$ on $L^2(X)$ is *a-compact* if $\tilde{K}^a = U_a K^a U_a^*$ is $\tilde{a}$−compact.

If $a$ contains no neutral clusters, which we assume to be true for all $a \neq a_{\max}$, $\tilde{K}^a$ is $\tilde{a}$-compact if and only if

$$\tilde{K}^a = 1_{\bar{Y}_a^c} \otimes \hat{K}^a, \tag{3.3.1}$$

where $\hat{K}^a$ is compact on $L^2(\hat{Y}_a^c \times X^a)$.

THEOREM 3.3.4. *Assume that (V1) (V2), (M1) and (Q2) hold. Then:*
*(i) for each $\lambda \in \mathbb{R} \setminus \tau_{\rm cm}$ and $\epsilon > 0$, there is an $\tilde{a}_{\max}$-compact operator $K^{\rm cm} = K^{\rm cm}(\epsilon)$ such that*

$$[\tilde{H}, \mathrm{i}A] \geq 2d_{\rm cm}(\lambda) - \epsilon + K^{\rm cm} \ \textit{at}\ \tilde{H} = \lambda; \tag{3.3.2}$$

*(ii) the set $\tau_{\rm cm}$ is closed and countable.*

PROOF OF THEOREM 3.3.2, GIVEN THEOREM 3.3.4. Let us assume first that $Q \neq 0$. Since $U_{\rm cm}$ preserves $A$, it suffices to prove *(i)–(ii)* with $H$ replaced by $\tilde{H}$. We may consider $\tilde{H}$, $A$ as operators acting only on $L^2(Y_{\rm cm}^{(2)}) \otimes L^2(X^{\rm cm})$. From (3.3.2) and Definition 3.3.3 we obtain that for any $\lambda \notin \tau_{\rm cm}$:

$$[\tilde{H}, \mathrm{i}A] \geq c(\lambda) + K^{\rm cm} \text{ at } \tilde{H} = \lambda, \tag{3.3.3}$$

where $c(\lambda) > 0$ and $K^{\rm cm}$ is compact on $L^2(Y_{\rm cm}^{(2)}) \otimes L^2(X^{\rm cm})$. Using also Lemma 3.3.1, we find that the conditions of Definition 3.2.5 are satisfied. Hence *(i)* and *(ii)* follow, respectively, from Lemma 3.2.6 and Lemma 3.1.4*(i)*.

For $Q = 0$, the argument is the same, except that (3.3.3) is replaced by

$$[\tilde{H}(\eta_{\rm cm}), \mathrm{i}A] \geq c(\lambda) + K(\eta_{\rm cm}) \text{ at } \tilde{H}(\eta_{\rm cm}) = \lambda,$$

for any $\lambda \notin \tau_{\rm cm}$ and $\eta_{\rm cm} \in Y'_{\rm cm}$, where $c(\lambda) > 0$ and $K(\eta_{\rm cm})$ is compact on $L^2(X^{\rm cm})$. □

PROOF OF THEOREM 3.3.4. The proof below is similar to that given by Froese-Herbst in [**FH1**] for standard $N-$body Hamiltonians; here we reorganized it and adapted it to our context. We will use the lattice structure on the set $\mathcal{A}$ of all cluster decompositions to prove by induction in $a \in \mathcal{A}$ that:

*(i')* for any $\lambda \in \mathbb{R}$ and $\epsilon > 0$ there is an $\tilde{a}$-compact operator $K^a = K^a(\epsilon)$ such that

$$[\tilde{H}^a, \mathrm{i}A^a] \geq 2d_a(\lambda) - \epsilon + K^a \text{ at } \tilde{H}^a = \lambda;$$

*(ii')* the set $\tau_a$ is closed and countable.

Here $A^a = \frac{1}{2}(\langle z^a, D_{z^a}\rangle + \langle D_{z^a}, z^a\rangle)$ is the generator of dilations along $Z^a$. This will imply the theorem, which states that *(i')*–*(ii')* hold for $a = a_{\max}$.

To start the induction, we note that

$$\tilde{H}_{a_{\min}} = \frac{1}{2}D_z^2 + \tilde{H}^{a_{\min}}$$

is unitarily equivalent to

$$H_{a_{\min}} = \frac{1}{2}D_z^2 + \sum_1^N \frac{1}{2m_i}(D_{y_i} - q_i J y_i)^2.$$

Hence the spectrum of $\tilde{H}^{a_{\min}}$ is purely discrete and consists of the Landau levels of noninteracting particles:

$$\sigma(\tilde{H}^{a_{\min}}) = \{\sum_1^M \frac{|q_i|}{m_i}(2n_i + 1)b,\ n_i \in \mathbb{N}\}.$$

Since $[\tilde{H}_{a_{\min}}, \mathrm{i}A] = D_z^2$, the conditions *(i')*–*(ii')* are trivially satisfied for $a = a_{\min}$.

Assume now that *(i')*–*(ii')* hold with $a$ replaced by $b$ for all $b < a$; we will prove that they hold for $a$. We first note that *(i')*–*(ii')* for all $b < a$ implies *(ii')* for $a$, since the threshold set $\tau_a$ of $\tilde{H}^a$ consists of eigenvalues of $\tilde{H}^b$ for $b < a$, which, by *(i')* for $b$ and Lemma 3.2.6*(ii)*, may accumulate only at $\tau_b$. Part *(i')* for $a$ is an immediate consequence of Lemmas 3.3.5 and 3.3.6 below. □

Let $b < a$. Since $X^b \subset X^a$, we may write

$$X^a =: X^b \oplus X_b^a.$$

Recall that $x_b^a$ is the orthogonal projection of $x^a \in X^a$ on $X_b^a$. Let also

$$H_b^a = H_b - \frac{1}{2}D_{z^a}^2.$$

LEMMA 3.3.5. *Assume that (V1) (V2), (M1) and (Q2) hold. Suppose that* $[\tilde{H}_b^a, iA^a] \geq c$ *at* $\tilde{H}_b^a = \lambda$ *for all* $b < a$, *where*

$$\tilde{H}_b^a = U_a H_b^a U_a^\star = U_a H_b U_a^\star - \frac{1}{2}D_{z^a}^2.$$

*Then for any $\epsilon > 0$ there is an $\tilde{a}$-compact operator $\tilde{K}^a = \tilde{K}^a(\epsilon)$ such that:*

$$[\tilde{H}^a, \mathrm{i}A^a] \geq c - \epsilon + \tilde{K}^a(\epsilon) \text{ at } \tilde{H}^a = \lambda.$$

PROOF. Let $\{q_b^a\}_{b \leq a}$ be a partition of unity as in Proposition 2.3.1, and let $q_{b,R}^a$ be the operator of multiplication by $q_b^a(\frac{x^a}{R})$. Then for any $f \in C_0^\infty(\mathbb{R})$ we have

$$\begin{aligned}
&f(\tilde{H}^a)[\tilde{H}^a, \mathrm{i}A^a]f(\tilde{H}^a) \\
&= \textstyle\sum_{b\leq a} q_{b,R}^a f(\tilde{H}_b^a)[\tilde{H}_b^a, \mathrm{i}A^a]f(\tilde{H}_b^a)q_{b,R}^a + o(R^0), \qquad (3.3.4)\\
&f^2(\tilde{H}^a) = \textstyle\sum_{b\leq a} q_{b,R}^a f^2(\tilde{H}_b^a)q_{b,R}^a + o(R^0),
\end{aligned}$$

as $R \to \infty$. Indeed, we have $\tilde{H}^a = \tilde{H}_b^a + I_b^a$, where $I_b^a = \sum_{(ij)\leq a, (ij)\not\leq b} V_{ij}$. Using *(V1)*, *(V2)*, we obtain as in Lemma 2.3.8:

$$q_{b,R}^a(f(\tilde{H}^a) - f(\tilde{H}_b^a)) = o(R^0),$$

$$[q_{b,R}^a, f(\tilde{H}_b^a)] = O(R^{-1}),$$

Hence the second identity of (3.3.4). To prove the first one, we write

$$\begin{aligned}
[\tilde{H}^a, \mathrm{i}A^a] &= [\tilde{H}_b^a, \mathrm{i}A^a] + [I_b^a, \mathrm{i}A^a] \\
&= D_{z^a}^2 - z^a\nabla_{z^a}V^b(x^a) - z^a\nabla_{z^a}I_b^a,
\end{aligned}$$

and argue as above, using *(M1)* and taking into account that $[\tilde{H}^a, \mathrm{i}A^a](\tilde{H}^a + \mathrm{i})^{-1}$ is bounded.

Pick $f \in C_0^\infty(\mathbb{R})$ such that $\operatorname{supp} f \subset [\lambda - 2\delta, \lambda + 2\delta]$ and $f \equiv 1$ on $[\lambda - \delta, \lambda + \delta]$. Given $\epsilon > 0$, the terms $o(R^0)$ in (3.3.4) are $< \epsilon/2$ if $R$ is sufficiently large. Hence the lemma will follow if we prove that $q_{a,R}^a f(\tilde{H}^a)$ is $\tilde{a}-$compact. Indeed, it follows from Kato's inequality and the form of $R_a$ given in (2.2.17) that

$$F(\frac{|x^a|}{R} \leq 1)(R_a(D_{\hat{y}_a^c}, \hat{y}_a^c, x^a) + \frac{1}{2}(D^a - \mathbf{A}^{a\,a}\, x^a)^2)^{-1}$$

is compact on $L^2(\hat{Y}_a^c \times X^a)$. If $a_{\max}$ is a charged cluster, the same argument shows that $q_{\mathrm{cm},R}^{\mathrm{cm}} f(\tilde{H})$ is $\tilde{a}_{\max}-$compact. If $a_{\max}$ is neutral, it is also straightforward to verify that

$$\begin{aligned}
\tilde{K}(\eta_{\mathrm{cm}}) = \ & F(\tfrac{|x^{\mathrm{cm}}|}{R} \leq 1)(\tfrac{1}{2}(\eta_{\mathrm{cm}} - 2(\mathbf{A}x)_{\mathrm{cm}})^2 \\
&+ \tfrac{1}{2}(D^{\mathrm{cm}} - \mathbf{A}^{\mathrm{cm}\,\mathrm{cm}}\, x^{\mathrm{cm}})^2 + 1)^{-1}
\end{aligned}$$

is compact on $L^2(X^{\mathrm{cm}})$, and that

$$\lim_{\eta_{\mathrm{cm}} \to \infty} \|\tilde{K}(\eta_{\mathrm{cm}})\| = 0.$$

This shows that $q_{\mathrm{cm},R}^{\mathrm{cm}} f(\tilde{H})$ is $\tilde{a}_{\max}-$compact if $a_{\max}$ is neutral and completes the proof of the lemma. □

LEMMA 3.3.6. *Assume that (i')–(ii') hold for a cluster decomposition* $b < a$. *Then for all* $\lambda \in \mathbb{R}$

$$[\tilde{H}_b^a, \mathrm{i}A^a] \geq_* 2d_a(\lambda) \text{ at } \tilde{H}_b^a = \lambda.$$

PROOF. We first claim that $\tilde{H}_b^a$ is unitarily equivalent to $\frac{1}{2}D_{z_b^a}^2 + \tilde{H}^b$. Indeed, this follows from the identities:

$$U_a H_b U_a^\star = \frac{1}{2}D_{z_a}^2 + \tilde{H}_b^a,$$

$$U_b H_b U_b^\star = \frac{1}{2}D_{z_b}^2 + \tilde{H}^b = \frac{1}{2}D_{z_a}^2 + \frac{1}{2}D_{z_b^a}^2 + \tilde{H}^b.$$

Moreover, $U_a$ and $U_b$ preserve the observables $z, D_z$, and hence $A^a$. It therefore suffices to prove the lemma with $\tilde{H}_a^b$ replaced by $\tilde{H}^b + \frac{1}{2}D_{z_b^a}^2$.

Fix $\lambda \in \mathbb{R}$ and let $f_\delta \in C_0^\infty((\lambda - \delta, \lambda + \delta))$ for $0 < \delta < 1$. We have

$$\begin{aligned} & f_\delta(\tfrac{1}{2}D_{z_b^a}^2 + \tilde{H}^b)[\tfrac{1}{2}D_{z_b^a}^2 + \tilde{H}^b, \mathrm{i}A^b]) f_\delta(\tfrac{1}{2}D_{z_b^a}^2 + \tilde{H}^b) \\ & = f_\delta(\tfrac{1}{2}D_{z_b^a}^2 + \tilde{H}^b)\big(D_{z_b^a}^2 + [\tilde{H}^b, \mathrm{i}A^b]\big) f_\delta(\tfrac{1}{2}D_{z_b^a}^2 + \tilde{H}^b) \qquad (3.3.5) \\ & = \int_{Z_b^{a\prime}}^{\oplus} f_\delta(\tfrac{1}{2}|\zeta_b^a|^2 + \tilde{H}^b)\big(|\zeta_b^a|^2 + [\tilde{H}^b, \mathrm{i}A^b]\big) f_\delta(\tfrac{1}{2}|\zeta_b^a|^2 + \tilde{H}^b) d\zeta_b^a. \end{aligned}$$

Since $\tilde{H}^b$ is bounded from below the integrand in (3.3.5) is supported in a compact set $\Omega = \{|\zeta_b^a| \leq C\}$, the same for all $\delta < 1$.

By *(i')–(ii')* for $\tilde{H}^b$, Lemma 3.1.4, and the virial relation, we have at $\tilde{H}^b = \nu$:

$$[\tilde{H}^b, \mathrm{i}A^b] \geq_* 2d_b(\nu) \geq 2d_a(\nu) \text{ if } \nu \notin \sigma_{\mathrm{pp}}(\tilde{H}^b),$$

$$[\tilde{H}^b, \mathrm{i}A^b] \geq_* 0 = 2d_a(\nu) \text{ if } \nu \in \sigma_{\mathrm{pp}}(\tilde{H}^b).$$

Hence $[\tilde{H}^b, \mathrm{i}A^b] \geq_* 2d_a(\lambda - \frac{1}{2}|\zeta_b^a|^2)$ at $\tilde{H}^b = \lambda - \frac{1}{2}|\zeta_b^a|^2$. By Lemma 3.1.3*(ii)*, this holds uniformly for $\zeta_b^a \in \Omega$. Therefore, given $\epsilon > 0$, for $\delta = \delta(\epsilon)$ sufficiently small the right hand side of (3.3.5) is bounded from below by

$$\begin{aligned} & \int_\Omega^\oplus f_\delta(\tfrac{1}{2}|\zeta_b^a|^2 + \tilde{H}^b)\big(|\zeta_b^a|^2 + 2d_a(\lambda - \tfrac{1}{2}|\zeta_b^a|^2)\big) f_\delta(\tfrac{1}{2}|\zeta_b^a|^2 + \tilde{H}^b) d\zeta_b^a \\ & \geq 2d_a(\lambda) \int_\Omega^\oplus f_\delta^2(\tfrac{1}{2}|\zeta_b^a|^2 + \tilde{H}^b) d\zeta_b^a \\ & \geq 2d_a(\lambda) f_\delta^2(\tfrac{1}{2}D_{z_b^a}^2 + \tilde{H}^b), \end{aligned}$$

where we used the triangle inequality:

$$d_a(\lambda - s) + s \geq d_a(\lambda) \text{ for } s \geq 0.$$

This proves the lemma. □

### 3.4. Analytically fibered operators

In this section we review the results of [**GN1**] on the Mourre theory for analytically fibered operators. We will later apply them to the "dispersive" scattering channels of 3-body magnetic Hamiltonians.

DEFINITION 3.4.1. ([**GN1**]) Let $H = \int_M^{\oplus} H(k) d\mu$ be a direct integral operator acting on a Hilbert space $\mathcal{H} = L^2(M, d\mu; \mathcal{H}')$. We will say that $H$ is *analytically decomposable* if:

*(i)* $M$ is a real analytic manifold and $d\mu$ is a $C^\infty$ 1−density on $M$;

*(ii)* the mapping $M \in k \mapsto (H(k) + \mathrm{i})^{-1}$ is analytic, with values in $B(\mathcal{H}')$;

*(iii)* $H(k)$ has only discrete spectrum;

*(iv)* the projection $p_{\mathbb{R}} : \Sigma \ni (\lambda, k) \mapsto \lambda \in \mathbb{R}$ is proper.

Recall that the energy-momentum set $\Sigma$ was defined as

$$\Sigma = \{(\lambda, k) \in \mathbb{R} \times M \mid \lambda \in \sigma(H(k))\}.$$

Note that since $k \mapsto (H(k) + \mathrm{i})^{-1}$ is continuous, $\Sigma$ is a closed subset of $\mathbb{R} \times M$.

The result of [**GN1**] that we will need is the following.

THEOREM 3.4.2. *Let $H$ be an analytically decomposable operator. Then there is a discrete set $\tilde{\tau} \subset \mathbb{R}$, called the* threshold set, *such that for any finite union of compact intervals $I$ disjoint from $\tilde{\tau}$ there is a self-adjoint operator $B_I$ satisfying the following conditions:*

*(i) $H \in C^1(B_I)$ and $[H, \mathrm{i}B_I]$ extends to a bounded operator on $\mathcal{H}$;*

*(ii) $[H, \mathrm{i}B_I] \geq c(\lambda)$ at $H = \lambda$, where $c(\lambda) \geq 0$ for all $\lambda \in \mathbb{R}$ and $c(\lambda) \geq c_I > 0$ for $\lambda \in I$;*

*(iii) we have:*

$$B_I = \phi_0(k) + \phi_1(k) D_k, \ [H, \mathrm{i}B_I] = \phi_2(k),$$

*where $\phi_i(k), i = 0, 1, 2$, are bounded operators on $\mathcal{H}'$ of the form $f_i(k, H(k))$ for $f_i \in C_0^\infty(M \times (\mathbb{R} \backslash \tilde{\tau}))$. Moreover, given $\epsilon > 0$, we may choose $B_I$ so that*

$$B_I = \mathbf{1}_{I_\epsilon}(H) B_I = B_I \mathbf{1}_{I_\epsilon}(H), \tag{3.4.1}$$

*where $I_\epsilon$ is the $\epsilon$-neighbourhood of $I$. (The constant $c_I$ in (ii) may depend on $\epsilon$.)*

COROLLARY 3.4.3. *Let $I_0 \subset \mathbb{R}$ be a finite union of compact intervals. Let $H$ satisfy the conditions of Definition 3.4.1, with (iii), (iv) replaced by:*

*(iii') $H(k)$ has only discrete spectrum in $I_0$;*

*(iv') the restriction of the projection $p_{\mathbb{R}} : \Sigma \ni (\lambda, k) \mapsto \lambda \in \mathbb{R}$ to $p_{\mathbb{R}}^{-1}(I_0)$ is proper.*

*Then there is a discrete set $\tilde{\tau} \subset I_0$ such that for any finite union of compact intervals $I \subset I_0 \backslash \tilde{\tau}$ there is a self-adjoint operator $B_I$ satisfying the conclusions (i)–(iii) of Theorem 3.4.2.*

We call a direct integral $H = \int_M H(k)dk$ satisfying the properties *(i)–(ii)* of Definition 3.4.1 and *(iii')–(iv')* of Corollary 3.4.3 for some $I_0 \subset \mathbb{R}$ *analytically decomposable over* $I_0$.

We now present the main argument in the construction of the conjugate operator in Theorem 3.4.2; a few technical parts are omitted. The full proof of the theorem is of course given in [**GN1**].

*Sketch of proof of Theorem 3.4.2.* We define the multiplicity function **mult** on $\mathbb{R} \times M$:

$$\mathbf{mult}(\lambda, k) = \dim(\operatorname{Ran} \mathbf{1}_{\{\lambda\}}(H(k))),$$

with values in nonnegative integers. Note that $\mathbf{mult}(\lambda, k) = 0$ if and only if $(\lambda, k) \notin \Sigma$.

A *stratification* of a real analytic manifold $N$ is a locally finite partition $\mathcal{S} = (S_i)_{i\in I}$ of $N$ such that:

- $S_i$ are connected analytic submanifolds of $N$;
- if $\overline{S_i} \cap S_j \neq \emptyset$, then $S_j \subset \overline{S_i}$.

The sets $S_i$ are called *strata* of the stratification $\mathcal{S}$. Using the analytic properties of $H(k)$ and a classical result from the theory of analytic singularities, one proves that there are stratifications $\mathcal{S}$ and $\mathcal{T}$ of $\mathbb{R}\times M$ and $\mathbb{R}$ respectively, such that:

- for all $S_i \in \mathcal{S}$, $p_{\mathbb{R}}(S_i) = T_j \in \mathcal{T}$ and $\operatorname{rank}(p_{\mathbb{R}}|_{S_i}) = \dim(T_j)$;
- $\mathbf{mult}(\lambda, k)$ is constant on each $S_i \in \mathcal{S}$.

We define the threshold set $\tilde{\tau}$ as

$$\tilde{\tau} := \bigcup_{\dim(T_j)=0} T_j.$$

The local finiteness of $\mathcal{T}$ implies that the set $\tilde{\tau}$ is discrete.

We can now explain the construction of the conjugate operator $B_I$. We begin by constructing "local" conjugate operators. Let $I'$ be a finite union of closed intervals disjoint from $\tilde{\tau}$ such that $I \subset \operatorname{Int} I'$. Pick functions $\chi, \chi_1 \in C_0^\infty(I')$ such that $\chi \equiv 1$ on $I$ and $\chi_1 \equiv 1$ on $\operatorname{supp}\chi$.

For any $k_0 \in M$, there are at most finitely many values of $\lambda \in \operatorname{supp}\chi_1$ such that $\lambda \in \sigma(H(k_0))$; we will label them $\lambda_i$, $i = 1, \dots, n$ $(n = n(k_0))$. Let $\lambda_0$ be one of these values, and let $S_0 \in \mathcal{S}$ be a stratum to which $(k_0, \lambda_0)$ belongs. We may choose an open neighbourhood $I_0 \times V_0 \subset \mathbb{R} \times M$, $I_0 \subset I'$, of $(k_0, \lambda_0)$ so that

$$\pi_0(k_0) = \mathbf{1}_{\{\lambda_0\}}(H(k_0)),$$

where we denote

$$\pi_0(k) = \mathbf{1}_{I_0}(H(k)).$$

Moreover, if $V_0$ is small enough, there is a system of local coordinates $(k', k'')$ on $V_0$ in which:

$$S_0 \cap (I_0 \times V_0) = \{(f_0(k', 0), (k', 0)) : (k', 0) \in V_0\},$$

for some real-analytic function $f_0(k)$ defined in a neighbourhood of $V_0$. We may also assume that $V_0$ is the same for all of the eigenvalues $\lambda_i$ of $H(k_0)$.

We define the local conjugate operator $B_{k_0,\lambda_0}$:

$$B_{k_0,\lambda_0} = \int_{V_0}^{\oplus} \pi_0(k) \circ \frac{1}{2}\left(\frac{\partial f_0}{\partial k'} D_{k'} + \text{h.c.}\right) \circ \pi_0(k) dk.$$

The commutator

$$F_{k_0,\lambda_0} := [H_{V_0}, iB_{k_0,\lambda_0}],$$

where $H_{V_0} = \int_{V_0}^{\oplus} H(k)dk$, may be written as a direct integral $F_{k_0,\lambda_0} = \int_{V_0}^{\oplus} F_{k_0,\lambda_0}(k)$, where

$$F_{k_0,\lambda_0}(k) = \pi_0(k)\frac{\partial f_0}{\partial k'}\frac{\partial}{\partial k'}(H(k)\pi_0(k))\pi_0(k)$$

extends to a bounded self-adjoint operator on $\mathcal{H}'$. Using the identities:

$$H(k',0)\pi_0(k',0) = f_0(k',0)\pi_0(k',0),$$

$$\pi_0(k)\left(\frac{\partial}{\partial k}\pi_0(k)\right)\pi_0(k) = 0$$

(the second one follows by differentiating $(\pi_0(k))^2 = \pi_0(k)$), we obtain that

$$F_{k_0,\lambda_0}(k) = \pi_0(k)\left(\frac{\partial f_0}{\partial k'}\right)^2 \pi_0(k).$$

Hence

$$F_{k_0,\lambda_0}(k) \geq 0 \text{ for all } k \in V_0. \tag{3.4.2}$$

Moreover, since $\lambda_0 \notin \tilde{\tau}$, we have $\frac{\partial f_0}{\partial k'}|_{k=k_0} \neq 0$, so that

$$F_{k_0,\lambda_0}(k_0) \geq c_0\pi_0(k_0), \ c_0 > 0. \tag{3.4.3}$$

Let

$$B_{k_0} = \sum_{i=1}^{n(k_0)} B_{k_0,\lambda_i(k_0)}.$$

Then

$$[H_{V_0}, \mathrm{i}B_{k_0}] = \int_{V_0}^{\oplus} F_{k_0}(k)dk,$$

where

$$F_{k_0}(k) = \sum_{i=1}^{n(k_0)} F_{k_0,\lambda_i(k_0)}(k).$$

By (3.4.2), we have

$$[H_{V_0}, \mathrm{i}B_{k_0}] \geq 0. \tag{3.4.4}$$

Moreover, (3.4.3) implies that

$$\chi_1(H(k_0))F_{k_0}(k_0)\chi_1(H(k_0)) \geq c_1\chi_1^2(H(k_0))$$

for some positive constant $c_1$. By continuity,

$$\chi_1(H(k))F_{k_0}(k)\chi_1(H(k)) \geq c_1\chi_1^2(H(k)) - c_1/2 \tag{3.4.5}$$

for $k \in U_0$, where $U_0 \subset V_0$ is a small enough neighbourhood of $k_0$. Multiplying (3.4.5) from both sides by $\chi(H(k))$, we obtain:

$$\chi(H(k))F_{k_0}(k)\chi(H(k)) \geq \frac{c_1}{2}\chi^2(H(k)),\ k \in U_0. \tag{3.4.6}$$

The set $K$ of $k \in M$ for which $\sigma(H(k)) \cap I' \neq \emptyset$ is compact, hence it can be covered by a finite number of open neighbourhoods $U_{k_i}$ chosen as above. Let $\{q_i\}$ be a smooth partition of unity associated with this covering: $q_i \in C_0^\infty(U_{k_i})$, $q_i \geq 0$, and $\sum q_i^2 \equiv 1$ on $K$. Let $B_{k_i}$ be the local conjugate operators constructed as above. Then we define

$$B_I := \sum q_i(k)B_{k_i}q_i(k).$$

Using (3.4.4), we obtain that

$$\begin{aligned}[H, \mathrm{i}B_I] &= \sum[H, q_i(k)\mathrm{i}B_{k_i}q_i(k)] \\ &= \sum q_i(k)[H_{V_i}, \mathrm{i}B_{k_i}]q_i(k) \geq 0;\end{aligned}$$

in particular, $[H, \mathrm{i}B_I] \geq 0$ at $H = \lambda$ for any $\lambda \in \mathbb{R}$. Next, from (3.4.6) we have

$$\chi(H(k))\Big(\sum q_i(k)F_{k_i}(k)q_i(k)\Big)\chi(H(k)) \geq \frac{c_1}{2}\chi^2(H(k))$$

for all $k \in K$. Hence

$$\chi(H)[H, \mathrm{i}B_I]\chi(H) \geq \frac{c_1}{2}\chi^2(H),$$

which implies *(ii)*.

To check that $B_I$ is self-adjoint one may use Nelson's commutator theorem (Lemma 1.2.5). Let $\Omega$ be an open neighbourhood of $\bigcup U_i$ in $M$ such that $\overline{\Omega}$ is compact and $\partial\Omega$ is smooth. Then there is a positive definite metric $g$ on $\overline{\Omega}$ such that $d\mu(k) = |\det g|^{1/2}(k)dk$. We define

$$N := \mathbf{1} - q\Delta_g q \text{ on } \mathcal{D}(N) = C_0^\infty(M; \mathcal{H}'),$$

where $\Delta_g$ is the Laplace-Beltrami operator on $\Omega$ associated to the metric $g$ and $q$ is a function in $C_0^\infty(\Omega)$ such that $q \equiv 1$ on $\bigcup \operatorname{supp} q_i$. Then one verifies that $N$ is essentially self-adjoint and that $B_i$ and $N$ satisfy the conditions of Nelson's theorem (Lemma 1.2.5*(i)*). Hence $B_I$ is essentially self-adjoint on $\mathcal{D}(N)$. Finally, since $[H, \mathrm{i}B_I]$ is bounded (this follows directly from the construction of $B_I$), by Lemma 3.2.2 we have $H \in C^1(B_I)$. The remaining properties of $B_I$ are easily seen from the construction.

## 3.5. Two-particle neutral systems

In this section we analyze 2-particle systems with total electric charge zero. The results presented here, apart from being of independent interest, will also be needed in the proof of the Mourre estimate for 3−particle systems containing a neutral pair (Sections 3.6, 3.7).

Throughout this section we consider Hamiltonians $H$ of the form

$$H = \int_{\mathcal{K}}^{\oplus} H(k)dk, \text{ acting on } \mathcal{H} = L^2(\mathcal{K}, dk; \mathcal{H}'). \tag{3.5.1}$$

If the space dimension $d = 3$, we assume that $\mathcal{H}' = L^2(X)$, $X = \mathbb{R}^2_y \times \mathbb{R}^1_z$, $x = (y, z)$, and

$$H(k) = \frac{1}{2}(k - \mathbf{A}_1 x)^2 + \frac{1}{2}(D_x - \mathbf{A}_2 x)^2 + V(x), \tag{3.5.2}$$

where $\mathbf{A}_1$ is a linear mapping $X \mapsto \mathcal{K}$ such that $\mathrm{Ker}(\mathbf{A}_1) = Z$ (i.e., $\mathbf{A}_1$ is bijective on $Y$), and $\mathbf{A}_2$ is a linear antisymmetric mapping $X \to X'$ such that $\mathbf{A}_2 z = 0$.

If $d = 2$, we assume that $\mathcal{H}' = L^2(Y)$, $Y = \mathbb{R}^2_y$, and

$$H(k) = \frac{1}{2}(k - \mathbf{A}_1 y)^2 + \frac{1}{2}(D_y - \mathbf{A}_2 y)^2 + V(y), \tag{3.5.3}$$

where $\mathbf{A}_1$ is a linear bijection $Y \mapsto \mathcal{K}$ and $\mathbf{A}_2$ is a linear antisymmetric mapping $Y \mapsto Y'$.

In both cases we assume that $\mathcal{K} = \mathbb{R}^2$, and that $V$ is a real multiplication operator (on $X$ or $Y$ respectively) satisfying *(V1), (V2), (M1), (M2)*. We will also denote:

$$H_0(k) = H(k) - V, \; H_0 = H - V = \int_{\mathcal{K}}^{\oplus} H_0(k)dk.$$

As demonstrated in Subsection 1.3.9, the Hamiltonian of a neutral pair ($N = 2$, $q_1 + q_2 = 0$, $q_i \neq 0$) has the above form. This is the application that the reader should keep in mind; however, the results presented here will also apply to more general operators $H$ which admit the given description.

**3.5.1. Three-dimensional case: basic results.** We first study the basic spectral properties of the fiber Hamiltonians $H(k)$ in the case $d = 3$. The case $d = 2$ is similar but simpler; it will be discussed briefly in the next subsection.

LEMMA 3.5.1. *Assume that (V1), (V2) hold. Let $T(k)$ be the unitary transformation implementing the symplectic transformation $\chi_T(k)$ given by:*

$$\chi_T(k) : (x, \xi) \to (\tilde{x}, \tilde{\xi}),$$

$$(\tilde{y}, \tilde{\eta}) = (y - y_0, \eta + \mathbf{A}_2\, y_0),$$

$$(\tilde{z}, \tilde{\zeta}) = (z, \zeta),$$

*where $y_0 \in Y$ solves $\mathbf{A}_1\, y_0 = k$. Let $H_T(k) = T(k)H(k)T(k)^*$. Then:*

*(i)* $H_T(k) = H_{0,T} + V(x + y_0)$, *where*

$$H_{0,T} = \frac{1}{2}(D_x - \mathbf{A}_2\, x)^2 + \frac{1}{2}(\mathbf{A}_1\, y)^2;$$

*(ii)* $H_T(k)$ *is self-adjoint for any* $k \in \mathcal{K}$, *with domain*

$$\mathcal{D}_0 = \{u \in L^2(X) : \ D^2 u, y^2 u \in L^2(X)\}.$$

*(iii)* $H_T(k) \to H_{0,T}$ *in the norm resolvent sense as* $|k| \to \infty$.

PROOF. *(i)* Applying $\chi_T(k)$ to the symbol $h(k)$ of $H(k)$, we obtain

$$h_T = \frac{1}{2}(\xi - \mathbf{A}_2 x)^2 + \frac{1}{2}(\mathbf{A}_1 y)^2 + V(x + y_0);$$

now *(i)* follows from Segal's theorem (see Subsection 1.2.5).

*(ii)* Let $N = C(D_x^2 + y^2 + 1)$, $C \gg 1$. Since $\mathbf{A}_2 x = \mathbf{A}_2 y$, $H_{0,T}$ is well defined as a symmetric operator on $\mathcal{D}(N)$. Moreover, the commutator $[H_{0,T}, N]$ is bounded on $\mathcal{D}(N)$. By Nelson's commutator theorem (Lemma 1.2.5), $H_{0,T}$ is essentially self-adjoint on $\mathcal{D}_0 = \mathcal{D}(N)$. Since $(\mathbf{A}_1 y)^2 \geq cy^2$, the graph norms for $H_{0,T}$ and for $N$ are equivalent, hence $H_{0,T}$ is self-adjoint on $\mathcal{D}_0$. By *(V1)* and Proposition 1.2.1, $V$ is $H_{0,T}$ bounded with relative bound 0. Hence $H_T(k)$ is self-adjoint on $\mathcal{D}_0$.

*(iii)* We need to prove that for $\operatorname{Im} z \neq 0$:

$$\begin{aligned} &\|(H_T(k) - z)^{-1} - (H_{0,T} - z)^{-1}\| \\ &= \|(H_T(k) - z)^{-1} V(\cdot + y_0)(H_{0,T} - z)^{-1}\| \to 0 \end{aligned}$$

as $|y_0| \to \infty$.

Note first that $\|(H_T(k) - z)^{-1}\| \leq |\operatorname{Im} z|^{-1}$ uniformly in $k$. Next, let $\epsilon > 0$. By *(V2)*, we may write $V = V_1 + V_2$, where $V_1$ is supported in $|x| \leq R$ for some $R > 0$, and $\|V_2(D^2 + 1)^{-1}\| \leq \epsilon$. By Proposition 1.2.1,

$$\|V_2(H_{0,T} - z)^{-1}\| < C_0 \epsilon, \tag{3.5.4}$$

where $C_0$ does not depend on $\epsilon$.

Since $\operatorname{supp} V_1 \subset \{|x| \leq R\}$, we may write $V_1 = V_1 \chi$, where $\chi$ is a function in $C_0^\infty(X)$ such that $\chi = 1$ for $|x| \leq R$. Using *(V1)*, we have:

$$\begin{aligned} &\|V_1(\cdot + y_0)(H_{0,T} - z)^{-1}\| \\ &= \|V_1(\cdot + y_0)\chi(\cdot + y_0)(H_{0,T} - z)^{-1}\| \\ &\leq 2\epsilon\|D^2 \chi(\cdot + y_0)(H_{0,T} - z)^{-1}\| + C_\epsilon \|\chi(\cdot + y_0)(H_{0,T} - z)^{-1}\|. \end{aligned} \tag{3.5.5}$$

By Proposition 1.2.1, the first norm on the right-hand side of (3.5.5) is bounded uniformly in $y_0$ and $\epsilon$. The second one can be estimated by

$$\begin{aligned} &\|\chi(\cdot + y_0)(H_{0,T} - z)^{-1}\| \\ &\leq \sup_{x \in X} |\chi(x + y_0)(|y|^2 + 1)^{-1}| \cdot \|(|y|^2 + 1)(H_{0,T} - z)^{-1}\| \\ &= O(|y_0|^{-2}), \end{aligned} \tag{3.5.6}$$

where we have used that, by *(ii)*, $(|y|^2+1)(H_{0,T}-z)^{-1}$ is bounded.

Collecting (3.5.4), (3.5.5), and (3.5.6), we obtain that for any $\epsilon > 0$:

$$\|V(\cdot + y_0)(H_{0,T}-z)^{-1}\| \le M\epsilon + O(|k|^{-1})$$

as $|k| \to \infty$, with $M$ independent of $\epsilon$. This proves *(iii)*. □

We recall that

$$H_0(k) = \frac{1}{2}(k - \mathbf{A}_1 y)^2 + \frac{1}{2}(D_x - \mathbf{A}_2 y)^2, \text{ acting on } L^2(X),$$

and set

$$H_0^{\perp}(k) = \frac{1}{2}(k - \mathbf{A}_1 y)^2 + \frac{1}{2}(D_y - \mathbf{A}_2 y)^2, \text{ acting on } L^2(Y).$$

LEMMA 3.5.2. *Assume (V1), (V2). Then for all $k \in \mathcal{K}$:*

*(i) $\sigma(H_0^{\perp}(k)) = \tau$ is a discrete subset of $\mathbb{R}$, independent of $k$, whose only accumulation point is $+\infty$;*

*(ii) $\sigma(H_0(k)) = \sigma_{\mathrm{ess}}(H_0(k)) = [\nu_0, \infty)$, where $\nu_0 = \inf \tau$.*

We remark that if $H$ is (as in our intended application) the magnetic Hamiltonian of a neutral pair of particles, $\tau$ is the set of the Landau levels:

$$\Lambda_n = \frac{|\vec{B}||q_1|}{2}(2n+1), \ n = 0, 1, 2, \ldots,$$

where $q_1$ and $q_2 = -q_1$ are the electric charges of the particles in question.

PROOF. Let $T(k)$ be as in Lemma 3.5.1. Then $T(k)H_0^{\perp}(k)T^*(k)$ is independent of $k$, hence so is

$$\sigma(H_0^{\perp}(k)) = \sigma(T(k)H_0^{\perp}(k)T^*(k)).$$

To prove that $\sigma(H_0^{\perp}(k))$ is a discrete set accumulating only at $+\infty$, we follow the same argument as in the proof of Lemma 3.5.1*(ii)*: in fact, $H_0^{\perp}(k)$ is self-adjoint with the domain

$$\{u \in L^2(Y) \mid D_y^2 u, y^2 u \in L^2(Y)\},$$

and therefore has compact resolvent. We also remark that if $H$ is the magnetic Hamiltonian of a neutral pair, *(i)* may be verified directly using the calculations in Subsection 1.3.9.

Finally, *(ii)* follows from *(i)* and the fact that $H_0(k) = D_z^2 \otimes \mathbf{1} + \mathbf{1} \otimes H_0^{\perp}(k)$. □

THEOREM 3.5.3. *Assume that (V1), (V2), (M1) hold. Then:*

*(i) $H(k)$ is self-adjoint with the domain $\mathcal{D}_0 = \{u \in L^2(X) : \ D^2 u, \ y^2 u \in L^2(X)\}$;*

*(ii) $\sigma_{\mathrm{ess}}(H(k)) = [\nu_0, \infty)$;*

*(iii) the eigenvalues of $H(k)$ may accumulate only at $\tau$;*

*(iv) if $E(k)$ is a discrete eigenvalue of $H(k)$, then $\lim_{|k|\to\infty} E(k) = \nu_0$;*

*(v) the map* $\mathcal{K} \ni k \mapsto H(k)$ *is analytic in the norm resolvent sense.*

In particular, Theorem 3.5.3 implies that $H$ is self-adjoint with domain $\mathcal{D}(H_0)$.

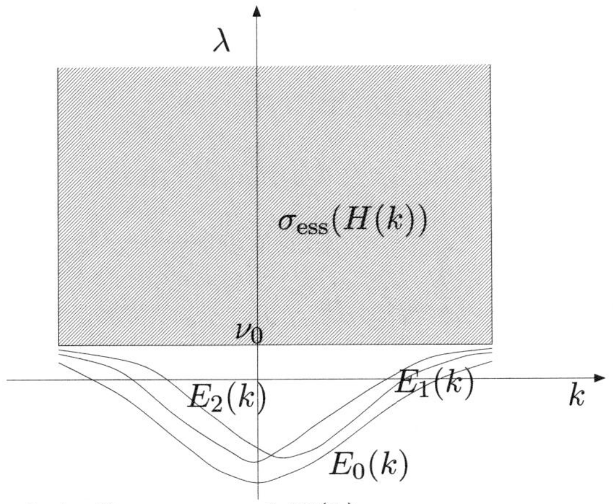


**Fig. 3.1.** Spectrum of $H(k)$

PROOF. *(i)* is an immediate consequence of Lemma 3.5.1*(ii)* and of the invariance of $\mathcal{D}_0$ under the transformations $T(k)$.

*(ii)* It suffices to prove that $(H(k)+\mathrm{i})^{-1}V(H_0(k)+\mathrm{i})^{-1}$ is compact; but this follows from *(i)* and Proposition 1.2.1.

*(iii)* follows from Lemma 3.5.7 (below) and Lemma 3.2.6.

*(iv)* follows from Lemma 3.5.1*(ii)*.

*(v)* is easily obtained by differentiating $(H(k)-z)^{-1}$ with respect to $k$ and using the characterization of the domain of $H(k)$ given in *(i)*. □

In the sequel we will need the following result on the exponential decay of the eigenfunctions belonging to discrete eigenvalues of $H(k)$. The proof follows a standard Combes-Thomas argument.

PROPOSITION 3.5.4. *Assume (V1), (V2). Let* $\chi \in C_0^\infty((-\infty, \nu_0))$. *Then for all* $\alpha$ *such that*

$$\sup(\operatorname{supp}\chi) + \frac{1}{2}\alpha^2 < \nu_0,$$

*and all multiindices* $\beta$, $|\beta| \geq 0$, *one has*

$$\|\mathrm{e}^{\alpha|x|}\partial_k^\beta \chi(H(k))\| \leq C_{\alpha,\beta} < \infty.$$

PROOF. For $R \gg 1$, let $H_R(k)$ be the operator obtained from $H(k)$ by replacing $V$ by $V_R := 1_{|x|\geq R}V$. From *(V2)* and Kato's inequality we have

$$\lim_{R\to\infty} \|V_R(H_0(k)+\mathrm{i})^{-1}\| = 0.$$

We choose $R$ to be large enough so that $\chi(H_R(k)) = 0$. Using (1.2.21), we obtain:

$$\begin{aligned}\chi(H(k)) = &\ \chi(H(k)) - \chi(H_R(k))\\ = &\ -\tfrac{\mathrm{i}}{2\pi}\textstyle\int_{\mathbb{C}} \partial_{\bar z}\tilde\chi(z)(z-H_R(k))^{-1}V_{1,R}(z-H(k))^{-1}dz\wedge d\bar z,\end{aligned}$$

for $V_{1,R} = 1_{|x|\leq R}V$, where $\tilde\chi$ is an almost analytic extension of $\chi$ (see Definition 1.2.8).

Pick a function $\phi(s) \in C^\infty(\mathbb{R}^+)$, satisfying:

$$\begin{aligned}&\phi(s) = s - 3/2 \text{ for } s\geq 2,\\ &\phi(s) = 0 \text{ for } s \leq 1,\\ &0 \leq \phi'(s) \leq 1.\end{aligned}$$

Let $\phi_R(x) := R\phi(\frac{|x|}{R})$ and $H_{R,\alpha}(k) := \mathrm{e}^{\alpha\phi_R}H_R\mathrm{e}^{-\alpha\phi_R}$. Since $\mathrm{e}^{\alpha\phi_R}V_{1,R} = V_{1,R}$, we have

$$\begin{aligned}&\mathrm{e}^{\alpha\phi_R}\chi(H(k))\\ = &\ -\tfrac{\mathrm{i}}{2\pi}\textstyle\int_{\mathbb{C}} \partial_{\bar z}\tilde\chi(z)(z-H_{R,\alpha}(k))^{-1}V_{1,R}(z-H(k))^{-1}dz\wedge d\bar z.\end{aligned} \tag{3.5.7}$$

Now

$$H_{R,\alpha}(k) - z = P(k) + \mathrm{i}Q,$$

where

$$\begin{aligned}&P(k) = H_R(k) - \tfrac{1}{2}\alpha^2|\nabla\phi_R|^2 - \mathrm{Re}\, z,\\ &Q = \tfrac{1}{2}\left(D_x\cdot\nabla\phi_R + \nabla\phi_R\cdot D_x\right) - \mathrm{Im}\, z.\end{aligned}$$

If $\mathrm{supp}\,\tilde\chi$ is sufficiently close to $\mathrm{supp}\,\phi$, we have $P \geq c > 0$, so that $P$ is invertible. Moreover, by Theorem 3.5.3 $Q$ is a relatively bounded perturbation of $P$ with the relative bound 0. Therefore

$$\|(H_{R,\alpha}(k)-z)^{-1}\| \leq \|P^{-\frac{1}{2}}\|^2\|(1-\mathrm{i}P^{-\frac{1}{2}}QP^{-\frac{1}{2}})^{-1}\| \leq C,$$

for $z\in\mathrm{supp}\tilde\chi$. Thus the left-hand side of (3.5.7) is bounded, which completes the proof of the lemma for $\beta = 0$. To obtain the bound for $\beta \neq 0$, we differentiate (3.5.7) with respect to $k$. We use that $\partial_k H_{R,\alpha}(k) = k - 2\mathbf{A}_1 x$ is bounded with respect to $P(k)^{\frac{1}{2}}$ and argue as above. □

**3.5.2. Three-dimensional case: the Mourre theory for $H(k)$.** In this subsection we prove a Mourre estimate for $H(k)$ in the 3−dimensional case, and obtain a number of related results which will be needed later. Let

$$A = \frac{1}{2}(\langle z, D_z\rangle + \langle D_z, z\rangle)$$

be the generator of dilations in the $z$ variables, acting on $L^2(X)$.

LEMMA 3.5.5. *Assume that (V1), (V2), (M1) hold. Let $f_\delta \in C_0^\infty(\mathbb{R})$ be a cutoff function supported in $[E-\delta, E+\delta]$ for some $E \in \mathbb{R}$, $\delta > 0$. Then*

$$\begin{aligned} &f_\delta(H(k))[H(k), \mathrm{i}A]f_\delta(H(k)) \\ &= f_\delta(H(k))D_z^2 f_\delta(H(k)) + K_1(k), \end{aligned} \tag{3.5.8}$$

*where:*

$$\begin{aligned} &K_1(k) \text{ is compact on } L^2(X), \\ &k \mapsto K_1(k) \text{ is norm continuous}, \\ &\|K_1(k)\| \to 0 \text{ as } k \to \infty. \end{aligned} \tag{3.5.9}$$

PROOF. We have

$$[H(k), iA] = D_z^2 - \langle z, \nabla_z V\rangle, \tag{3.5.10}$$

hence (3.5.8) holds with

$$K_1(k) = -f_\delta(H(k))\langle z, \nabla_z V\rangle f_\delta(H(k)).$$

This operator is compact on $L^2(X)$ by *(M1)* and Theorem 3.5.3 *(i)*. The norm continuity of $K_1(k)$ follows from Theorem 3.5.3*(v)* and Proposition 1.2.1. To prove the last statement in (3.5.9), we write

$$\begin{aligned} &T(k)K_1(k)T(k)^\star \\ &= -f_\delta(H_T(k))z\nabla_z V(x+y_0)f_\delta(H_T(k)) \end{aligned}$$

for $y_0 = \mathbf{A}_1^{-1}k$, and proceed as in the proof of Lemma 3.5.1*(iii)*. □

We define

$$d(E) := \begin{cases} \text{dist}\,(E, \tau), & E \geq \nu_0, \\ 0, & E < \nu_0, \end{cases} \tag{3.5.11}$$

where $\tau$ is the *threshold set* defined in Lemma 3.5.2.

LEMMA 3.5.6. *Assume (V1), (V2), (M1), and let $E \in \mathbb{R}$. Then for any $\epsilon > 0$ there is a $\delta > 0$ such that for any function $f_\delta \in C_0^\infty((E-\delta, E+\delta))$ we have*

$$f_\delta(H(k))D_z^2 f_\delta(H(k)) \geq (2d(E) - \epsilon)f_\delta^2(H(k)) + K_2(k), \tag{3.5.12}$$

*where $K_2(k)$ are as in (3.5.9).*

PROOF. Since $D_z^2 = 2(H_0(k) - H_0^{\perp}(k))$, $D_z$ commutes with $H_0^{\perp}(k)$, and $\sigma(H_0^{\perp}(k)) = \tau$, for sufficiently small $\delta > 0$ and all $k$ we have

$$f_\delta(H_0(k))D_z^2 f_\delta(H_0(k)) \geq 2d(E) f_\delta(H_0(k))^2. \tag{3.5.13}$$

Writing $V$ as $VF(|x| \leq R) + VF(|x| \geq R)$, using the hypotheses *(V1)*, *(V2)*, and letting $R \to +\infty$, we obtain that

$$(z - H_0(k))^{-1} - (z - H(k))^{-1} = -(z - H_0(k))^{-1} V (z - H(k))^{-1}$$

is compact with norm $O(|\mathrm{Im}\, z|^{-2})$. This implies that $f(H_0(k)) - f(H(k))$ is compact on $L^2(X)$ for any $f \in C_0^\infty(\mathbb{R})$, hence (3.5.12) holds. The properties (3.5.9) of $K_2$ are verified as in Lemma 3.5.5. □

As a consequence, we obtain the following Mourre estimate for $H(k)$.

LEMMA 3.5.7. *Assume (V1), (V2), (M1). Then:*

*(i)* $H(k) \in C^1(A)$ *for all* $k \in \mathcal{K}$;

*(ii) for any* $k \in \mathcal{K}$, $E \in \mathbb{R}$, *and* $\epsilon > 0$, *there is an operator* $K_\epsilon(k)$, *compact on* $L^2(X)$, *such that*

$$[H(k), \mathrm{i}A] > 2d(E) - \epsilon + K_\epsilon(k) \text{ at } H(k) = E.$$

PROOF. Using (3.5.10) and that $\mathrm{e}^{\mathrm{i}sA}$ preserves $\mathcal{D}_0$, we find that the conditions of Proposition 3.2.1 are satisfied, which yields *(i)*. Part *(ii)* is an immediate consequence of of Lemmas 3.5.5 and 3.5.6. □

Later on, we will also need the following two lemmas.

LEMMA 3.5.8. *Assume (V1), (V2), (M1).*

*(i) Let* $E \in \mathbb{R}$. *Then there is a constant* $C > 0$ *such that*

$$[H(k), iA] \geq_* 2d(E) \text{ at } H(k) = E \tag{3.5.14}$$

*uniformly for* $|k| > C$.

*(ii) If* $I \subset \mathbb{R}$ *is a compact interval, for any* $\epsilon > 0$ *there is a* $C > 0$ *such that (3.5.14) holds uniformly for* $|k| > C$ *and* $E \in I$.

PROOF. *(i)* Choose a $\delta > 0$ such that (3.5.12) holds for all $f_\delta \in C_0^\infty([E-\delta, E+\delta])$, and a $f_\delta$ such that $f_\delta \equiv 1$ on $[E - \delta/2, E + \delta/2]$. Applying Lemmas 3.5.5 and 3.5.6 to this function, we find a constant $C > 0$ such that $\|K_1(k)\| < \epsilon$ and $\|K_2(k)\| < \epsilon$ for $|k| > C$. Then

$$\begin{aligned}
&f_\delta(H(k))[H(k), iA] f_\delta(H(k)) \\
&\geq f_\delta(H(k)) D_z^2 f_\delta(H(k)) - \epsilon \\
&\geq (2d(E) - \epsilon) f_\delta^2(H(k)) - 2\epsilon
\end{aligned}$$

for all $|k| > C$, which implies *(i)*.

*(ii)* Let $I \subset \mathbb{R}$ be a compact interval, and let $\epsilon > 0$. Since $d(\lambda)$ is continuous, it is uniformly continuous on $I$. We may therefore find $\bar{\delta} > 0$ such that

$$|d(\lambda) - d(\lambda')| < \epsilon/2 \text{ if } |\lambda - \lambda'| < \bar{\delta}, \ \lambda, \lambda' \in I.$$

By *(i)*, for each $\lambda \in I$ there are constants $\delta(\lambda) > 0$ and $C(\lambda)$ such that

$$f(H(k))[H(k), \mathrm{i}A]f(H(k)) \geq (2d(\lambda) - \epsilon/2)f^2(H(k))$$

for any function $f \in C_0^\infty(I_\lambda)$, where $I_\lambda = (\lambda - \delta(\lambda), \lambda + \delta(\lambda))$, and for all $|k| > C(\lambda)$. Moreover, we can assume that $\delta(\lambda) < \bar{\delta}$ for all $\lambda$.

The intervals $I_\lambda$, $\lambda \in I$, cover $I$; since $I$ is compact, we may choose a finite subcovering $I_{\lambda_1}, \ldots, I_{\lambda_N}$. Let $\delta > 0$ be the Lebesgue number for this covering, i.e., a number $\delta$ such that for any $\lambda \in I$ the interval $(\lambda - \delta, \lambda + \delta)$ is entirely contained in one of the $I_{\lambda_j}$'s. Then any function $f$ supported in an interval $(\lambda - \delta, \lambda + \delta)$ is also supported in $I_{\lambda_j}$ for some $j$, so that:

$$\begin{aligned} f(H(k))[H(k), \mathrm{i}A]f(H(k)) &\geq (2d(\lambda_j) - \epsilon/2)f^2(H(k)) \\ &\geq (2d(\lambda) - \epsilon)f^2(H(k)), \end{aligned}$$

for all $|k| > C = \max(C(\lambda_1), \ldots, C(\lambda_n))$. □

LEMMA 3.5.9. *Assume (V1), (V2), (M1). Let*

$$N = (D_k^2 + k^2) \otimes \mathbf{1} + \mathbf{1} \otimes (D_x^2 + x^2)$$

*be the harmonic oscillator on* $\mathcal{H}$. *Let* $\chi \in C^\infty(\mathbb{R})$ *satisfy* $\chi \equiv 1$ *on* $[c, \infty)$ *for some* $c > 0$. *Then* $\chi(H)A\chi(H)$ *is symmetric on* $\mathcal{D}(N)$ *and:*

*(i)* $\|\chi(H)A\chi(H)u\| \leq C\|Nu\|$, $u \in \mathcal{D}(N)$,

*(ii)* $|(\chi(H)A\chi(H)u, Nu) - (Nu, \chi(H)A\chi(H)u)| \leq C\|N^{\frac{1}{2}}u\|^2$, $u \in \mathcal{D}(N)$.

PROOF. It is easy to see that Lemmas 1.2.3 and 1.2.4 extend to operators $H$ of the form considered in this section. Since $H$ is bounded from below and $\chi \equiv 1$ near $+\infty$, we may write $\chi(H) = \mathbf{1} + \chi_1(H)$, where $\chi_1 \in C_0^\infty(\mathbb{R})$; hence by Lemma 1.2.4 $\chi(H)$ preserves $\mathcal{D}(N)$. It follows that $\chi(H)A\chi(H)$ is well defined and symmetric on $\mathcal{D}(N)$, and that *(i)* holds.

We now prove *(ii)*. By Remark 3.2.4, we may apply the Leibniz rule to $[\chi(H)A\chi(H), N]$, hence it suffices to prove that:

$$N^{-1/2}A[\chi(H), N]N^{-1/2} \text{ is bounded,} \tag{3.5.15}$$

$$N^{-1/2}\chi(H)[A, N]\chi(H)N^{-1/2} \text{ is bounded.} \tag{3.5.16}$$

We first prove (3.5.16), which is easier. We have

$$[A, \mathrm{i}N] = 2D_z^2 - 2z^2,$$

hence $N^{-1/2}[A, \mathrm{i}N]N^{-1/2}$ is bounded. Using also that, by Lemma 1.2.4 and interpolation, $\chi(H)$ preserves $\mathcal{D}(N^{1/2})$, we obtain (3.5.16).

We now turn to (3.5.15). As in the proof of *(i)*, we may replace $\chi$ by $\chi_1$. We must prove that

(3.5.17) $$N^{-1/2}A[\chi_1(H), D_x^2+x^2]N^{-1/2} \text{ is bounded,}$$

(3.5.18) $$N^{-1/2}[\chi(H), D_k^2+k^2]N^{-1/2} = N^{-1/2}[\chi(H), D_k^2]N^{-1/2} \text{ is bounded.}$$

We begin with (3.5.17). By explicit computation, we have:

$$\begin{aligned}&[\chi_1(H), D_x^2+x^2]\\ &= -\chi_1(H)B_1 + B_1\chi_1(H) + \chi_1(H)B_2 - B_2\chi_1(H) + [\chi_1(H), (\mathbf{A}_2x)^2+x^2],\end{aligned}$$

where

$$B_1 = (D_x - \mathbf{A}_2x)^2, \ B_2 = \langle D_x - \mathbf{A}_2x, D_x\rangle + \text{h.c.}.$$

We thus need to prove that:

(3.5.19) $$N^{-1/2}A\chi_1(H)B_iN^{-1/2} \text{ is bounded, } i=1,2,$$

(3.5.20) $$N^{-1/2}AB_i\chi_1(H)N^{-1/2} \text{ is bounded, } i=1,2,$$

(3.5.21) $$N^{-1/2}A[\chi_1(H), (\mathbf{A}_2x)^2+x^2]N^{-1/2} \text{ is bounded.}$$

Clearly,

$$N^{-1/2}\langle x\rangle, \ N^{-1/2}\langle D_x\rangle, \ \langle z\rangle^{-1}\langle D_z\rangle^{-1}A, \ \langle z\rangle^{-1}A\langle D_z\rangle^{-1}, \ \langle D_z\rangle^{-1}A\langle z\rangle^{-1}$$

are bounded; moreover, it is easy to verify that so is $\langle D_z\rangle^s\chi_1(H)\langle D_x - \mathbf{A}_2x\rangle^s$ for $0\le s\le 1$. From this we easily obtain (3.5.19).

The proof of (3.5.20) is slightly more difficult. We begin with $B_1$. Since $N^{-1/2}A\langle z\rangle^{-1}$ is bounded, it suffices to prove that

$$\langle z\rangle(D_x - \mathbf{A}_2x)^2\chi_1(H)N^{-1/2} \text{ is bounded;}$$

using also that $[\langle z\rangle, (D_x-\mathbf{A}_2x)^2]\chi_1(H)$ is bounded by direct computation, we reduce this to:

(3.5.22) $$(D_x - \mathbf{A}_2x)^2\langle z\rangle\chi_1(H)N^{-1/2} \text{ is bounded.}$$

We claim now that

(3.5.23) $$(H+\mathrm{i})[\langle z\rangle, \chi_1(H)] \text{ is bounded.}$$

Indeed, by (1.2.21) it suffices to prove this with $\chi_1(H)$ replaced by $(z-H)^{-1}$. By Lemma 1.2.3 and Proposition 3.2.1, $H\in C^1(\langle z\rangle)$; now (3.5.23) follows by Proposition 3.2.3*(iii)* and direct computation.

By (3.5.23), (3.5.22) is equivalent to:

$$\Big((D_x - \mathbf{A}_2x)^2\chi_1(H)\Big)\Big(\langle z\rangle N^{-1/2}\Big) \text{ is bounded.}$$

But this is clear, since both factors above are bounded. This proves (3.5.20) for $i=1$.

Using that $N^{-1/2}D_x$ is bounded, we reduce the verification of (3.5.20) for $i=2$ to checking that

$$\langle z\rangle\langle D_z\rangle\langle D_x-\mathbf{A}_2x\rangle\chi_1(H)N^{-1/2} \text{ is bounded,}$$

or equivalently,

$$\langle D_z\rangle\langle D_x-\mathbf{A}_2x\rangle\langle z\rangle\chi_1(H)N^{-1/2} \text{ is bounded.}$$

Now the rest of the proof is similar to that for $i=1$.

It remains to prove (3.5.21). By (1.2.21), we have

$$[\chi_1(H),(\mathbf{A}_2x)^2+x^2]=\frac{\mathrm{i}}{2\pi}\int_{\mathbb{C}}\partial_{\bar z}\tilde\chi_1(z)[(z-H)^{-1},(\mathbf{A}_2x)^2+x^2]dz\wedge d\bar z.$$

Since by Lemma 1.2.3 $(z-H)^{-1}$ preserves $\mathcal{D}(N)$, the identity

$$[(z-H)^{-1},(\mathbf{A}_2x)^2+x^2]=(z-H)^{-1}[H,(\mathbf{A}_2x)^2+x^2](z-H)^{-1}$$

is justified. Next, we have

$$\mathrm{i}[H,(\mathbf{A}_2x)^2+x^2]=\langle{}^t\mathbf{A}_2(D_x-\mathbf{A}_2x),\mathbf{A}_2x\rangle+\langle D_x-\mathbf{A}_2x,x\rangle+\text{h.c.}$$

Using also Lemma 1.2.3, we obtain

$$\begin{aligned}&\|\langle D_z\rangle(z-H)^{-1}[H,(\mathbf{A}_2x)^2+x^2](z-H)^{-1}\langle x\rangle^{-1}\|\\&\le C\|\langle D_z\rangle(z-H)^{-1}\langle D_x-\mathbf{A}_2x\rangle\|\|\langle x\rangle(z-H)^{-1}\langle x\rangle^{-1}\|\\&\le C|\mathrm{Im}\, z|^{-3}.\end{aligned}$$

Hence $\langle D_z\rangle[\chi_1(H),(\mathbf{A}_2x)^2+x^2]N^{-1/2}$ is bounded. Since $N^{-1/2}A\langle D_z\rangle^{-1}$ is bounded, this implies (3.5.21). This also completes the proof of (3.5.17).

Finally, we prove (3.5.18). By (1.2.21), we have

$$\begin{aligned}&[\chi_1(H),D_k^2]\\=\;&\tfrac{\mathrm{i}}{2\pi}\textstyle\int_{\mathbb{C}}\partial_{\bar z}\tilde\chi_1(z)(z-H)^{-1}(\langle D_k,k-\mathbf{A}_1(x)\rangle+\text{h.c.})(z-H)^{-1}dz\wedge d\bar z.\end{aligned}$$

Since

$$D_k(z-H)^{-1}=(z-H)^{-1}D_k-\mathrm{i}(z-H)^{-1}(k-\mathbf{A}_1x)(z-H)^{-1},$$

we find that $(z-H)^{-1}$ preserves $\mathcal{D}(D_k)$ with norm $O(|\mathrm{Im}\, z|)^{-2})$, and that

$$\|(H+\mathrm{i})D_k(z-H)^{-1}\langle D_k\rangle^{-1}\|=O(|\mathrm{Im}\, z|^{-2}).$$

Arguing as before, we deduce that

$$\langle D_z\rangle[\chi_1(H),D_k^2]N^{-1/2},\;\; N^{-1/2}A[\chi_1(H),D_k^2]N^{-1/2}$$

are bounded. This completes the proof of (3.5.18) and of (3.5.15). □

**3.5.3. Two-dimensional case: basic results.** The following results are the $d = 2$ analogues of the results of Subsection 3.5.1. The proofs are similar but easier; we leave them as an exercise for the reader.

THEOREM 3.5.10. *Assume that (V1), (V2) hold. Then:*

*(i)* $H(k)$ *is self-adjoint with domain*

$$\mathcal{D}_0 := \{u \in L^2(Y) : \ D_y^2 u, \ y^2 u \in L^2(Y)\};$$

*(ii) the set* $\tau := \sigma(H_0(k))$ *is a discrete subset of* $\mathbb{R}$, *independent of* $k$, *and may accumulate only at* $+\infty$;

*(iii)* $\sigma_{\mathrm{ess}}(H(k)) = \tau$ *for all* $k$;

*(iv) the mapping* $\mathcal{K} \ni k \mapsto H(k)$ *is analytic in the norm resolvent sense;*

*(v)* $\sigma(H(k))$ *converges to* $\tau$ *as* $|k| \to \infty$ *(i.e., for any* $\epsilon > 0$ *there is* $R > 0$ *such that* $\sigma(H(k))$ *is contained in the* $\epsilon$*-neighbourhood of* $\tau$ *for all* $|k| > R$*).*

PROPOSITION 3.5.11. *Assume that (V1), (V2) hold. Let* $\chi \in C_0^\infty(\mathbb{R} \backslash \tau)$. *Then for all* $\alpha$ *such that*

$$\mathrm{dist}\,(\mathrm{supp}\,\chi, \tau) > \frac{1}{2}\alpha^2,$$

*and for all multiindices* $\beta$, $|\beta| \geq 0$, *we have*

$$\|\mathrm{e}^{\alpha|y|}\partial_k^\beta \chi(H(k))\| \leq C_{\alpha,\beta} < \infty.$$

## 3.6. Dispersive channels

We now apply the results of Section 3.5 to the following situation. Let $N = 3$, and assume (in addition to *(Q1)*) that the system contains at least one neutral pair. Without loss of generality, we may assume that the electric charges $q_1$, $q_2$, $q_3$ of the particles satisfy

$$q_i \neq 0, \ i = 1, 2, 3; \ q_1 + q_2 = 0. \tag{3.6.1}$$

Let $a$ be the cluster decomposition $a = \{\{1, 2\}, \{3\}\}$. The purpose of this section is to construct a conjugate operator and prove a Mourre estimate for the channel Hamiltonian $H_a$.

We define the *threshold set* for $\tilde{H}^{a,\mathrm{n}}$ as the set of Landau levels of the pair $\{1, 2\}$:

$$\tau^a = \{\sum_1^2 q_i|\vec{B}|(n_i + \frac{1}{2}), \ n_i \in \mathbb{N}\}. \tag{3.6.2}$$

(This set was denoted by $\tau$ in Section 3.5.) The Landau levels of the third particle are

$$\Lambda_j := q_3|\vec{B}|(j + \frac{1}{2}), j \in \mathbb{N}. \tag{3.6.3}$$

**3.6.1. Two-dimensional case.** We first consider the (simpler) case $d = 2$. Let $a = \{\{1,2\},\{3\}\}$, and assume that (3.6.1) holds. As we saw in Section 2.2, the Hamiltonian $H_a$ is unitarily equivalent to

$$U_a H_a U_a^\star \equiv \tilde{H}_a = \tilde{H}^{a,\mathrm{n}} + \tilde{H}^{a,\mathrm{c}}, \tag{3.6.4}$$

where:

$$\begin{aligned} \tilde{H}^{a,\mathrm{n}} &= \int_{Y_a^{\mathrm{n}\prime}}^{\oplus} \tilde{H}^{a,\mathrm{n}}(\eta_a^{\mathrm{n}}) d\eta_a^{\mathrm{n}}, \\ \tilde{H}^{a,\mathrm{n}}(\eta_a^{\mathrm{n}}) &= \tfrac{1}{2}(\eta_a^{\mathrm{n}} - 2A_a{}^a y^a)^2 + \tfrac{1}{2}(D_{y^{a,\mathrm{n}}} - A^{a\,a}\, y^{a,\mathrm{n}})^2 + V^a(y^a), \end{aligned} \tag{3.6.5}$$

and

$$\tilde{H}^{a,\mathrm{c}} = R_a^{\mathrm{w}}(D_{\hat{y}_a^{\mathrm{c}}}, \hat{y}_a^{\mathrm{c}}). \tag{3.6.6}$$

We consider $\tilde{H}^{a,\mathrm{n}}(\eta_a^{\mathrm{n}})$ and $\tilde{H}^{a,\mathrm{c}}$ as operators on $L^2(Y^a)$ and on $L^2(\hat{Y}_a^{\mathrm{c}})$ respectively. (Observe that, since the charged cluster in $a$ is a single particle, we have $Y^{a,\mathrm{c}} = \{0\}$, so that no internal variables $y^{a,\mathrm{c}}$ are present.) $\tilde{H}^{a,\mathrm{n}}$ is an operator of the form considered in Section 3.5 with $d = 2$; in particular, all results of Subsection 3.5.3 apply to it. Moreover, we have

$$\sigma(\tilde{H}^{a,\mathrm{c}}) = \{\Lambda_j : j \in \mathbb{N}\},$$

where $\Lambda_j$ are given by (3.6.3).

We will use the results of Section 3.4 to prove a Mourre estimate for $\tilde{H}^{a,\mathrm{n}}$, and then deduce from it a Mourre estimate for $\tilde{H}_a$.

Let $I$ be a compact interval such that $I \subset \mathbb{R} \setminus \tau^a$. Then $\tilde{H}^{a,\mathrm{n}}$ is analytically decomposable over $I$ (i.e., satisfies the conditions *(i)–(ii)* of Definition 3.4.1 and *(iii')–(iv')* of Corollary 3.4.3). Indeed, *(i)* is obvious, and *(ii)*, *(iii')*, *(iv')* follow from Theorem 3.5.10*(iv)*, *(iii)*, *(v)* respectively. From Corollary 3.4.3 we obtain the following.

PROPOSITION 3.6.1. *Assume that (V1), (V2) hold. There is a discrete set $\tilde{\tau}^a \subset \mathbb{R}$ such that for any compact interval $I \subset \mathbb{R} \setminus (\tau^a \cup \tilde{\tau}^a)$ there is a self-adjoint operator $B_I^a$ such that:*

*(i) $\tilde{H}^{a,\mathrm{n}} \in C^1(B_I^a)$ and $[\tilde{H}^{a,\mathrm{n}}, \mathrm{i}B_I^a]$ extends to a bounded operator on $\tilde{\mathcal{H}}_a^{\mathrm{n}} := \int_{Y_a^{\mathrm{n}\prime}}^{\oplus} L^2(Y^a) d\eta_a^{\mathrm{n}}$;*

*(ii) $[\tilde{H}^{a,\mathrm{n}}, \mathrm{i}B_I^a] \geq c(\lambda)$ at $\tilde{H}^{a,\mathrm{n}} = \lambda$, where $c(\lambda) \geq 0$ for all $\lambda \in \mathbb{R}$ and $c(\lambda) \geq c_I > 0$ for $\lambda \in I$;*

*(iii) we have*

$$B_I^a = \phi_0(\eta_a^{\mathrm{n}}) + \phi_1(\eta_a^{\mathrm{n}}) D_{\eta_a^{\mathrm{n}}}, \quad [\tilde{H}^{a,\mathrm{n}}, \mathrm{i}B_I^a] = \phi_2(\eta_a^{\mathrm{n}}),$$

*where $\phi_i(\eta_a^{\mathrm{n}})$, $i = 0,1,2$, are bounded operators on $L^2(Y^a)$ of the form $f_i(\eta_a^{\mathrm{n}}, H(\eta_a^{\mathrm{n}}))$ for $f_i \in C_0^\infty(Y_a^{\mathrm{n}\prime} \times (\mathbb{R}\setminus(\tau^a \cup \tilde{\tau}^a))$. Moreover, for any $\epsilon > 0$ we may choose $B_I^a$ so that*

$$B_I^a = \mathbf{1}_{I_\epsilon}(\tilde{H}^{a,\mathrm{n}})\, B_I^a\, \mathbf{1}_{I_\epsilon}(\tilde{H}^{a,\mathrm{n}}), \tag{3.6.7}$$

*where $I_\epsilon$ is the $\epsilon$-neighbourhood of $I$.*

PROOF. The proposition follows immediately from Corollary 3.4.3. Indeed, we may write

$$\mathbb{R} \setminus \tau^a = \bigcup_{i=1}^{\infty} I_0^{(i)},$$

where each $I_0^{(i)}$ is a finite union of compact intervals and $I_0^{(i)} \subset I_0^{(i+1)}$. Let

$$\tilde{\tau}^a = \bigcup_{i=1}^{\infty} \tilde{\tau}_i^a, \tag{3.6.8}$$

where $\tilde{\tau}_i^a$ is the threshold set for $I_0^{(i)}$ (as in Corollary 3.4.3). Then $\tilde{\tau}^a$ is discrete in $\mathbb{R} \setminus \tau^a$ as an increasing union of discrete sets. (We will call it the *secondary threshold set.*)

Let $I$ be a finite union of compact intervals disjoint from $\tau^a \cup \tilde{\tau}^a$. Then $I \subset I_0^{(i)}$ for some $i$ large enough. Hence Corollary 3.4.3 (with $I_0$ replaced by $I_0^{(i)}$) yields the existence of a conjugate operator $B_I^a$ with the required properties. □

Later on we will need the following properties of $B_I^a$:

LEMMA 3.6.2. *Assume that (V1), (V2) hold. Let*

$$N_a^{\mathrm{n}} = (D_{\eta_a^{\mathrm{n}}}^2 + |\eta_a^{\mathrm{n}}|^2) \otimes \mathbf{1} + \mathbf{1} \otimes (D_{y^a}^2 + |y^a|^2)$$

*be the harmonic oscillator on $\tilde{\mathcal{H}}_a^{\mathrm{n}}$. Then $\mathcal{D}(N_a^{\mathrm{n}}) \subset \mathcal{D}(B_I^a)$ and:*

*(i)* $\|B_I^a u\| \leq C\|N_a^{\mathrm{n}} u\|$, $u \in \mathcal{D}(N_a^{\mathrm{n}})$,

*(ii)* $|(B_I^a u, N_a^{\mathrm{n}} u) - (N_a^{\mathrm{n}} u, B_I^a u)| \leq C\|(N_a^{\mathrm{n}})^{1/2} u\|^2$, $u \in \mathcal{D}(N_a^{\mathrm{n}})$.

PROOF. Part *(i)* follows from Proposition 3.6.1*(iii)*. Let now $\phi_0(k)$, $\phi_1(k)$ be as in Proposition 3.6.1*(ii)*. By Proposition 3.5.11, the operators $\langle y\rangle^2 \partial_k^\alpha \phi_i$, $\langle D_y^2\rangle \partial_k^\alpha \phi_i$ are bounded for all $\alpha$; hence Lemma A.5 in the Appendix implies *(ii)*. □

The main result of this subsection is the following proposition.

PROPOSITION 3.6.3. *Assume that (V1), (V2) hold and let $\tau_a = (\tau^a \cup \tilde{\tau}^a) + \sigma(\tilde{H}^{a,\mathrm{c}})$. For any compact interval $I \subset \mathbb{R} \setminus \tau_a$ there is a self-adjoint operator $\tilde{G}_{a,I}$, equal to a finite sum of operators $B_{I_j}^a$ as in Proposition 3.6.1, such that:*

*(i) $\tilde{H}_a$ is in $C^1(\tilde{G}_{a,I})$, and $[\tilde{H}_a, \mathrm{i}\tilde{G}_{a,I}]$ extends to a bounded operator on $\tilde{\mathcal{H}}_a := L^2(Y_a^{\mathrm{c}}) \otimes \int_{Y_a^{\mathrm{n}\prime}}^{\oplus} L^2(Y^a) d\eta_a^{\mathrm{n}}$;*

*(ii) $[\tilde{H}_a, \mathrm{i}\tilde{G}_{a,I}] \geq c(\lambda)$ at $\tilde{H}_a = \lambda$, where $c(\lambda) \geq 0$ for all $\lambda \in \mathbb{R}$ and $c(\lambda) \geq c_I > 0$ for $\lambda \in I$.*

PROOF. Let $N_0 \in \mathbb{N}$ be large enough so that

$$\inf(\sigma(\tilde{H}^{a,\mathrm{n}})) + \Lambda_j \geq 1 + \sup I \text{ for } j > N_0.$$

Since $I$ is disjoint from $\tau_a$, the set

$$\bigcup_{j=1}^{N_0} I_j, \ I_j = I - \Lambda_j,$$

is disjoint from $\tau^a \cup \tilde{\tau}^a$. We then define:

$$\tilde{G}_{a,I} = \sum_{j=1}^{N_0} B^a_{I_j},$$

where $B^a_{I_j}$ are as in Proposition 3.6.1. We may of course consider $\tilde{G}_{a,I}$ as an operator on $\tilde{\mathcal{H}}_a$.

Part *(i)* of the proposition follows immediately from Proposition 3.6.1*(i)*. Let us prove *(ii)*. Let $\lambda \in \mathbb{R}$ and let $f \in C_0^\infty(\mathbb{R})$ be supported in a sufficiently small neighbourhood of $\lambda$. We denote by $P_j, j \in \mathbb{N}$ the spectral projections on the eigenspaces of $\mathbf{1} \otimes \tilde{H}^{a,\mathrm{c}}$ corresponding to the eigenvalues $\Lambda_j$. Then

$$f(\tilde{H}_a) = \sum_{j=1}^{N(\lambda)} P_j f(\tilde{H}^{a,\mathrm{n}} + \Lambda_j).$$

Hence

$$\begin{aligned} & f(\tilde{H}_a)[\tilde{H}_a, \mathrm{i}\tilde{G}_{a,I}]f(\tilde{H}_a) \\ = \ & \textstyle\sum_{j=1}^{N(\lambda)} P_j f(\tilde{H}^{a,\mathrm{n}} + \Lambda_j)[\tilde{H}^{a,\mathrm{n}}, \mathrm{i}\sum_{k=1}^{N_0} B^a_{I_k}]f(\tilde{H}^{a,\mathrm{n}} + \Lambda_j)P_j. \end{aligned} \tag{3.6.9}$$

By Proposition 3.6.1*(ii)*,

$$f(\tilde{H}^{a,\mathrm{n}} + \Lambda_j)[\tilde{H}^{a,\mathrm{n}}, \mathrm{i}B^a_{I_k}]f(\tilde{H}^{a,\mathrm{n}} + \Lambda_j) \geq 0$$

for any $\lambda \in \mathbb{R}$ and any $1 \leq k \leq N_0$, $1 \leq j \leq N(\lambda)$, provided that $\operatorname{supp} f$ is sufficiently small. Hence the right-hand side of (3.6.9) is nonnegative. Let now $\lambda \in I$; then $\lambda - \Lambda_j \in I_j$, $j = 1, \ldots, N_0$. Moreover, if $\operatorname{supp} f$ is small enough, we have

$$f(\tilde{H}^{a,\mathrm{n}} + \Lambda_j) = 0 \text{ for } j > N_0.$$

Applying Proposition 3.6.1*(ii)* again, we obtain from (3.6.9) that

$$\begin{aligned} & f(\tilde{H}_a)[\tilde{H}_a, \mathrm{i}\tilde{G}_{a,I}]f(\tilde{H}_a) \\ & \geq \textstyle\sum_{j=1}^{N_0} P_j f(\tilde{H}^{a,\mathrm{n}} + \Lambda_j)[\tilde{H}^{a,\mathrm{n}}, \mathrm{i}B^a_{I_j}]f(\tilde{H}^{a,\mathrm{n}} + \Lambda_j)P_j \\ & \geq \textstyle\sum_{j=1}^{N_0} c_{I_j} P_j f(\tilde{H}^{a,\mathrm{n}} + \Lambda_j)^2 P_j \\ & \geq c_I f(\tilde{H}_a)^2, \end{aligned}$$

where $c_I = \min_{1 \leq j \leq N_0} c_{I_j} > 0$. □

**3.6.2. Three-dimensional case.** We now consider the case $d = 3$. Assume that (3.6.1) holds, and let $a = \{\{1,2\},\{3\}\}$. Then the channel Hamiltonian $H_a$ is unitarily equivalent to

$$U_a H_a U_a^\star = \tilde{H}_a = \frac{1}{2}D_{z_a}^2 + \tilde{H}^{a,\mathrm{n}} + \tilde{H}^{a,\mathrm{c}}, \tag{3.6.10}$$

where:
(3.6.11)

$$\tilde{H}^{a,\mathrm{n}} = \int_{Y_a^{\mathrm{n}\prime}}^{\oplus} \tilde{H}^{a,\mathrm{n}}(\eta_a^{\mathrm{n}})d\eta_a^{\mathrm{n}},$$

$$\tilde{H}^{a,\mathrm{n}}(\eta_a^{\mathrm{n}}) = \tfrac{1}{2}(\eta_a^{\mathrm{n}} - 2A_a{}^a\, y^a)^2 + \tfrac{1}{2}(D_{x^{a,\mathrm{n}}} - A^{a\,a}\, y^{a,\mathrm{n}})^2 + V^{a,\mathrm{n}}(x^a),$$

and

$$\tilde{H}^{a,\mathrm{c}} = R_a^{\mathrm{w}}(D_{\hat{y}_a^{\mathrm{c}}}, \hat{y}_a^{\mathrm{c}}). \tag{3.6.12}$$

We consider $\tilde{H}^{a,\mathrm{n}}(\eta_a^{\mathrm{n}})$ and $\tilde{H}^{a,\mathrm{c}}$ as operators on $L^2(X^a)$ and on $L^2(\hat{Y}_a^{\mathrm{c}})$ respectively. As in the case $d = 2$, we have $X^{a,\mathrm{c}} = \{0\}$.

We will assume that:

($S$) For all $\eta_a^{\mathrm{n}} \in Y_a^{\mathrm{n}\prime}$, the point spectrum of $\tilde{H}^{a,\mathrm{n}}(\eta_a^{\mathrm{n}})$ is purely discrete.

It follows from Theorem 3.5.3 that $\tilde{H}^{a,\mathrm{n}}$ is analytically decomposable over any compact interval $I_0 \subset (-\infty, \nu_0)$. By Corollary 3.4.3 $\tilde{H}^{a,\mathrm{n}}$ satisfies a Mourre estimate with a suitable conjugate operator $B_{I_0}$ on any such interval. On the other hand, the results of Subsection 3.5.2 imply that at $\tilde{H}^{a,\mathrm{n}} = \lambda$, $\lambda \in (\nu_0, \infty) \setminus \tau^a$, the Mourre estimate holds with the conjugate operator

$$A^a = \frac{1}{2}(\langle z^a, D_{z^a}\rangle + \langle D_{z^a}, z^a\rangle). \tag{3.6.13}$$

We will combine these results to construct a "composite" conjugate operator $\tilde{G}_{a,I}$ for $\tilde{H}_a$.

Let us turn to the details. We write $(-\infty, \nu_0) = \bigcup_{i=1}^{\infty} I_0^{(i)}$, where $I_0^{(i)}$ are compact intervals, and define the *secondary threshold set* $\tilde{\tau}^a$ as

$$\tilde{\tau}^a := \bigcup_{i=1}^{\infty} \tilde{\tau}_i^a, \tag{3.6.14}$$

where $\tilde{\tau}_i^a$ is the threshold set for $\tilde{H}^{a,\mathrm{n}}$ on the interval $I_0^{(i)}$, as defined in Corollary 3.4.3. It follows that $\tilde{\tau}^a$ is closed in $(-\infty, \nu_0)$ and discrete. Applying Corollary 3.4.3 to $\tilde{H}^{a,\mathrm{n}}$, we obtain the following proposition.

PROPOSITION 3.6.4. *Assume that (V1), (V2), (M1), (S) hold. Let $I$ be a compact interval such that*

$$I \subset ((-\infty, \nu_0) \setminus \tilde{\tau}^a).$$

*Let $B_I^a$ be the conjugate operator for $\tilde{H}^{a,\mathrm{n}}$ constructed in Corollary 3.4.3. Then:*

*(i) $B_I^a$ is self-adjoint, $\tilde{H}^{a,\mathrm{n}} \in C^1(B_I^a)$, $[\tilde{H}^{a,\mathrm{n}}, \mathrm{i}B_I^a]$ extends to a bounded operator on $\mathcal{H}_a^{\mathrm{n}} := \int_{Y_a^{\mathrm{n}\prime}}^{\oplus} L^2(Y^a)d\eta_a^{\mathrm{n}}$;*

*(ii)* $[\tilde{H}^{a,\mathrm{n}}, \mathrm{i}B_I^a] \geq c(\lambda)$ *at* $\tilde{H}^{a,\mathrm{n}} = \lambda$, *where* $c(\lambda) \geq 0$ *for all* $\lambda \in \mathbb{R}$ *and* $c(\lambda) \geq c_I > 0$ *for* $\lambda \in I$;

*(iii) we have:*

$$B_I^a = \phi_0(\eta_a^{\mathrm{n}}) + \phi_1(\eta_a^{\mathrm{n}}) D_{\eta_a^{\mathrm{n}}}, \ [\tilde{H}^{a,\mathrm{n}}, \mathrm{i}B_I^a] = \phi_2(\eta_a^{\mathrm{n}}),$$

*where* $\phi_i(\eta_a^{\mathrm{n}})$, $i = 0, 1, 2$, *are bounded operators on* $L^2(X^a)$ *of the form* $f_i(\eta_a^{\mathrm{n}}, H(\eta_a^{\mathrm{n}}))$ *for* $f_i \in C_0^\infty(Y_a^{\mathrm{n}\prime} \times (\mathbb{R} \backslash (\tau^a \cup \tilde{\tau}^a)))$. *Moreover, for any* $\epsilon > 0$ *we may choose* $B_I^a$ *so that*

$$B_I^a = \mathbf{1}_{I_\epsilon}(\tilde{H}^{a,\mathrm{n}})\, B_I^a\, \mathbf{1}_{I_\epsilon}(\tilde{H}^{a,\mathrm{n}}), \tag{3.6.15}$$

*where* $I_\epsilon$ *is the* $\epsilon$*-neighbourhood of* $I$.

We will also need a technical lemma.

LEMMA 3.6.5. *Assume (V1), (V2), (M1), (S). Let* $I$ *be a compact interval such that* $I \subset ((-\infty, \nu_0) \setminus \tilde{\tau}^a)$, *and let* $B_I^a$ *be the operator defined in Proposition 3.6.4. Let*

$$N_a^{\mathrm{n}} = (D_{\eta_a^{\mathrm{n}}}^2 + |\eta_a^{\mathrm{n}}|^2) \otimes \mathbf{1} + \mathbf{1} \otimes (D_{x^a}^2 + |x^a|^2) \tag{3.6.16}$$

*be the harmonic oscillator on* $\tilde{\mathcal{H}}_a^{\mathrm{n}}$. *Then* $\mathcal{D}(N_a^{\mathrm{n}}) \subset \mathcal{D}(B_I^a)$ *and:*

*(i)* $\|B_I^a u\| \leq C\|N_a^{\mathrm{n}} u\|$, $u \in \mathcal{D}(N_a^{\mathrm{n}})$,

*(ii)* $|(B_I^a u, N_a^{\mathrm{n}} u) - (N_a^{\mathrm{n}} u, B_I^a u)| \leq C\|(N_a^{\mathrm{n}})^{\frac{1}{2}} u\|^2$, $u \in \mathcal{D}(N_a^{\mathrm{n}})$.

PROOF. The proof is identical to the proof of Lemma 3.6.2, except that Proposition 3.5.4 is used instead of Proposition 3.5.11. □

From Lemma 3.5.9, we obtain the following.

LEMMA 3.6.6. *Assume (V1), (V2), (M1). Let*

$$\tilde{G}_0^a := \chi_a(\tilde{H}^{a,\mathrm{n}}) A^a \chi_a(\tilde{H}^{a,\mathrm{n}}),$$

*where* $\chi_a \in C^\infty(\mathbb{R})$ *satisfies*

$$\chi_a \equiv 0 \text{ on } (-\infty, \nu_0 - 2\epsilon), \ \chi_a \equiv 1 \text{ on } [\nu_0 - \epsilon, \infty). \tag{3.6.17}$$

*Let* $N_a^{\mathrm{n}}$ *be defined by (3.6.16). Then:*

*(i)* $\|\tilde{G}_0^a u\| \leq C\|N_a^{\mathrm{n}} u\|$, $u \in \mathcal{D}(N_a^{\mathrm{n}})$,

*(ii)* $|(\tilde{G}_0^a u, N_a^{\mathrm{n}} u) - (N_a^{\mathrm{n}} u, \tilde{G}_0^a u)| \leq C\|(N_a^{\mathrm{n}})^{\frac{1}{2}} u\|^2$, $u \in \mathcal{D}(N_a^{\mathrm{n}})$.

PROPOSITION 3.6.7. *Assume that (V1), (V2), (M1), (S) hold. Let* $\tilde{G}_0^a$ *be the operator defined in Lemma 3.6.6. Then:*

*(i)* $\tilde{G}_0^a$ *is essentially self-adjoint on* $\mathcal{D}(N_a^{\mathrm{n}})$ *and* $\tilde{H}^{a,\mathrm{n}} \in C^1(\tilde{G}_0^a)$;

*(ii)* $[\tilde{H}^{a,\mathrm{n}}, \mathrm{i}\tilde{G}_0^a] \geq_* c(\lambda)$ *at* $\tilde{H}^{a,\mathrm{n}} = \lambda$, *where* $c(\lambda) \geq 0$ *for all* $\lambda \in \mathbb{R}$ *and* $c(\lambda) \geq \mathrm{dist}\,(\lambda, \tau^a)$ *for* $\lambda \geq \nu_0$.

PROOF. By Lemma 3.6.6, $\tilde{G}_0^a$, $N_a^{\rm n}$, and $\tilde{H}^{a,{\rm n}}$ satisfy the hypotheses of Lemmas 1.2.5 and 3.2.2. This yields *(i)*.

We now prove *(ii)*. All commutators in the proof below will be considered as quadratic forms on $\mathcal{D}(N_a^{\rm n})$. By Theorem 3.5.3, *(S)* is equivalent to:

$$\sigma_{\rm pp}(\tilde{H}^{a,{\rm n}}(\eta_a^{\rm n})) \cap (\nu_0, \infty) = \emptyset \text{ for all } \eta_a^{\rm n} \in Y_a^{\rm n\prime}.$$

By Lemma 3.5.7, Lemma 3.1.4*(i)*, and the virial relation (3.2.5), we have for all $\lambda \in \mathbb{R}$

$$[\tilde{H}^{a,{\rm n}}(\eta_a^{\rm n}), {\rm i}A^a] \geq_* d^a(\lambda) \text{ at } \tilde{H}^{a,{\rm n}}(\eta_a^{\rm n}) = \lambda,$$

where

$$d^a(\lambda) = \begin{cases} \operatorname{dist}(\lambda, \tau^a), & \lambda \geq \nu_0, \\ 0, & \lambda \leq \nu_0. \end{cases}$$

By Lemma 3.5.8 we have

$$[\tilde{H}^{a,{\rm n}}(\eta_a^{\rm n}), {\rm i}A^a] \geq d^a(\lambda) - \epsilon \text{ at } \tilde{H}^{a,{\rm n}}(\eta_a^{\rm n}) = \lambda,$$

uniformly for $|\eta_a^{\rm n}| \geq C$. Using Lemma 3.1.3, we obtain that for all $\lambda \in \mathbb{R}$

$$[\tilde{H}^{a,{\rm n}}(\eta_a^{\rm n}), {\rm i}A^a] \geq_* d^a(\lambda) \text{ at } \tilde{H}^{a,{\rm n}}(\eta_a^{\rm n}) = \lambda,$$

uniformly for $\eta_a^{\rm n} \in Y_a^{\rm n\prime}$. Hence

$$[\tilde{H}^{a,{\rm n}}, {\rm i}A^a] \geq_* d^a(\lambda) \text{ at } \tilde{H}^{a,{\rm n}} = \lambda.$$

Let $\chi_a \in C^\infty(\mathbb{R})$ be as in the statement of the proposition. Since $\chi_a(\tilde{H}^{a,{\rm n}})$ preserves $\mathcal{D}(N_a^{\rm n})$, by Remark 3.2.4 we have the equality

$$[\tilde{H}^{a,{\rm n}}, i\tilde{G}_0^a] = \chi_a(\tilde{H}^{a,{\rm n}})[\tilde{H}^{a,{\rm n}}, {\rm i}A^a]\chi_a(\tilde{H}^{a,{\rm n}})$$

between the quadratic forms on $\mathcal{D}(N_a^{\rm n})$. Hence for any $\lambda \in \mathbb{R}$

$$[\tilde{H}^{a,{\rm n}}, {\rm i}\tilde{G}_0^a] \geq_* d^a(\lambda) \text{ at } \tilde{H}^{a,{\rm n}} = \lambda. \tag{3.6.18}$$

Furthermore,

$$(\tilde{H}^{a,{\rm n}} + {\rm i})^{-1}[\tilde{H}^{a,{\rm n}}, {\rm i}\tilde{G}_0^a](\tilde{H}^{a,{\rm n}} + {\rm i})^{-1}$$

is bounded as a quadratic form on $\mathcal{D}(N_a^{\rm n})$. By Lemma 3.2.2, (3.6.18) extends to $\mathcal{D}(\tilde{H}^{a,{\rm n}}) \cap \mathcal{D}(\tilde{G}_0^a)$. This completes the proof of the proposition. □

After these preliminaries, we begin the proof of the main result of this section – the Mourre estimate for

$$\tilde{H}_a = \frac{1}{2}D_{z_a}^2 + \tilde{H}^{a,{\rm n}} + \tilde{H}^{a,{\rm c}}.$$

We first construct the conjugate operator. Let

$$\tau_a = (\tau^a \cup \tilde{\tau}^a) + \sigma(\tilde{H}^{a,{\rm c}}), \tag{3.6.19}$$

and fix a compact interval $I \subset (\mathbb{R} \setminus \tau_a)$. Recall that

$$\sigma(\tilde{H}^{a,{\rm c}}) = \{\Lambda_j : j = 1, 2, \dots\},$$

where $\Lambda_j$ are given by (3.6.3); in particular, we have

$$\Lambda_j < \Lambda_{j+1}, \ \lim_{j\to\infty} \Lambda_j = \infty. \tag{3.6.20}$$

Let $P_j$ be the eigenprojections on the eigenspaces of $\mathbf{1}\otimes\mathbf{1}\otimes\tilde{H}^{a,\mathrm{c}}$ corresponding to $\Lambda_j$. Let $I'$ be a compact interval such that $I' \subset (\mathbb{R}\setminus\tau_a)$ and $I \subset \mathrm{Int}\, I'$. We define for $j = 1, 2, \ldots$:

$$I_j = I - \Lambda_j, \ I'_j = I' - \Lambda_j.$$

Then $I'_j$, $j = 1, 2, \ldots$, are disjoint from $\mathbb{R}\setminus(\tau^a\cup\tilde{\tau}^a)$. By (3.6.20), there are integers $1 \le N_1 \le N_2 < \infty$ such that:

$$\begin{aligned} &I'_j \subset \big((\nu_0,\infty)\setminus\tau^a\big), \ j < N_1,\\ &I'_j \subset \big((-\infty,\nu_0)\setminus\tilde{\tau}^a\big), \ j \ge N_1,\\ &I'_j \subset \big((-\infty, \inf(\sigma(\tilde{H}^{a,\mathrm{n}})) - 1\big), \ j \ge N_2. \end{aligned} \tag{3.6.21}$$

We define

$$\tilde{G}_{a,I} := A_a + \tilde{G}^a_0 + \tilde{G}^a_I, \tag{3.6.22}$$

where

$$A_a = \frac{1}{2}(\langle z_a, D_{z_a}\rangle + \langle D_{z_a}, z_a\rangle),$$

$\tilde{G}^a_0$ is defined in Lemma 3.6.6, and

$$\tilde{G}^a_I = \sum_{j=N_1}^{N_2-1} \tilde{G}^a_j,$$

where $\tilde{G}^a_j = B^a_{I'_j}$, $j = N_1, \ldots, N_2 - 1$, are conjugate operators for $\tilde{H}^{a,\mathrm{n}}$ over the intervals $I'_j$, defined in Proposition 3.6.4. (If $N_2 = N_1$, we put $\tilde{G}^a_I = 0$.) Moreover, we may assume that the $\epsilon$ in (3.6.15) and (3.6.17) is small enough, so that

$$\chi_a(\tilde{H}^{a,\mathrm{n}})\tilde{G}^a_I = 0. \tag{3.6.23}$$

$\tilde{G}_{a,I}$ is defined on $\mathcal{D}(N_a)$, where

$$N_a = (D^2_{z_a} + z^2_a)\otimes\mathbf{1} + \mathbf{1}\otimes N^{\mathrm{n}}_a \tag{3.6.24}$$

is the harmonic oscillator on $\tilde{\mathcal{H}}_a = L^2(Z_a)\otimes\int^{\oplus}_{Y^{\mathrm{n}\prime}_a} L^2(X^a)d\eta^{\mathrm{n}}_a$.

PROPOSITION 3.6.8. *Assume (V1), (V2), (M1), (S). Then:*

*(i)* $\tilde{H}_a \in C^1(\tilde{G}_{a,I})$,

*(ii)* $[\tilde{H}_a, \mathrm{i}\tilde{G}_{a,I}] \ge_* c(\lambda)$ *at* $\tilde{H}_a = \lambda$, *where* $c(\lambda) \ge 0$ *for all* $\lambda\in\mathbb{R}$ *and* $c(\lambda) \ge c_I > 0$ *for* $\lambda \in I$.

PROOF. By Lemmas 3.6.5 and 3.6.6 for $\tilde{G}^a_j$ and $\tilde{G}^a_0$, and by direct computation for $A_a$, $\tilde{G}_{a,I}$ and $N_a$ satisfy the hypotheses of Nelson's commutator theorem, hence $\tilde{G}_{a,I}$ is essentially self-adjoint on $\mathcal{D}(N_a)$. Moreover, we have for $u \in \mathcal{D}(N_a)$:

$$
\begin{aligned}
&|(\tilde{G}_{a,I}u, \tilde{H}_a u) - (\tilde{H}_a u, \tilde{G}_{a,I}u)| \\
&\leq |(u, D^2_{z_a}u)| + |(\tilde{G}_{a,I}u, \tilde{H}^{a,\mathrm{n}}u) - (\tilde{H}^{a,\mathrm{n}}u, \tilde{G}_{a,I}u)| \\
&\leq C(\|D^2_{z_a}u\|^2 + \|\tilde{H}^{a,\mathrm{n}}u\|^2 + \|u\|^2).
\end{aligned}
\tag{3.6.25}
$$

By Lemma 1.2.3, $(z - \tilde{H}_a)^{-1}$ preserves $\mathcal{D}(N_a)$. We may therefore apply Lemma 3.2.2 to $\tilde{H}_a, \tilde{G}_{a,I}, N_a$ and conclude that $\tilde{H}_a \in C^1(\tilde{G}_{a,I})$. We also see that *(ii)* will follow if we prove that

$$D^2_{z_a} \otimes \mathbf{1} + \mathbf{1} \otimes [\tilde{H}_a, \mathrm{i}\tilde{G}_{a,I}] \geq c_0 > 0 \text{ at } \tilde{H}_a = \lambda,\ \lambda \in I,$$

on $\mathcal{D}(N_a)$. Let $\lambda \in \mathbb{R}$, and let $f \in C_0^\infty(\mathbb{R})$ be supported in a sufficiently small neighbourhood of $\lambda$. We will abbreviate

$$f_j = f(\frac{1}{2}{\zeta_a}^2 + \tilde{H}^{a,\mathrm{n}} + \Lambda_j).$$

Using that $\tilde{H}^{a,\mathrm{n}}$ is bounded from below, we obtain that

$$
\begin{aligned}
f(\tilde{H}_a) &= \textstyle\sum_{j=1}^{N(\lambda)} P_j f(\frac{1}{2}D^2_{z_a} + \tilde{H}^{a,\mathrm{n}} + \Lambda_j) \\
&= \textstyle\sum_{j=1}^{N(\lambda)} \int^{\oplus}_{U(\lambda)} P_j f_j \, d\zeta_a,
\end{aligned}
$$

and that on $\mathcal{D}(N_a)$

$$
\begin{aligned}
&f(\tilde{H}_a)[\tilde{H}_a, \mathrm{i}\tilde{G}_{a,I}]f(\tilde{H}_a) \\
&= \textstyle\sum_{j=1}^{N(\lambda)} \int^{\oplus}_{U(\lambda)} P_j f_j(\zeta_a^2 + [\tilde{H}^{a,\mathrm{n}}, \mathrm{i}\tilde{G}^a_0 + \mathrm{i}\tilde{G}^a_I]) f_j P_j \, d\zeta_a,
\end{aligned}
\tag{3.6.26}
$$

where $U(\lambda)$ is a compact subset of $Z'_a$.

We first prove that for any $\lambda \in \mathbb{R}$,

$$[\tilde{H}_a, \mathrm{i}\tilde{G}_{a,I}] \geq_* 0 \text{ at } \tilde{H}_a = \lambda. \tag{3.6.27}$$

Indeed, by Proposition 3.6.4*(ii)* and 3.6.7*(ii)*:

$$[\tilde{H}^{a,\mathrm{n}}, \mathrm{i}\tilde{G}^a_0 + \mathrm{i}\tilde{G}^a_I] \geq 0 \text{ at } \tilde{H}^{a,\mathrm{n}} = \lambda - \frac{1}{2}\zeta_a^2 - \Lambda_j$$

for any $\lambda \in \mathbb{R}$, $\zeta \in U(\lambda)$, $j \leq N(\lambda)$. Hence (3.6.27) follows from (3.6.26) and Lemma 3.1.3*(ii)*.

Assume now that $\lambda \in I$, and that $\operatorname{supp} f$ is small enough. Then

$$f_j \equiv f(\frac{1}{2}\zeta_a^2 + \tilde{H}^{a,\mathrm{n}} + \Lambda_j) = 0 \text{ for } j \geq N_2,$$

and there is a constant $M = M(I)$ such that

$$f_j \equiv f(\frac{1}{2}\zeta_a^2 + \tilde{H}^{a,\mathrm{n}} + \Lambda_j) = 0 \text{ for } |\zeta_a| > M.$$

Hence (3.6.26) becomes:

$$\begin{aligned} & f(\tilde H_a)[\tilde H_a, \mathrm{i}\tilde G_{a,I}]f(\tilde H_a) \\ (3.6.28)\qquad & = \textstyle\sum_{j=1}^{N_2-1}\int^{\oplus}_{|\zeta|\le M} P_j f_j(\zeta_a^2 + [\tilde H^{a,\mathrm{n}}, \mathrm{i}\tilde G_0^a + \sum_{k=N_1}^{N_2-1}\mathrm{i}G_k^a]) f_j P_j\, d\zeta_a \\ & = I_1 + I_2 + I_3 + I_4, \end{aligned}$$

where

$$\begin{aligned} I_1 &= \textstyle\sum_{j<N_1}\int^{\oplus}_{|\zeta_a|\le M} P_j f_j(\zeta_a^2 + [\tilde H^{a,\mathrm{n}}, \mathrm{i}\tilde G_0^a]) f_j P_j\, d\zeta_a, \\ I_2 &= \textstyle\sum_{j=N_1}^{N_2-1}\int^{\oplus}_{|\zeta_a|\le M} P_j f_j(\zeta_a^2 + [\tilde H^{a,\mathrm{n}}, \mathrm{i}\tilde G_j^a]) f_j P_j\, d\zeta_a, \\ I_3 &= \textstyle\sum_{j<N_1}\int^{\oplus}_{|\zeta_a|\le M} P_j f_j[\tilde H^{a,\mathrm{n}}, \mathrm{i}\sum_{k=N_1}^{N_2-1}\tilde G_k^a]) f_j P_j\, d\zeta_a, \\ I_4 &= \textstyle\sum_{j=N_1}^{N_2-1}\int^{\oplus}_{|\zeta_a|\le M} P_j f_j\left[\tilde H^{a,\mathrm{n}}, \mathrm{i}\tilde G_0^a + \sum_{N_1\le k\le N_2-1, k\ne j}\mathrm{i}\tilde G_k^a\right] f_j P_j\, d\zeta_a. \end{aligned}$$

By Propositions 3.6.4*(ii)*, 3.6.7*(ii)*, and Lemma 3.1.3*(ii)*, for any $\epsilon > 0$ we have

$$I_3 + I_4 \ge -\epsilon,$$

provided that the support of $f$ is sufficiently small. It therefore suffices to prove that there is a constant $c_0 > 0$ such that

$$I_1 + I_2 \ge c_0 f^2(\tilde H_a), \tag{3.6.29}$$

whenever supp $f$ is small enough. By Lemma 3.1.3*(ii)*, we only need to prove that there is a constant $c_0 > 0$ such that for any $\lambda \in I$, $j = 1, \dots, N_2 - 1$, and $|\zeta_a| \le M$:

$$\zeta_a^2 + [\tilde H^{a,\mathrm{n}}, \mathrm{i}\tilde G_{j'}^a] \ge c_0 > 0 \text{ at } \tilde H^{a,\mathrm{n}} = \lambda - \Lambda_j - \frac{1}{2}\zeta_a^2, \tag{3.6.30}$$

where

$$j' = \begin{cases} 0, & j < N_1, \\ j, & N_1 \le j \le N_2 - 1. \end{cases}$$

(3.6.30) is equivalent to

$$\zeta_a^2 + [\tilde H^{a,\mathrm{n}}, \mathrm{i}\tilde G_{j'}^a] \ge c_0 > 0 \text{ at } \tilde H^{a,\mathrm{n}} = \lambda_j - \frac{1}{2}\zeta_a^2, \tag{3.6.31}$$

for $j = 1, \dots, N_2 - 1$, for any $\lambda_j \in I_j := I - \Lambda_j$, and any $|\zeta_a| \le M$.

By Propositions 3.6.4*(ii)* and 3.6.7*(ii)*, we have

$$[\tilde H^{a,\mathrm{n}}, \mathrm{i}\tilde G_{j'}^a] \ge_* c(s) \text{ at } \tilde H^{a,\mathrm{n}} = s, \tag{3.6.32}$$

where $c(s) \ge 0$ for all $s \in \mathbb{R}$, and $c(s) \ge c_j > 0$ for $s \in I_j'$.

Let $\theta > 0$ be small enough so that

$$0 < \theta < \frac{1}{2}\min(\mathrm{dist}\,(I', \tau_a), \mathrm{dist}\,(I, \partial I')). \tag{3.6.33}$$

Let $\lambda_j \in I_j$ and $|\zeta_a| \le M$.

Assume first that $\zeta_a^2 \geq \theta$. By (3.6.32) we have

$$[\tilde{H}^{a,\mathrm{n}}, \mathrm{i}\tilde{G}^a_{j'}] \geq -\frac{\theta}{2} \text{ at } \tilde{H}^{a,\mathrm{n}} = \lambda_j - \frac{1}{2}\zeta_a^2, \tag{3.6.34}$$

hence (3.6.31) holds with $c_0 = \theta/2$.

Suppose now that $\zeta_a^2 \leq \theta$. Then, by (3.6.33), $s := \lambda_j - \frac{1}{2}\zeta_a^2 \in I'_j$, so that (3.6.32) holds with $c(s) \geq c_j > 0$. Hence

$$[\tilde{H}^{a,\mathrm{n}}, \mathrm{i}\tilde{G}^a_{j'}] \geq \frac{c_j}{2} \text{ at } \tilde{H}^{a,\mathrm{n}} = s. \tag{3.6.35}$$

Combining (3.6.34) and (3.6.35), we obtain that (3.6.31) holds with $c_0 = \min(\theta/2, c_j/2)$. □

## 3.7. Three-dimensional dispersive systems

The purpose of this section is to obtain a Mourre estimate for the magnetic Hamiltonian $H$ of a three-body system satisfying *(Q1)* and containing at least one neutral pair. (The case when no neutral pairs are present is considered in Section 3.3 for $d = 3$ and in Section 6.5 for $d = 2$). Observe that such a system may actually contain more than one neutral pair: for instance, if $q_1 = 1$, $q_2 = q_3 = -1$, then both pairs $(1, 2)$ and $(1, 3)$ are neutral. We also observe that, since we are assuming all particles to be charged ($q_i \neq 0$), the presence of a neutral pair in a 3-body system implies that the total charge $Q = q_1 + q_2 + q_3$ is nonzero.

The Mourre estimate for $H$ will be obtained from the "channel Mourre estimates" proved in Section 3.6. The main new difficulty that arises at this time is that we are not allowed to restrict ourselves to just one scattering channel (as we did in Section 3.6). Instead, we have to construct conjugate operators $M_a$ corresponding to different scattering channels, and then glue them together in order to obtain a Mourre estimate that holds for all channels simultaneously.

The proof of the Mourre estimate with the "composite" conjugate operator $M$, and the verification of the additional properties of $M$ which will be needed later on, are unfortunately rather long and technical. The reader who wishes to move on as quickly as possible to the rest of our proof of asymptotic completeness might want to read only the first three subsections. The Mourre estimate and other results that will be used in the sequel are stated in Subsection 3.7.1, and (with the exception of Proposition 3.7.5) proved in Subsections 3.7.2 and 3.7.3. In Subsection 3.7.4 we prove Proposition 3.7.5.

Throughout this section we assume that *(V1)*, *(V4)*, *(M1)*, *(S)* hold.

**3.7.1. The main results.** We begin by constructing the conjugate operator $M$ and stating the main results of this section. The statement of the Mourre estimate itself is given in Theorem 3.7.3; we will also prove that $M$ is self-adjoint (Proposition 3.7.1) and that $H \in C^{1+\epsilon}(M)$ for some $\epsilon > 0$ (Proposition 3.7.5). Additional properties of $M$, which will be used in the proof of the Mourre estimate and of the *minimal velocity estimates* in Chapter 5, are given in Lemmas 3.7.2 and 3.7.4.

We first define the threshold set. Let

$$\tau_{a_{\min}} := \sigma(\tilde{H}^{a_{\min}}),$$

$$\tau_a = \tau_{a_{\min}} \cup \sigma(\tilde{H}^a),\ a \in \mathcal{A}^{\mathrm{c}}, \sharp a = 2,$$

and recall that the set $\tau_a$ for $a \in \mathcal{A}^{\mathrm{n}}$, $\sharp a = 2$, was defined in Section 3.6. Then the threshold set for $H$ is

$$\tau := \cup_{a < a_{\max}} \tau_a. \tag{3.7.1}$$

The set $\tau$ is closed and discrete.

Fix an interval $I \subset \mathbb{R} \setminus \tau$. Let

$$\mathbf{1} = \sum_{a \in \mathcal{A}} q_a^2(x^{\mathrm{cm}})$$

be an $N$-body partition of unity as in Proposition 2.3.4, with the constants $\delta_a, \epsilon_a$ sufficiently small (to be determined later). We also define for $a \in \mathcal{A}^{\mathrm{n}}$:

$$F_a = F\left(\frac{|y_a^{\mathrm{c}}|}{\langle x_a^{\mathrm{cm}} \rangle} \leq c_0\right), \tag{3.7.2}$$

where $c_0 > 0$ is small enough. We note that

$$|D_x^\alpha F_a| \leq C_\alpha \langle x_a^{\mathrm{cm}} \rangle^{-|\alpha|},\ |\alpha| > 0;$$

this follows by explicit computation and by observing that $\nabla F_a$ is supported in the region where $|y_a^{\mathrm{c}}| \sim \langle x_a^{\mathrm{cm}} \rangle$. On $\operatorname{supp} q_a$ we have $|x^a| \leq \delta_a |x^{\mathrm{cm}}|$, hence $\langle x_a^{\mathrm{cm}} \rangle \geq C \langle x^{\mathrm{cm}} \rangle$, provided that $\delta_a$ is small enough. Thus

$$|D_x^\alpha F_a| \leq C_\alpha \langle x^{\mathrm{cm}} \rangle^{-|\alpha|},\ |\alpha| \geq 0, \text{ on } \operatorname{supp} q_a. \tag{3.7.3}$$

Let $U_a$, $a \in \mathcal{A}$, denote the reducing transformations defined in Section 2.2. Let $A$ be the generator of dilations on $L^2(X)$:

$$A = \frac{1}{2}(\langle z, D_z \rangle + \langle D_z, z \rangle);$$

we will also write

$$A_a = \frac{1}{2}(\langle z_a, D_{z_a} \rangle + \langle D_{z_a}, z_a \rangle),\ A^a = \frac{1}{2}(\langle z^a, D_{z^a} \rangle + \langle D_{z^a}, z^a \rangle).$$

For each $a \in \mathcal{A}^{\mathrm{n}}$, let $\tilde{G}_0^a$ and $\tilde{G}_I^a$ be the operators constructed in Section 3.6 for the interval $I$; recall that

$$\tilde{G}_0^a = \chi_a(\tilde{H}^{a,\mathrm{n}}) A^a \chi_a(\tilde{H}^{a,\mathrm{n}}),$$

where $\chi_a$ is a function in $C^\infty(\mathbb{R})$ equal to 1 on a neighbourhood of $+\infty$, and that $\tilde{G}_I^a$ is a finite sum of conjugate operators as in Proposition 3.6.4. Since the interval $I$ will be fixed throughout this section, we may simplify the notation by dropping the subscripts $I$ and write $\tilde{G}^a$ instead of $\tilde{G}_I^a$.

We define:

$$
\begin{aligned}
G_0^a &= U_a^* \tilde{G}_0^a U_a = \chi_a(H^{a,\mathrm{n}}) A^a \chi_a(H^{a,\mathrm{n}}), \\
G^a &= U_a^* \tilde{G}^a U_a.
\end{aligned} \tag{3.7.4}
$$

The auxiliary conjugate operators $M_a$ are:

$$
M_a = \begin{cases} A, \ a \in \mathcal{A}^{\mathrm{c}}; \\ A_a + G_0^a + F_a G^a F_a, \ a \in \mathcal{A}^{\mathrm{n}}. \end{cases} \tag{3.7.5}
$$

For $a \in \mathcal{A}^{\mathrm{n}}$, $M_a$ is defined on $\mathcal{D}(N)$, where

$$
N = D_x^2 + x^2 \tag{3.7.6}
$$

is the harmonic oscillator on $L^2(X)$. We then define the conjugate operator $M$:

$$
M := \sum_{a < a_{\max}} q_a M_a q_a. \tag{3.7.7}
$$

Note that the channel conjugate operators $M_a$, $a \in \mathcal{A}^{\mathrm{n}}$, are very similar to

$$
G_a = A_a + \chi_a(H^{a,\mathrm{n}}) A^a \chi_a(H^{a,\mathrm{n}}) + G^a = U_a^* \tilde{G}_{a,I} U_a, \tag{3.7.8}
$$

where $\tilde{G}_{a,I}$ is the conjugate operator for $\tilde{H}_a$ constructed in Proposition 3.6.8. This will allow us to use the results of the previous section in our proof of the Mourre estimate for $H$. The only difference between $M_a$ and $G_a$ is the presence of the cutoffs $F_a$; the Mourre estimate could actually be proved without these cutoffs, but we will need them later on to obtain a minimal velocity estimate for $H$.

The main results of this section are as follows.

PROPOSITION 3.7.1. *Assume (V1), (V4), (M1), (S). Then $M$ satisfies the hypotheses of Nelson's commutator theorem (Lemma 1.2.5) with the comparison operator $N$; in particular, $M$ is essentially self-adjoint on $\mathcal{D}(N)$.*

We will continue to use $M$ to denote the unique self-adjoint extension of the operator $M$ defined above.

LEMMA 3.7.2. *Assume (V1), (V4), (M1), (S), and let $c > -\inf \sigma(H) + 1$. Then the quadratic form $(H+c)^{-\frac{1}{2}}[H, \mathrm{i}M](H+c)^{-\frac{1}{2}}$ extends to a bounded operator on $L^2(X)$.*

THEOREM 3.7.3. *Assume (V1), (V4), (M1), (S). Then:*

*(i) $H \in C^1(M)$;*

*(ii) the eigenvalues of $H$ may accumulate only at $\tau$. Consequently, $\tau \cup \sigma_{\mathrm{pp}}(H)$ is a closed countable set;*

*(iii) for any $\lambda \in I \backslash \tau \cup \sigma_{\mathrm{pp}}(H)$ and $F \in C_0^\infty(\mathbb{R}^2)$, there is a $c(\lambda) > 0$ such that*

$$
F(K_{\mathrm{cm}})[H, \mathrm{i}M]F(K_{\mathrm{cm}}) \geq c(\lambda) \ \textit{at} \ H = \lambda.
$$

LEMMA 3.7.4. *Assume (V1), (V4), (M1), (S).*

*(i) We have:*

$$H \in C^1(\langle x^{\rm cm}\rangle); \tag{3.7.9}$$

$$\langle x^{\rm cm}\rangle \in C^1(M); \tag{3.7.10}$$

$$[M, \langle x^{\rm cm}\rangle]\langle x^{\rm cm}\rangle^{-1} \text{ is bounded;} \tag{3.7.11}$$

$$\langle x^{\rm cm}\rangle^{-1} M(H+\mathrm{i})^{-1} \text{ is bounded;} \tag{3.7.12}$$

$$-C\langle x^{\rm cm}\rangle \le \chi(H)M\chi(H) \le C\langle x^{\rm cm}\rangle \text{ for } \chi \in C_0^\infty(\mathbb{R}). \tag{3.7.13}$$

*(ii) Let $I(t,x)$ be a time-dependent potential such that*

$$|\partial_x^\alpha I(t,x)| \le C_\alpha(\langle x\rangle + t)^{-1-\epsilon}, \ |\alpha| \ge 0. \tag{3.7.14}$$

*Then the commutator $[M, I(t,x)]$, defined as a quadratic form on $\mathcal{D}(M)$, extends to a bounded operator with norm bounded by $Ct^{-\epsilon}$.*

PROPOSITION 3.7.5. *Assume that (V1), (V4), (M1), (S) hold. Then there is an $\epsilon > 0$ such that $H \in C^{1+\epsilon}(M)$.*

In Subsection 3.7.2 we will prove Proposition 3.7.1 and Lemmas 3.7.2 and 3.7.4, along with a number of other technical results. Theorem 3.7.3 is proved in Subsection 3.7.3. Finally, in Subsection 3.7.4 we will prove Proposition 3.7.5.

**3.7.2. A few technical lemmas.** Throughout this subsection, $a$ denotes a cluster decomposition with $\sharp a \ge 2$, and *(V1), (V4), (M1), (S)* continue to hold.

We define the class of functions

$$S^0_{\rm cm}(X) := \{f \in C^\infty(X) : \ |\partial_x^\alpha f| \le C_\alpha \langle x^{\rm cm}\rangle^{-|\alpha|}, \ |\alpha| \ge 0\}.$$

The functions in $S^0_{\rm cm}(X)$ have essentially the same properties as those in $S^0_{\rm cl}(X^{\rm cm})$, except that they may also depend on $x_{\rm cm}$.

We will use $q'_a, q''_a$ to denote functions in $S^0_{\rm cm}(X)$ supported in

$$\{x \in X : \ |x^a| \le \delta'_a, \ |x^b| \ge \epsilon'_a \ \forall b \not\le a\},$$

where $\delta'_a, \epsilon'_a$ are small enough. In particular, we may put $q'_a = q''_a = q_a$. We may occasionally abuse notation and use "$B$ + h.c." to denote "$B$ + the adjoint of $B$ with $q'_a$ and $q''_a$ interchanged". We also observe that by (3.7.3), $q'_a F_a \in S^0_{\rm cm}(X)$. We first state for future reference the following properties of $M_a$, which follow directly from the construction.

LEMMA 3.7.6. *(i) We have:*

$$G^a = \phi_0(D_{y^{\rm n}_a} + (\mathbf{A}x)^{\rm n}_a) + \langle \phi_1(D_{y^{\rm n}_a} + (\mathbf{A}x)^{\rm n}_a), y^{\rm n}_a\rangle, \tag{3.7.15}$$

$$[H^{a,{\rm n}}, {\rm i}G^a] = \phi_2(D_{y^{\rm n}_a} + (\mathbf{A}x)^{\rm n}_a), \tag{3.7.16}$$

*where $\phi_i$, $i = 0, 1, 2$, are operator-valued functions $Y^{{\rm n}\prime}_a \to B(L^2(X^a))$ such that:*

$$\phi_i(\eta^{\rm n}_a) = f_i(\eta^{\rm n}_a, \tilde{H}^{a,{\rm n}}(\eta^{\rm n}_a)),\ \ f_i \in C_0^\infty(Y^{{\rm n}\prime}_a \times (-\infty, \inf \tau^a)), \tag{3.7.17}$$

$$(\langle x^a\rangle^k + D^2_{x^a})\partial^\alpha_{\eta^{\rm n}_a}\phi_i(\eta^{\rm n}_a) \in B(L^2(X)),\ \ |\alpha| \geq 0,\ i = 0, 1, 2,\ k \geq 0; \tag{3.7.18}$$

*(ii)* $\chi_a(H^{a,{\rm n}})G^a = 0$;

*(iii)* $\mathcal{D}(q'_a M_a q'_a) \subset \mathcal{D}(N)$.

PROOF. *(i)* follows from Proposition 3.6.4*(iii)* and Proposition 3.5.4. *(ii)* is obtained from (3.6.23). Finally, we observe that by (1.2.19) $U_a$ and multiplication by $q'_a$ preserve $\mathcal{D}(N)$, which implies *(iii)*. □

The main purpose of this subsection is to prove Proposition 3.7.1, Lemmas 3.7.2 and 3.7.4, The next three technical lemmas will also be important throughout the rest of this section.

The first lemma establishes the "pseudodifferential" properties of the operator $q'_a M_a q''_a$.

LEMMA 3.7.7. *Let $c > -\inf \sigma(H) + 1$. Then:*

*(i)* $\langle x^{\rm cm}\rangle^{-1} q'_a M_a (H + c)^{-\frac{1}{2}}$, $\langle x^{\rm cm}\rangle^{-1} M_a q'_a (H + c)^{-\frac{1}{2}}$ *are bounded.*

*(ii) (pseudolocality) If $q_1, q_2 \in S^0_{\rm cm}(X)$ satisfy* $\operatorname{supp} q_1 \cap \operatorname{supp} q_2 = \emptyset$, *then*

$$q_1 q'_a M_a q''_a q_2 \langle x^{\rm cm}\rangle \text{ is bounded.} \tag{3.7.19}$$

*(iii) Let $I(x^{\rm cm})$ be a real-valued function satisfying:*

$$I = I^{\rm s} + I^{\rm l},$$

$$I^{\rm s}\langle x^{\rm cm}\rangle^2 (H + c)^{-\frac{1}{2}} \text{ is bounded,} \tag{3.7.20}$$

$$|\partial^\alpha_{x^{\rm cm}} I^{\rm l}| \leq C_\alpha \langle x^{\rm cm}\rangle^{-\epsilon - |\alpha|}, |\alpha| \geq 0, \epsilon > 0.$$

*Then*

$$\langle x^{\rm cm}\rangle^{\min(1,\epsilon)}(H + c)^{-\frac{1}{2}}[I, {\rm i}q'_a M_a q''_a](H + c)^{-\frac{1}{2}} \text{ is bounded.} \tag{3.7.21}$$

*(iv) Suppose that $I(x^{\rm cm})$ satisfies (3.7.20) with $I^{\rm s} \equiv 0$. Then*

$$\langle x^{\rm cm}\rangle^{\min(1,\epsilon)}[I, {\rm i}q'_a M_a q''_a] \text{ is bounded.} \tag{3.7.22}$$

*(v)* $(H + c)^{-\frac{1}{2}}[H_a, {\rm i}q'_a M_a q''_a](H + c)^{-\frac{1}{2}}$ *is bounded.*

The next lemma states, roughly speaking, that $[H_a, {\rm i}M_a]$ is similar to $[H_a, {\rm i}G_a]$ modulo "small" errors which will be estimated appropriately later on. This will be used to deduce a Mourre estimate for $H$ from the results of Section 3.6.

LEMMA 3.7.8. *Let $a \in \mathcal{A}^{\mathrm{n}}$, and let $G_a$ be defined by (3.7.8). Then:*

*(i) we have*

$$[H_a, \mathrm{i}M_a] = [H_a, \mathrm{i}G_a] + R_a, \tag{3.7.23}$$

*where*

$$R_a = F\Big(\frac{|y_a^{\mathrm{c}}|}{\langle x_a^{\mathrm{cm}}\rangle} \geq c_1\Big) R_{a,1} + R_{a,2} F\Big(\frac{|y_a^{\mathrm{c}}|}{\langle x_a^{\mathrm{cm}}\rangle} \geq c_1\Big) \tag{3.7.24}$$

*for some $c_1 > 0$, and $(H+c)^{-1/2} q_a' R_{a,1} q_a''$, $q_a' R_{a,2} q_a'' (H+c)^{-1/2}$ are bounded;*

*(ii) we have for any $c > -\inf \sigma(H) + 1$:*

$$(H+c)^{-1/2}[H_a, \mathrm{i}G_a](H+c)^{-1/2} \text{ is bounded,} \tag{3.7.25}$$

$$(H+c)^{-1/2}[[H_a, \mathrm{i}G_a], q_a'](H+c)^{-1/2}\langle x^{\mathrm{cm}}\rangle \text{ is bounded.} \tag{3.7.26}$$

Finally, we will need a lemma to the effect that the local conjugate operators $M_a$ are similar in the regions where the cutoff functions $q_a$ overlap; this will be the key to combining "local" Mourre estimates into a "global" one.

LEMMA 3.7.9. *For $a \in \mathcal{A}^{\mathrm{n}}$, let*

$$T_a = T_a(q_a', q_q'') := q_a'(M_a - A) q_a''.$$

*Then for any function $q \in S^0_{\mathrm{cm}}(X)$ such that*

$$\langle x^a \rangle \geq c \langle x^{\mathrm{cm}} \rangle \text{ on } \operatorname{supp} q, \tag{3.7.27}$$

*we have*

$$T_a\, q \langle x^{\mathrm{cm}} \rangle,\ q T_a \langle x^{\mathrm{cm}} \rangle \text{ are bounded.} \tag{3.7.28}$$

We now begin the proof of the results stated above. First, we prove a simple geometrical lemma.

LEMMA 3.7.10. *For all $a \in \mathcal{A}^{\mathrm{n}}$:*

$$|y_a^{\mathrm{n}}| \leq C \langle x^{\mathrm{cm}} \rangle \text{ on } \operatorname{supp} q_a' F_a,$$

*provided that $\delta_a'$, $c_0$ have been chosen small enough.*

PROOF. We have:

$$x^{\mathrm{cm}} = \pi^{\mathrm{cm}} y_a^{\mathrm{n}} + \pi^{\mathrm{cm}} y_a^{\mathrm{c}} + \pi^{\mathrm{cm}} y^a + z^{\mathrm{cm}},$$

$$|y^a| \leq \delta_a \langle x^{\mathrm{cm}} \rangle \text{ on } \operatorname{supp} q_a',$$

$$|y_a^{\mathrm{c}}| \leq c_0 \langle x^{\mathrm{cm}} \rangle \text{ on } \operatorname{supp} F_a.$$

Hence on $\operatorname{supp} q_a' F_a$

$$|\pi^{\mathrm{cm}} y_a^{\mathrm{n}}| \leq C \langle x^{\mathrm{cm}} \rangle,$$

provided that $\delta_a$, $c_0$ are small enough. Since $\pi^{\mathrm{cm}}$ is invertible on $Y_a^{\mathrm{n}}$, this proves the lemma. □

LEMMA 3.7.11. *Let $a \in \mathcal{A}^n$, and let $N$ be defined by (3.7.6). Then:*
*(i)* $\|G^a u\| \le C\|Nu\|$ *for* $u \in \mathcal{D}(N)$*;*
*(ii)* $|(G^a u, Nu) - (Nu, G^a u)| \le C\|N^{1/2}u\|^2$ *for* $u \in \mathcal{D}(N)$*;*
*(iii)* $G^a F_a q'_a \langle x^{\mathrm{cm}}\rangle^{-1}$ *and* $q'_a F_a G^a \langle x^{\mathrm{cm}}\rangle^{-1}$ *are bounded.*

PROOF. *(i)* and *(ii)* are immediate consequences of Lemma 3.6.5. (Alternatively, they can be proved directly using Lemma 3.7.6*(i)*.)

*(iii)* By Lemma 3.7.6, $G^a \langle y_a^{\mathrm{n}}\rangle^{-1}$ is bounded. This together with Lemma 3.7.10 implies the first part of *(iii)*; moreover, to prove the second part it suffices to prove that

$$q'_a F_a [G^a, \langle x^{\mathrm{cm}}\rangle^{-1}] \text{ is bounded.}$$

But this also easily follows from Lemmas 3.7.6 and 3.7.10. □

LEMMA 3.7.12. *Let $B_i = \phi_i(D_{y_a^{\mathrm{n}}} + (\mathbf{A}x)_a^{\mathrm{n}})$, $i = 0, 1, 2$, where $\phi_i$ are as in Lemma 3.7.6. Then:*
*(i)* $\langle x^a\rangle^k B_i$ *preserves* $\mathcal{D}(\langle x^{\mathrm{cm}}\rangle^k)$, $k \ge 1$*;*
*(ii)* $[B_i, q]\langle x^{\mathrm{cm}}\rangle^{1-\epsilon}$ *is bounded for any* $q \in S^\epsilon(X^{\mathrm{cm}})$*;*
*(iii) if* $q_1, q_2 \in S^0_{\mathrm{cm}}(X)$ *satisfy* $\operatorname{supp} q_1 \cap \operatorname{supp} q_2 = \emptyset$, *then*

$$q_1 B_i q_2 \langle x^{\mathrm{cm}}\rangle^2 \textit{ is bounded;}$$

*(iv) let $I(t,x)$ satisfy $|\partial_x^\alpha I(t,x)| \le C_\alpha(t + \langle x^{\mathrm{cm}}\rangle)^{-1-\epsilon}$, $\epsilon > 0$, for all $\alpha$, then*

$$[I(t,x), B_i]((\langle t\rangle + \langle x^{\mathrm{cm}}\rangle))^{1+\epsilon} \textit{ is bounded;}$$

*(v)* $[B_i, N]$ *is bounded on* $\mathcal{D}(N^{1/2})$.

PROOF. Consider the unitary transformation

$$Uu(x) = \mathrm{e}^{\mathrm{i}\langle x_a^{\mathrm{n}}, \mathbf{A}x\rangle} u(x).$$

Then $U$ preserves the position operator $x$, and

$$U^{-1} B_i U = \phi_i(D_{y_a^{\mathrm{n}}}).$$

(Observe that the reducing transformation $U_a$ does not have this property because of its action on $Y^{a,\mathrm{c}}$, and therefore cannot be used here instead of $U$).

It suffices to prove the lemma with $B_i$ replaced by $U^{-1}B_iU$. But this follows from Lemmas A.4 and A.5, with $x_1 = y_a^{\mathrm{n}}$ and $x_2 = x^a$. (Observe that Lemma A.5 may be applied to $U^{-1}NU$, since by Segal's theorem (1.2.19) we have $U^{-1}NU = n(x, D_x)$, where $n(x,\xi)$ is a second-order polynomial satisfying $n(x,\xi) \ge c_0(x^2 + \xi^2) - c_1$.) □

LEMMA 3.7.13. *(i) Let $q_\epsilon \in S_{\mathrm{cl}}^{-\epsilon}(X^{\mathrm{cm}})$, $\epsilon \ge 0$. Then*

$$[q'_a F_a G^a q''_a F_a, q_\epsilon]\langle x^{\mathrm{cm}}\rangle^\epsilon \textit{ is bounded.}$$

*(ii) Let* $q_1, q_2 \in S^0_{\mathrm{cm}}(X)$ *satisfy* $\operatorname{supp} q_1 \cap \operatorname{supp} q_2 = \emptyset$. *Then*

$$q_1 q'_a F_a G^a q''_a F_a q_2 \langle x^{\mathrm{cm}}\rangle \textit{ is bounded.}$$

PROOF. Since $\epsilon \geq 0$, $q_\epsilon$ preserves $\mathcal{D}(N)$, so that $[G^a, q_\epsilon]$ and $q_1 G^a q_2$ are bounded operators on $\mathcal{D}(N)$. Now *(i)–(ii)* follow from Lemma 3.7.6*(i)*, Lemma 3.7.12*(ii)–(iii)*, and Lemma 3.7.10. □

Before we proceed further, we make the following simple observation, which will be used repeatedly throughout this section.

REMARK 3.7.14. Let $a \in \mathcal{A}^{\mathrm{n}}$. Since $\chi_a(s) \equiv 1$ for $s$ large enough, and since $H^{a,\mathrm{n}}$ is bounded from below, we may write

$$\chi_a(H^{a,\mathrm{n}}) = \mathbf{1} - \chi_{a,1}(H^{a,\mathrm{n}}),\ \chi_{a,1} \in C_0^\infty(\mathbb{R}). \tag{3.7.29}$$

In particular, Lemmas 1.2.4 and 2.3.9 may be applied to $\chi_a(H^{a,\mathrm{n}})$, even though $\chi_a$ is not compactly supported.

We also need to take care of a technicality.

LEMMA 3.7.15. *Assume (V1), (V4), and let* $c > -\inf \sigma(H) + 1$. *Then:*

$$\langle x^{\mathrm{cm}}\rangle (H+c)^{-1/2} \langle x^{\mathrm{cm}}\rangle^{-1} (H+c)^{1/2} \textit{ is bounded;} \tag{3.7.30}$$

$$\langle x^{\mathrm{cm}}\rangle^{-1} (H+c)^{-1/2} \langle x^{\mathrm{cm}}\rangle (H+c)^{1/2} \textit{ is bounded;} \tag{3.7.31}$$

$$(H+c)^{-1/2} q (H+c)^{1/2} \textit{ is bounded for any } q \in S^0_{\mathrm{cm}}(X). \tag{3.7.32}$$

PROOF. We will only prove (3.7.30); the proof of (3.7.31) and (3.7.32) is almost identical. Since $H + c \geq 0$, we may use the functional calculus formula:

$$(H+c)^{-\frac{1}{2}} = \pi^{-1} \int_0^{+\infty} s^{-\frac{1}{2}} (H+c+s)^{-1} ds. \tag{3.7.33}$$

From (3.7.33) it follows that for any bounded operator $B$ we have

$$\begin{aligned}(H+c)^{-\frac{1}{2}} B &= B (H+c)^{-\frac{1}{2}} \\ &-\pi^{-1} \textstyle\int_0^{+\infty} s^{-\frac{1}{2}} (H+c+s)^{-1} [H, B] (H+c+s)^{-1} ds.\end{aligned} \tag{3.7.34}$$

Let $B = \langle x^{\mathrm{cm}}\rangle^{-1}$. By (3.7.9), $H \in C^1(\langle x^{\mathrm{cm}}\rangle)$, hence we may use Proposition 3.2.3*(iii)* to write

$$\begin{aligned}&\langle x^{\mathrm{cm}}\rangle (s+c+H)^{-1} \\ &= (s+c+H)^{-1} [H, \langle x^{\mathrm{cm}}\rangle] (s+c+H)^{-1} + (s+c+H)^{-1} \langle x^{\mathrm{cm}}\rangle,\end{aligned} \tag{3.7.35}$$

for $s > 0$. (The proof of (3.7.9), which will be given shortly, does not use Lemma 3.7.15.) Combining (3.7.34) and (3.7.35), we obtain:

$$\begin{aligned}
&\langle x^{\rm cm}\rangle(c+H)^{-\frac{1}{2}}\langle x^{\rm cm}\rangle^{-1}(c+H)^{\frac{1}{2}}\\
= \ & \mathbf{1} - \tfrac{1}{\pi}\int_0^{+\infty} s^{-\frac{1}{2}}\langle x^{\rm cm}\rangle(s+c+H)^{-1}[H,\langle x^{\rm cm}\rangle^{-1}]\\
&\times(s+c+H)^{-1}(c+H)^{\frac{1}{2}}ds\\
= \ & \mathbf{1} - \tfrac{1}{\pi}\int_0^{+\infty} s^{-\frac{1}{2}}(s+c+H)^{-1}\langle x^{\rm cm}\rangle\\
&\times[H,\langle x^{\rm cm}\rangle^{-1}](s+c+H)^{-1}(c+H)^{\frac{1}{2}}ds\\
- \ & \tfrac{1}{\pi}\int_0^{+\infty} s^{-\frac{1}{2}}(s+c+H)^{-1}[H,\langle x^{\rm cm}\rangle](s+c+H)^{-1}\\
&\times[H,\langle x^{\rm cm}\rangle^{-1}](s+c+H)^{-1}(c+H)^{\frac{1}{2}}ds.
\end{aligned}$$

To bound the integral terms, we use that

$$(H+c)^{-\frac{1}{2}}\langle x^{\rm cm}\rangle[H,\langle x^{\rm cm}\rangle^{-1}],\ (H+c)^{-\frac{1}{2}}[H,\langle x^{\rm cm}\rangle]$$

are bounded, and that

$$\begin{aligned}
&\|(s+c+H)^{-1}\| \le C\langle s\rangle^{-1},\\
&\|(H+c)^{\frac{1}{2}}(s+c+H)^{-1}\| \le C\langle s\rangle^{-\frac{1}{2}}.
\end{aligned}$$

We obtain that the integrands are bounded in norm by $Cs^{-\frac{1}{2}}\langle s\rangle^{-1}$ and $Cs^{-\frac{1}{2}}\langle s\rangle^{-3/2}$ respectively. Hence the integrals are convergent in norm and (3.7.30) holds. □

Proof of Proposition 3.7.1. By Lemmas 3.7.11 and 3.6.6, $\|Mu\| \le \|Nu\|$ for $u \in \mathcal{D}(N)$; in particular, $\mathcal{D}(N) \subset \mathcal{D}(M)$. Since $q_a$ and $F_a$ preserve $\mathcal{D}(N)$, by the argument given in Remark 3.2.4 we may apply the Leibniz rule to expand the commutator $[q_aF_aG^aF_aq_a, N]$. We obtain that

$$[q_aM_aq_a, N] = I_1 + I_2 + \cdots + I_6,$$

where:

$$\begin{aligned}
&I_1 = [q_a, N]A_aq_a + \text{ h.c.}, && I_4 = q_a[G_0^a, N]q_a,\\
&I_2 = q_a[A_a, N]q_a, && I_5 = [q_aF_a, N]G^aF_aq_a + \text{ h.c.},\\
&I_3 = [q_a, N]G_0^aq_a + \text{ h.c.}, && I_6 = q_aF_a[G^a, N]F_aq_a.
\end{aligned}$$

Using that $q_a, F_a \in S^0_{\rm cm}(X)$, one easily obtains that the terms $I_1$, $I_2$, $I_3$, $I_5$ are form-bounded on $\mathcal{D}(N^{1/2})$. The term $I_4$ is bounded on $\mathcal{D}(N^{1/2})$ by exactly the same arguments as in the proof of Lemma 3.5.9, and $I_6$ – by Lemma 3.7.11. This shows that $[M, N]$ is bounded on $\mathcal{D}(N^{1/2})$, and ends the proof of the proposition. □

PROOF OF LEMMA 3.7.4. To prove (3.7.9), we will verify that $H$ and $\langle x^{\mathrm{cm}}\rangle$ satisfy the conditions *(i)–(ii)* of Proposition 3.2.1. Indeed, by Lemma 1.2.3 $(z-H)^{-1}$ preserves $\mathcal{D}(\langle x^{\mathrm{cm}}\rangle)$, hence $H$ and $\langle x^{\mathrm{cm}}\rangle$ satisfy *(ii)*. Moreover, we have the identity

$$[H, \mathrm{i}\langle x^{\mathrm{cm}}\rangle] = \frac{1}{2}(\langle \frac{x^{\mathrm{cm}}}{\langle x^{\mathrm{cm}}\rangle}, (D-\mathbf{A}x)^{\mathrm{cm}}\rangle + \mathrm{h.c.}),$$

where both sides are understood as quadratic forms on $\mathcal{D}(H)\cap\mathcal{D}(\langle x^{\mathrm{cm}}\rangle)$. Thus *(i)* is also satisfied; by Proposition 3.2.1, this implies (3.7.9).

Next, we will prove (3.7.11). It suffices to prove that

$$[q_a M_a q_a, \langle x^{\mathrm{cm}}\rangle]\langle x^{\mathrm{cm}}\rangle^{-1} \text{ is bounded} \tag{3.7.36}$$

as a quadratic form on $\mathcal{D}(N)$. Clearly,

$$[q_a A q_a, \langle x^{\mathrm{cm}}\rangle]\langle x^{\mathrm{cm}}\rangle^{-1} \text{ is bounded,} \tag{3.7.37}$$

which proves (3.7.36) for $a\in\mathcal{A}^{\mathrm{c}}$. Assume now that $a\in\mathcal{A}^{\mathrm{n}}$. By Remark 3.7.14 and Lemma 1.2.4,

$$\chi_a(H^{a,\mathrm{n}}) \text{ preserves } \mathcal{D}(N), \mathcal{D}(\langle x^{\mathrm{cm}}\rangle). \tag{3.7.38}$$

Moreover, $q_a$ and $F_a$ clearly preserve $\mathcal{D}(\langle x^{\mathrm{cm}}\rangle)$. Hence by Remark 3.2.4 we may use the Leibniz rule to compute the commutator in (3.7.36). We write:

$$\begin{aligned}
&[M_a, \langle x^{\mathrm{cm}}\rangle] = R_1+R_2+R_3+R_4,\\
&R_1 = [A_a, \langle x^{\mathrm{cm}}\rangle],\\
&R_2 = [\chi_a(H^{a,\mathrm{n}}), \langle x^{\mathrm{cm}}\rangle]A^a\chi_a(H^{a,\mathrm{n}}) + \chi_a(H^{a,\mathrm{n}})A^a[\chi_a(H^{a,\mathrm{n}}), \langle x^{\mathrm{cm}}\rangle],\\
&R_3 = \chi_a(H^{a,\mathrm{n}})[A^a, \langle x^{\mathrm{cm}}\rangle]\chi_a(H^{a,\mathrm{n}}),\\
&R_4 = F_a[G^a, \langle x^{\mathrm{cm}}\rangle]F_a.
\end{aligned}$$

We first observe that

$$[A_a, \langle x^{\mathrm{cm}}\rangle]\langle x^{\mathrm{cm}}\rangle^{-1},\ [A^a, \langle x^{\mathrm{cm}}\rangle]\langle x^{\mathrm{cm}}\rangle^{-1} \text{ are bounded.}$$

Using also (3.7.38) we obtain that

$$R_1\langle x^{\mathrm{cm}}\rangle^{-1},\ R_3\langle x^{\mathrm{cm}}\rangle^{-1} \text{ are bounded.} \tag{3.7.39}$$

Next we check that

$$R_2\langle x^{\mathrm{cm}}\rangle^{-1} \text{ is bounded.} \tag{3.7.40}$$

To prove that $[\chi_a(H^{a,\mathrm{n}}), \langle x^{\mathrm{cm}}\rangle]A^a\chi_a(H^{a,\mathrm{n}})\langle x^{\mathrm{cm}}\rangle^{-1}$ is bounded, by (3.7.38) it suffices to check that

$$[\chi_a(H^{a,\mathrm{n}}), \langle x^{\mathrm{cm}}\rangle]A^a\langle x^{\mathrm{cm}}\rangle^{-1}\chi_a(H^{a,\mathrm{n}}) \text{ is bounded.}$$

Since $(H^{a,\mathrm{n}}+c)^{-1}A^a\langle x^{\mathrm{cm}}\rangle^{-1}\chi_a(H^{a,\mathrm{n}})$ is bounded, we only need to prove that

$$[\chi_a(H^{a,\mathrm{n}}), \langle x^{\mathrm{cm}}\rangle](H^{a,\mathrm{n}}+c) \text{ is bounded.} \tag{3.7.41}$$

Similarly, to prove that $\chi_a(H^{a,\mathrm{n}})A^a[\chi_a(H^{a,\mathrm{n}}), \langle x^{\mathrm{cm}}\rangle]\langle x^{\mathrm{cm}}\rangle^{-1}$ is bounded it suffices to verify that

$$\chi_a(H^{a,\mathrm{n}})A^a\langle x^{\mathrm{cm}}\rangle^{-1}[\chi_a(H^{a,\mathrm{n}}), \langle x^{\mathrm{cm}}\rangle] \text{ is bounded;}$$

since $\chi_a(H^{a,\mathrm{n}})A^a\langle x^{\mathrm{cm}}\rangle^{-1}(H^{a,\mathrm{n}}+c)^{-1}$ is bounded, this again follows from (3.7.41).

We now verify (3.7.41). By Remark 3.7.14 we may write $\chi_a(H^{a,\mathrm{n}}) = \mathbf{1} - \chi_{a,1}(H^{a,\mathrm{n}})$, where $\chi_{a,1} \in C_0^\infty(\mathbb{R})$. Then

$$[\chi_a(H^{a,\mathrm{n}}), \langle x^{\mathrm{cm}}\rangle] = -[\chi_{a,1}(H^{a,\mathrm{n}}), \langle x^{\mathrm{cm}}\rangle].$$

Applying (1.2.21) to $\chi_{a,1}$ and using that $[(H^{a,\mathrm{n}}-z)^{-1}, \langle x^{\mathrm{cm}}\rangle]$ is bounded with norm $O(|\mathrm{Im}\, z|^{-1})$, we obtain (3.7.41).

Finally, we need to prove that

$$q_aR_4q_a\langle x^{\mathrm{cm}}\rangle^{-1} = q_aF_a[G^a, \langle x^{\mathrm{cm}}\rangle]F_aq_a\langle x^{\mathrm{cm}}\rangle^{-1} \text{ is bounded.} \tag{3.7.42}$$

By Lemma 3.7.10, it suffices to prove that

$$[G^a, \langle x^{\mathrm{cm}}\rangle]\langle y_a^{\mathrm{n}}\rangle^{-1} \text{ is bounded.} \tag{3.7.43}$$

Using the explicit form of $G^a$ given in Lemma 3.7.6, we are reduced to proving that

$$[\phi_i(D_{y_a^{\mathrm{n}}} + (\mathbf{A}x)_a^{\mathrm{n}}), \langle x^{\mathrm{cm}}\rangle] \text{ is bounded, } i = 0, 1.$$

But this follows from Lemma 3.7.12*(ii)*.

Combining (3.7.39), (3.7.40), (3.7.42), we obtain (3.7.11) for $a \in \mathcal{A}^{\mathrm{n}}$.

(3.7.12) is an immediate consequence of Lemma 3.7.7*(i)*, the proof of which will be given shortly.

To prove (3.7.10), we use Lemma 3.2.2 with $N$ defined by (3.7.6) and with $H = H_0 = \langle x^{\mathrm{cm}}\rangle$, $A = M$. We first check that $\langle x^{\mathrm{cm}}\rangle$ and $M$ satisfy the hypotheses of Lemma 1.2.5: for $\langle x^{\mathrm{cm}}\rangle$ it is obvious, and for $M$ it follows from Proposition 3.7.1. The condition (3.2.4) of Lemma 3.2.2 follows from (3.7.11). Hence (3.7.10) holds.

To prove (3.7.13), we have to verify that

$$\langle x^{\mathrm{cm}}\rangle^{-\frac{1}{2}}\chi(H)M\chi(H)\langle x^{\mathrm{cm}}\rangle^{-\frac{1}{2}}$$

is bounded; this can be done using the same arguments as in the proof of (3.7.11). The proof of the part *(ii)* of the lemma is also similar, except that we now use Lemma 3.7.12*(iv)*. The details are left to the reader. □

Proof of Lemma 3.7.7. For $a \in \mathcal{A}^{\mathrm{c}}$, we need to prove *(i)*–*(iv)* with $M_a = A$; for $a \in \mathcal{A}^{\mathrm{n}}$ we write:

$$M_a = A_a + G_0^a + F_aG^aF_a,$$

and consider each term separately.

*(i)* It is clear that

$$\langle x^{\mathrm{cm}}\rangle^{-1}q_a'A(H+c)^{-1/2},\ \langle x^{\mathrm{cm}}\rangle^{-1}Aq_a'(H+c)^{-1/2}$$

are bounded for any $a \in \mathcal{A}$, and that the same is true with $A$ replaced by $A_a$.

By (3.7.32), to prove *(i)* with $M_a$ replaced by $G_0^a$ it suffices to prove that

$$\langle x^{\mathrm{cm}}\rangle^{-1} G_0^a (H+c)^{-1/2} \text{ is bounded.}$$

By Lemma 1.2.4, $\chi_a(H^{a,\mathrm{n}})$ preserves $\mathcal{D}(\langle x^{\mathrm{cm}}\rangle)$. Hence

$$\begin{aligned}\langle x^{\mathrm{cm}}\rangle^{-1} G_0^a (H+c)^{-1/2} &= \langle x^{\mathrm{cm}}\rangle^{-1} \chi_a(H^{a,\mathrm{n}}) A^a \chi_a(H^{a,\mathrm{n}})(H+c)^{-1/2} \\ &= \left(\langle x^{\mathrm{cm}}\rangle^{-1} \chi_a(H^{a,\mathrm{n}})\langle x^{\mathrm{cm}}\rangle\right)\left(\langle x^{\mathrm{cm}}\rangle^{-1} A^a (H^{a,\mathrm{n}}+i)^{-1/2}\right) \\ &\quad \left((H^{a,\mathrm{n}}+i)^{1/2}\chi_a(H^{a,\mathrm{n}})(H+c)^{-1/2}\right)\end{aligned}$$

is bounded.

Finally, *(i)* with $M_a$ replaced by $F_a G^a F_a$ follows from Lemma 3.7.11*(iii)*.

*(ii)* (3.7.19) is trivial with $M_a$ replaced by $A$ or $A_a$, since $q_1 A q_2 = q_1 A_a q_2 = 0$.

Assume now that $a \in \mathcal{A}^{\mathrm{n}}$, and pick $q_3, q_4 \in S^0_{\mathrm{cl}}(X^{\mathrm{cm}})$ satisfying:

$$\begin{aligned}&q_3 \equiv 1 \text{ on } \operatorname{supp} q_1,\ q_3 \equiv 0 \text{ on } \operatorname{supp} q_2, \\ &q_4 \equiv 0 \text{ on } \operatorname{supp} q_1,\ q_4 \equiv 1 \text{ on } \operatorname{supp} q_2, \\ &\operatorname{supp} q_3 \cap \operatorname{supp} q_4 = \emptyset.\end{aligned}$$

By Remark 3.7.14 we may apply Lemma 2.3.9*(iii)* to $\chi_a(H^{a,\mathrm{n}})$, hence

$$\begin{aligned}&q_1 \chi_a(H^{a,\mathrm{n}})(1-q_3)\langle x^{\mathrm{cm}}\rangle^2 (H^{a,\mathrm{n}}+c), \\ &(1-q_4)\chi_a(H^{a,\mathrm{n}}) q_2 \langle x^{\mathrm{cm}}\rangle^2 (H^{a,\mathrm{n}}+c),\end{aligned}$$

are bounded. Using also that $\chi_a(H^{a,\mathrm{n}})$ preserves $\mathcal{D}(\langle x^{\mathrm{cm}}\rangle)$ and that

$$\langle x^{\mathrm{cm}}\rangle^{-1} A^a (H^{a,\mathrm{n}}+c)^{-\frac{1}{2}}$$

is bounded, we obtain:

$$\begin{aligned}&q_1 \chi_a(H^{a,\mathrm{n}}) A^a \chi_a(H^{a,\mathrm{n}}) q_2 \langle x^{\mathrm{cm}}\rangle \\ &= q_1 \chi_a(H^{a,\mathrm{n}}) q_3 A^a q_4 \chi_a(H^{a,\mathrm{n}}) q_2 \langle x^{\mathrm{cm}}\rangle + R,\end{aligned}$$

where $R$ is bounded. But $q_3 A^a q_4 = 0$; hence the last line is equal to $R$. This proves (3.7.19) for $G_0^a$.

We conclude the proof of *(ii)* by observing that (3.7.19) for $q_a' F_a G^a F_a q_a''$ follows from Lemma 3.7.13*(ii)*.

*(iii)* We may assume that $0 < \epsilon \le 1$. We will only consider the case $a \in \mathcal{A}^{\mathrm{n}}$; the proof for $a \in \mathcal{A}^{\mathrm{c}}$ is similar but simpler. We write $I$ in (3.7.21) as $I^{\mathrm{s}} + I^{\mathrm{l}}$, and consider the commutators $[I^{\mathrm{s}}, q_a' M_a q_a'']$ and $[I^{\mathrm{l}}, q_a' M_a q_a'']$ separately.

We first estimate $[I^{\mathrm{s}}, q_a' M_a q_a'']$. It suffices to prove that:

$$\langle x^{\mathrm{cm}}\rangle (H+c)^{-1/2} I^{\mathrm{s}} q_a' M_a q_a'' (H+c)^{-1/2} \text{ is bounded,} \tag{3.7.44}$$

$$\langle x^{\mathrm{cm}}\rangle(H+c)^{-1/2}q_a'M_aq_a''I^{\mathrm{s}}(H+c)^{-1/2} \text{ is bounded.} \tag{3.7.45}$$

We begin with (3.7.44). We write

$$\begin{aligned}&\langle x^{\mathrm{cm}}\rangle(H+c)^{-1/2}I^{\mathrm{s}}q_a'M_aq_a''(H+c)^{-1/2}\\ &= \Big(\langle x^{\mathrm{cm}}\rangle(H+c)^{-1/2}\langle x^{\mathrm{cm}}\rangle^{-1}(H+c)^{1/2}\Big)\\ &\Big((H+c)^{-1/2}\langle x^{\mathrm{cm}}\rangle^2I^{\mathrm{s}}\Big)\Big(\langle x^{\mathrm{cm}}\rangle^{-1}q_a'M_aq_a''(H+c)^{-1/2}\Big).\end{aligned} \tag{3.7.46}$$

The first term on the right-hand side of (3.7.46) is bounded by (3.7.30), the second one – by (3.7.20), and the last one – by part *(i)* of the lemma.

We now turn to (3.7.45). We write

$$\begin{aligned}&\langle x^{\mathrm{cm}}\rangle(H+c)^{-1/2}q_a'M_aq_a''I^{\mathrm{s}}(H+c)^{-1/2}\\ &= \Big(\langle x^{\mathrm{cm}}\rangle(H+c)^{-\frac12}\langle x^{\mathrm{cm}}\rangle^{-1}(H+c)^{\frac12}\Big)\\ &\Big((H+c)^{-\frac12}\langle x^{\mathrm{cm}}\rangle q_a'M_aq_a''\langle x^{\mathrm{cm}}\rangle^{-2}\Big)\Big(\langle x^{\mathrm{cm}}\rangle^2I^s(H+c)^{-\frac12}\Big).\end{aligned} \tag{3.7.47}$$

The first term on the right-hand side of (3.7.47) is bounded by (3.7.30), and the third one – by (3.7.20). Hence it suffices to verify that

$$(H+c)^{-\frac12}\langle x^{\mathrm{cm}}\rangle q_a'M_aq_a''\langle x^{\mathrm{cm}}\rangle^{-2} \text{ is bounded.} \tag{3.7.48}$$

It is clear that $(H+c)^{-\frac12}\langle x^{\mathrm{cm}}\rangle q_a'A_aq_a''\langle x^{\mathrm{cm}}\rangle^{-2}$ is bounded. Next, we write

$$\begin{aligned}&(H+c)^{-\frac12}\langle x^{\mathrm{cm}}\rangle q_a'G_0^aq_a''\langle x^{\mathrm{cm}}\rangle^{-2}\\ &= \Big((H+c)^{-\frac12}\langle x^{\mathrm{cm}}\rangle q_a'\chi_a(H^{a,\mathrm{n}})\langle x^{\mathrm{cm}}\rangle^{-1}(H^{a,\mathrm{n}}+c)\Big)\times\\ &\Big((H^{a,\mathrm{n}}+c)^{-1}\langle x^{\mathrm{cm}}\rangle A^a\langle x^{\mathrm{cm}}\rangle^{-2}\Big)\Big(\langle x^{\mathrm{cm}}\rangle^2\chi_a(H^{a,\mathrm{n}})q_a''\langle x^{\mathrm{cm}}\rangle^{-2}\Big).\end{aligned} \tag{3.7.49}$$

The second term on the right-hand side of (3.7.49) is bounded by direct computation. The last one is bounded because $\chi_a(H^{a,\mathrm{n}})$ preserves $\mathcal{D}(\langle x^{\mathrm{cm}}\rangle^2)$. To prove that the first one is bounded, it suffices to verify that

$$\langle x^{\mathrm{cm}}\rangle[\chi_a(H^{a,\mathrm{n}}),\langle x^{\mathrm{cm}}\rangle^{-1}](H^{a,\mathrm{n}}+c)$$

is bounded. But this follows by applying Lemma 2.3.9*(ii)* to $\chi_a(H^{a,\mathrm{n}})$.

Finally, from Lemma 3.7.12*(i)* and Lemma 3.7.10 we obtain that

$$\langle x^{\mathrm{cm}}\rangle q_a'F_aG^aF_aq_a''\langle x^{\mathrm{cm}}\rangle^{-2} \text{ is bounded.}$$

This completes the proof of (3.7.48) and of (3.7.45).

It remains to prove that

$$\langle x^{\mathrm{cm}}\rangle^{\epsilon}(H+c)^{-\frac12}[I^{\mathrm{l}},q_a'M_aq_a''](H+c)^{-\frac12} \text{ is bounded.} \tag{3.7.50}$$

We rewrite (3.7.50) as

$$\Big(\langle x^{\mathrm{cm}}\rangle^{\epsilon}(H+c)^{-\frac12}\langle x^{\mathrm{cm}}\rangle^{-\epsilon}\Big)\Big(\langle x^{\mathrm{cm}}\rangle^{\epsilon}[I_{\mathrm{l}},q_a'M_aq_a''](H+c)^{-\frac12}\Big).$$

Since by Lemma 1.2.4 $(H+c)^{-1/2}$ preserves $\mathcal{D}(\langle x^{\mathrm{cm}}\rangle)$ the first term is bounded. It therefore suffices to show that

$$\langle x^{\mathrm{cm}}\rangle^{\epsilon}[I^{\mathrm{l}}, q_a' M_a q_a''] \text{ is bounded.} \tag{3.7.51}$$

Observe that this also proves *(iv)*.

By direct computation, $\langle x^{\mathrm{cm}}\rangle^{\epsilon}[I^{\mathrm{l}}, q_a' A_a q_a'']$ is bounded. Letting $q_\epsilon = I^{\mathrm{l}}$ in Lemma 3.7.13*(i)*, we also obtain that

$$\langle x^{\mathrm{cm}}\rangle^{\epsilon}[I^{\mathrm{l}}, q_a' F_a G^a F_a q_a''] \text{ is bounded.}$$

It remains to estimate the term

$$\begin{aligned} &\langle x^{\mathrm{cm}}\rangle^{\epsilon}[I^{\mathrm{l}}, q_a' G_0^a q_a''] \\ = \ &\langle x^{\mathrm{cm}}\rangle^{\epsilon} q_a' [I^{\mathrm{l}}, \chi_a(H^{a,\mathrm{n}})] A^a \chi_a(H^{a,\mathrm{n}}) q_a'' \\ &+\langle x^{\mathrm{cm}}\rangle^{\epsilon} q_a' \chi_a(H^{a,\mathrm{n}}) [I^{\mathrm{l}}, A^a] \chi_a(H^{a,\mathrm{n}}) q_a'' \\ &+\langle x^{\mathrm{cm}}\rangle^{\epsilon} q_a' \chi_a(H^{a,\mathrm{n}}) A^a [I^{\mathrm{l}}, \chi_a(H^{a,\mathrm{n}})] q_a''. \end{aligned} \tag{3.7.52}$$

By Remark 3.7.14 and Lemma 1.2.4, $\chi_a(H^{a,\mathrm{n}})$ preserves the domain of $\langle x^{\mathrm{cm}}\rangle^{\epsilon}$. Hence it suffices to prove that:

$$\langle x^{\mathrm{cm}}\rangle^{\epsilon}[I^{\mathrm{l}}, \chi_a(H^{a,\mathrm{n}})] A^a, \tag{3.7.53}$$

$$\langle x^{\mathrm{cm}}\rangle^{\epsilon}[I^{\mathrm{l}}, A^a], \tag{3.7.54}$$

$$\langle x^{\mathrm{cm}}\rangle^{\epsilon} A^a [I^{\mathrm{l}}, \chi_a(H^{a,\mathrm{n}})], \tag{3.7.55}$$

are bounded. (3.7.54) is easily seen to be bounded by direct computation. (3.7.53) can be rewritten as

$$\Big(\langle x^{\mathrm{cm}}\rangle^{\epsilon}[I^{\mathrm{l}}, \chi_a(H^{a,\mathrm{n}})]\langle x^{\mathrm{cm}}\rangle (H^{a,\mathrm{n}}+c)\Big)\Big((H^{a,\mathrm{n}}+c)^{-1}\langle x^{\mathrm{cm}}\rangle^{-1} A^a\Big);$$

the first term is bounded by Lemma 2.3.9*(ii)*, and the second one – by commuting $\langle x^{\mathrm{cm}}\rangle^{-1}$ through $A^a$. Finally, we rewrite (3.7.55) as

$$\Big(\langle x^{\mathrm{cm}}\rangle^{\epsilon} A^a (H^{a,\mathrm{n}}+c)^{-1}\langle x^{\mathrm{cm}}\rangle^{-1-\epsilon}\Big)\Big(\langle x^{\mathrm{cm}}\rangle^{1+\epsilon}(H^{a,\mathrm{n}}+c)[I^{\mathrm{l}}, \chi_a(H^{a,\mathrm{n}})]\Big).$$

The first term is bounded by direct computation, and the second one – by Lemma 2.3.9*(ii)*.

*(iv)* was proved above (see the proof of (3.7.51)).

*(v)* Suppose first that $a \in \mathcal{A}^{\mathrm{c}}$, in which case $M_a = A$. We have:

$$[H_a, \mathrm{i}q_a' A q_a''] = [H_0, \mathrm{i}q_a' A q_a''] + [V^a, \mathrm{i}q_a' A q_a''].$$

Observe first that:

$$\begin{aligned} &(H+c)^{-\frac{1}{2}}[H_0, \mathrm{i}q_a' A q_a''](H+c)^{-\frac{1}{2}} \\ &= \Big((H+c)^{-\frac{1}{2}}(D-\mathbf{A}x)\Big)\Big([D-\mathbf{A}x, \mathrm{i}q_a' A q_a''](H+c)^{-\frac{1}{2}}\Big) - \text{h.c.} \end{aligned} \tag{3.7.56}$$

The commutator $[D - \mathbf{A}x, \mathrm{i}q'_a A q''_a]$ can be computed explicitly and is easily seen to have the form

$$B_1 + \langle B_2, D_z\rangle,$$

where $B_1, B_2$ are bounded. Hence (3.7.56) is bounded.

Next, since $[V^a, A_a] = 0$, we have

$$\begin{aligned}[V^a, \mathrm{i}q_a A q_a] &= [V^a, \mathrm{i}q''_a A^a\, q''_a] \\ &= [V^{a,\mathrm{s}}, \mathrm{i}q'_a A^a q''_a] + [V^{a,\mathrm{l}}, \mathrm{i}q'_a A^a\, q''_a].\end{aligned} \tag{3.7.57}$$

The second term on the right-hand side of (3.7.57) is bounded by *(V4)*. To estimate the first term, we write

$$\begin{aligned}&(H+c)^{-\frac{1}{2}} q'_a [V^{a,\mathrm{s}}, \mathrm{i}A^a] q''_a (H+c)^{-\frac{1}{2}} \\ &= (H+c)^{-\frac{1}{2}} q'_a \mathrm{i}(V^{a,\mathrm{s}} z^a D_{z^a} - z^a D_{z^a} V^{a,\mathrm{s}}) q''_a (H+c)^{-\frac{1}{2}},\end{aligned}$$

which is also easily seen to be bounded by *(V4)*. This proves *(v)* for $a \in \mathcal{A}^{\mathrm{c}}$.

Assume now that $a \in \mathcal{A}^{\mathrm{n}}$. By Remark 3.2.4, we have the identity between quadratic forms on $\mathcal{D}(N)$:

$$[H_a, \mathrm{i}q'_a M_a q''_a] = q'_a [H_a, \mathrm{i}M_a] q''_a + [H_a, \mathrm{i}q'_a] M_a q''_a + \text{ h.c.} \tag{3.7.58}$$

We first observe that

$$\begin{aligned}&(H+c)^{-1/2} [H_a, \mathrm{i}q'_a] M_a q''_a (H+c)^{-1/2} \\ &= \Big((H+c)^{-1/2} [H_a, \mathrm{i}q'_a] \langle x^{\mathrm{cm}}\rangle\Big) \Big(\langle x^{\mathrm{cm}}\rangle^{-1} M_a q''_a (H+c)^{-1/2}\Big);\end{aligned}$$

the first term is easily seen to be bounded by an explicit computation, and the second one is bounded by part *(i)* of the lemma.

It remains to prove that

$$(H+c)^{-1/2} q'_a [H_a, \mathrm{i}M_a] q''_a (H+c)^{-1/2} \text{ is bounded.} \tag{3.7.59}$$

To this end, we write

$$\begin{aligned}[H_a, iM_a] &= D_{z^a}^2 + \chi_a(H^{a,\mathrm{n}}) [H^{a,\mathrm{n}}, \mathrm{i}A^a] \chi_a(H^{a,\mathrm{n}}) \\ &+ \Big([H_a, \mathrm{i}F_a] G^a F_a + \text{ h.c.}\Big) + F_a [H^{a,\mathrm{n}}, \mathrm{i}G^a] F_a.\end{aligned}$$

The term

$$(H+c)^{-1/2} q'_a\, D_{z^a}^2\, q''_a (H+c)^{-1/2} \tag{3.7.60}$$

is seen to be bounded by commuting a $D_{z_a}$ through $q'_a, q''_a$. Next, we verify that

$$(H+c)^{-1/2} q'_a \chi_a(H^{a,\mathrm{n}}) [H^{a,\mathrm{n}}, \mathrm{i}A^a] \chi_a(H^{a,\mathrm{n}})\, q''_a (H+c)^{-1/2} \tag{3.7.61}$$

is bounded. Using (3.7.32) and that

$$(H+c)^{-1/2} (H^{a,\mathrm{n}} + c)^{1/2}$$

is bounded, we see that it suffices to check that

$$(H^{a,\mathrm{n}} + c)^{-1/2} \chi_a(H^{a,\mathrm{n}})\, [H^{a,\mathrm{n}}, \mathrm{i}A^a]\, \chi_a(H^{a,\mathrm{n}}) (H^{a,\mathrm{n}} + c)^{-1/2}$$

is bounded. But this again follows by explicit calculation. The term

$$(H+c)^{-1/2}q_a' F_a[H^{a,\mathrm{n}}, \mathrm{i}G^a]F_a q_a''(H+c)^{-1/2} \tag{3.7.62}$$

is bounded by Proposition 3.6.4*(i)*. Finally, we need to consider the term

$$(H+c)^{-1/2}q_a'[H^{a,\mathrm{n}}, \mathrm{i}F_a]G^a F_a q_a''(H+c)^{-1/2}. \tag{3.7.63}$$

By (3.7.3) $\nabla F_a = O(\langle x^{\mathrm{cm}}\rangle^{-1})$ on $\operatorname{supp} q_a$, so that

$$(H+c)^{-1/2}q_a'[H^{a,\mathrm{n}}, \mathrm{i}F_a]\langle x^{\mathrm{cm}}\rangle \text{ is bounded.} \tag{3.7.64}$$

Hence it suffices to prove that

$$\langle x^{\mathrm{cm}}\rangle^{-1}G^a F_a q_a''(H+c)^{-1/2}$$

is bounded. But this follows from Lemma 3.7.6. □

PROOF OF LEMMA 3.7.2. We write

$$[H, \mathrm{i}q_a M_a q_a] = [H_a, \mathrm{i}q_a M_a q_a] + [I_a, \mathrm{i}q_a M_a q_a]. \tag{3.7.65}$$

The first term on the right-hand side of (3.7.65) is form-bounded on $\mathcal{D}((H+c)^{\frac{1}{2}})$ by Lemma 3.7.7*(v)*. To estimate the second term, we observe that using *(V4)* and the support properties of $q_a$ (see Proposition 2.3.4) we may write

$$[I_a, \mathrm{i}q_a M_a q_a] = [I_{a,1}, \mathrm{i}q_a M_a q_a],$$

where $I_{a,1}$ satisfies (3.7.20). By (3.7.21), the second term is also form-bounded on $\mathcal{D}((H+c)^{\frac{1}{2}})$. □

PROOF OF LEMMA 3.7.8. *(i)* By (3.7.5) and (3.7.8), we have:

$$\begin{aligned} R_a := \; & [H_a, \mathrm{i}M_a] - [H_a, \mathrm{i}G_a] \\ = \; & [H_a, \mathrm{i}F_a]G^a F_a + F_a G^a[H_a, \mathrm{i}F_a] \\ & -(1-F_a)[H^{a,\mathrm{n}}, \mathrm{i}G^a]F_a - [H^{a,\mathrm{n}}, \mathrm{i}G^a](1-F_a). \end{aligned}$$

The last two terms are of the form (3.7.24), since by Proposition 3.6.4*(i)* $[H^{a,\mathrm{n}}, iG^a]$ is bounded and since $1-F_a$ is supported in $|y_a^{\mathrm{c}}| \geq c_1\langle x_a^{\mathrm{cm}}\rangle$. Consider now the first term; the second one is similar. Since $\nabla F_a$ is supported in $|y_a^{\mathrm{c}}| \geq c_1\langle x_a^{\mathrm{cm}}\rangle$, we may write

$$q_a'[H_a, iF_a]G^a F_a = F\Big(\frac{|y_a^{\mathrm{c}}|}{\langle x_a^{\mathrm{cm}}\rangle} \geq c_1\Big)q_a'[H_a, iF_a]G^a F_a.$$

By (3.7.3), it remains to verify that

$$\langle x^{\mathrm{cm}}\rangle^{-1}G^a F_a q_a' \text{ is bounded.}$$

But this follows from Lemma 3.7.11*(iii)*.

*(ii)* We write

$$[H_a, \mathrm{i}G_a] = [H_a, \mathrm{i}A_a] + \chi_a(H^{a,\mathrm{n}})[H^{a,\mathrm{n}}, \mathrm{i}A^a]\chi_a(H^{a,\mathrm{n}}) + [H^{a,\mathrm{n}}, \mathrm{i}G^a]. \tag{3.7.66}$$

(3.7.25) follows directly from (3.7.66), *(V4)*, and Lemma 3.7.6*(i)*, if we write

$$[H^{a,\mathrm{n}}, \mathrm{i}A^a] = D^2_{z^a} - z^a \nabla_{z^a} V^{a,\mathrm{l}} + \mathrm{i}V^{a,\mathrm{s}} A^a - \mathrm{i}A^a V^{a,\mathrm{s}}. \tag{3.7.67}$$

It remains to prove (3.7.26). Since $\langle x^{\mathrm{cm}} \rangle \nabla q_a'$ is bounded, we see by direct computation that

$$(H+c)^{1/2}[[H_a, \mathrm{i}A_a], q_a'] \langle x^{\mathrm{cm}} \rangle \text{ is bounded.}$$

Similarly, $(H^{a,\mathrm{n}} + c)^{1/2}[[H^{a,\mathrm{n}}, \mathrm{i}A^a], q_a'] \langle x^{\mathrm{cm}} \rangle$ is bounded. Using also that $\chi_a(H^{a,\mathrm{n}})$ preserves $\mathcal{D}(\langle x^{\mathrm{cm}} \rangle)$ and that $(H+c)^{-1}(H^{a,\mathrm{n}} + c)^{1/2}$ is bounded, we obtain that

$$(H+c)^{-1/2} \chi_a(H^{a,\mathrm{n}})[[H^{a,\mathrm{n}}, \mathrm{i}A^a], q_a'] \chi_a(H^{a,\mathrm{n}}) \langle x^{\mathrm{cm}} \rangle \text{ is bounded.}$$

Next, by Lemma 3.7.12*(ii)* we have:

$$[[H^{a,\mathrm{n}}, \mathrm{i}G^a], q_a'] \langle x^{\mathrm{cm}} \rangle \text{ is bounded.}$$

It remains to prove that

$$(H+c)^{-1/2}(R + R^*)(H+c)^{-1/2} \langle x^{\mathrm{cm}} \rangle \text{ is bounded,} \tag{3.7.68}$$

where

$$R = [\chi_a(H^{a,\mathrm{n}}), q_a'][H^{a,\mathrm{n}}, \mathrm{i}A^a] \chi_a(H^{a,\mathrm{n}}).$$

We will prove that $(H+c)^{-1/2} R (H+c)^{-1/2} \langle x^{\mathrm{cm}} \rangle$ is bounded; the proof for $R^*$ is similar but simpler. By Remark 3.7.14 and Lemma 2.3.9*(ii)* applied to $H^{a,\mathrm{n}}$,

$$[\chi_a(H^{a,\mathrm{n}}), q_a'] \langle x^{\mathrm{cm}} \rangle (H^{a,\mathrm{n}} + c) \text{ is bounded.}$$

From (3.7.67) and *(V4)* we have

$$(H^{a,\mathrm{n}} + c)^{-1} \langle x^{\mathrm{cm}} \rangle^{-1} [H^{a,\mathrm{n}}, \mathrm{i}A^a] \langle x^{\mathrm{cm}} \rangle (H^{a,\mathrm{n}} + c)^{-1/2} \text{ is bounded.}$$

Hence (3.7.68) will follow if we prove that

$$(H^{a,\mathrm{n}} + c)^{1/2} \langle x^{\mathrm{cm}} \rangle^{-1} \chi_a(H^{a,\mathrm{n}})(H+c)^{-1/2} \langle x^{\mathrm{cm}} \rangle \tag{3.7.69}$$

is bounded. We rewrite (3.7.69) as $I_1 + I_2$, where:

$$I_1 = (H^{a,\mathrm{n}} + c)^{1/2} \chi_a(H^{a,\mathrm{n}}) \langle x^{\mathrm{cm}} \rangle^{-1} (H+c)^{-1/2} \langle x^{\mathrm{cm}} \rangle,$$

$$I_2 = (H^{a,\mathrm{n}} + c)^{1/2} [\langle x^{\mathrm{cm}} \rangle^{-1}, \chi_a(H^{a,\mathrm{n}})] \langle x^{\mathrm{cm}} \rangle^2 \langle x^{\mathrm{cm}} \rangle^{-2} (H+c)^{-1/2} \langle x^{\mathrm{cm}} \rangle.$$

By Lemma 2.3.9*(ii)*,

$$(H^{a,\mathrm{n}} + c)[\langle x^{\mathrm{cm}} \rangle^{-1}, \chi_a(H^{a,\mathrm{n}})] \langle x^{\mathrm{cm}} \rangle^2 \text{ is bounded;}$$

using also (3.7.31), we find that $I_2$ is bounded. Finally,

$$I_1 = (H^{a,\mathrm{n}} + c)^{1/2} \chi_a(H^{a,\mathrm{n}})(H+c)^{-1/2}(H+c)^{1/2} \langle x^{\mathrm{cm}} \rangle^{-1} (H+c)^{-1/2} \langle x^{\mathrm{cm}} \rangle$$

is bounded by (3.7.30). This completes the proof of (3.7.26) and of Lemma 3.7.8. □

PROOF OF LEMMA 3.7.9. By (3.7.27) to show that $T_a\, q\langle x^{\mathrm{cm}}\rangle$ is bounded it suffices to prove that

$$T_a\, q\langle x^a\rangle \text{ is bounded.} \tag{3.7.70}$$

We have

$$M_a - A = (\mathbf{1} - \chi_a(H^{a,\mathrm{n}}))A^a + \chi_a(H^{a,\mathrm{n}})\, A^a(\mathbf{1} - \chi_a(H^{a,\mathrm{n}})) + F_a G^a F_a.$$

By Remark 3.7.14, $\mathbf{1} - \chi_a(H^{a,\mathrm{n}}) = \chi_{a,1}(H^{a,\mathrm{n}})$, where $\chi_{a,1} \in C_0^\infty(\mathbb{R})$. Moreover, since $\chi_a \equiv 1$ on a neighbourhood of $[\inf \tau_a, +\infty)$, $\chi_{a,1}$ is supported in $(-\infty, \inf \tau_a)$. By Proposition 3.5.4,

$$\chi_{a,1}(\tilde{H}^{a,\mathrm{n}})\langle x^a\rangle^2 \text{ is bounded;}$$

since $U_a$ commutes with $x^a$, it follows that

$$\chi_{a,1}(H^{a,\mathrm{n}})\langle x^a\rangle^2 \text{ is bounded.} \tag{3.7.71}$$

Using that by Lemma 1.2.4 $\chi_{a,1}(H^{a,\mathrm{n}})$ preserves the domain of $\langle x^a\rangle$, we obtain from (3.7.71) that

$$\chi_{a,1}(H^{a,\mathrm{n}})A^a\langle x^a\rangle \text{ is bounded,}$$

$$\chi_a(H^{a,\mathrm{n}})A^a\chi_{a,1}(H^{a,\mathrm{n}})\langle x^a\rangle \text{ is bounded.}$$

It remains to prove that

$$q_a' F_a G^a F_a q_a'' q\langle x^a\rangle \text{ is bounded.}$$

By Proposition 3.6.4, we have $G^a = G^a\chi_{a,2}(H^{a,\mathrm{n}})$, where $\chi_{a,2}$ is a $C_0^\infty$ function supported in $(-\infty, \inf \tau_a)$. We write

$$\begin{aligned} G^a F_a q_a'' q\langle x^a\rangle &= G^a\chi_{a,2}(H^{a,\mathrm{n}})F_a q_a'' q\langle x^a\rangle \\ &= G^a\langle y_a^{\mathrm{n}}\rangle^{-1}\,\langle y_a^{\mathrm{n}}\rangle\chi_{a,2}(H^{a,\mathrm{n}})F_a q_a'' q\langle x^a\rangle. \end{aligned}$$

By Lemma 3.7.6, $G^a\langle y_a^{\mathrm{n}}\rangle^{-1}$ is bounded. Since $\chi_{a,2}(H^{a,\mathrm{n}})$ preserves $\mathcal{D}(x)$, it suffices to prove that

$$\chi_{a,2}(H^{a,\mathrm{n}})\langle y_a^{\mathrm{n}}\rangle F_a q_a'' q\langle x^a\rangle \text{ is bounded.}$$

By Lemma 3.7.10 $\langle y_a^{\mathrm{n}}\rangle \le C\langle x^{\mathrm{cm}}\rangle$ on $\operatorname{supp} q_a'' F_a$, and by (3.7.27) $\langle x^{\mathrm{cm}}\rangle \le C\langle x^a\rangle$ on $\operatorname{supp} q$. Hence it suffices to prove that

$$\chi_{a,2}(H^{a,\mathrm{n}})\langle x^a\rangle^2 \text{ is bounded.}$$

But this again follows from Proposition 3.5.4.

To prove that $qT_a\langle x^{\mathrm{cm}}\rangle$ is bounded, we write

$$qT_a\langle x^{\mathrm{cm}}\rangle = qT_a q'\langle x^{\mathrm{cm}}\rangle + \text{ bounded,}$$

where $q'$ also satisfies (3.7.27), possibly with a different constant. We may do so by Lemma 3.7.7*(ii)* for $q_a' M_a q_a''$, and by direct computation for $q_a' A_a q_a''$. Now the claim follows from (3.7.70). □

**3.7.3. Proof of the Mourre estimate.** In this section we prove the main result of this section – the Mourre estimate for $H$. Our Theorem 3.7.3 (the *strict* Mourre estimate) will be obtained from Theorem 3.7.17, which states that $H$ satisfies a Mourre estimate at all energies $\notin \tau$ modulo a "small" error. Since we do not separate the center of mass, this error will not be compact on $L^2(X)$; instead, it will belong to the class $\mathcal{R}$ defined below.

DEFINITION 3.7.16. We define

$$\mathcal{R} = \{R \in B(L^2(X)) : F(K_{\mathrm{cm}})RF(K_{\mathrm{cm}}) \text{ is compact for all } F \in C_0^\infty(\mathbb{R}^2)\}.$$

The class $\mathcal{R}$ is similar to the class of $a_{\max}$-compact operators (Definition 3.3.3), except that we do not require that $R$ commute with $K_{\mathrm{cm}}$.

THEOREM 3.7.17. *Assume that (V1), (V4), (M1), and (S) hold. Then:*
*(i)* $H \in C^1(M)$;
*(ii) for any* $\lambda \notin \tau$, *there are* $c(\lambda) > 0$ *and* $R \in \mathcal{R}$ *such that*

$$[H, \mathrm{i}M] \geq c(\lambda) + R, \ \text{at } H = \lambda.$$

Before we begin the proof of Theorem 3.7.17, we prove two technical lemmas which will be used to estimate the error terms.

LEMMA 3.7.18. *Let* $R \in B(L^2(X))$ *be such that one of the operators:*

$$\langle x^{\mathrm{cm}}\rangle^{\epsilon}(H+c)^{\epsilon'}R,\ (H+c)^{\epsilon'}\langle x^{\mathrm{cm}}\rangle^{\epsilon}R,$$

$$R\langle x^{\mathrm{cm}}\rangle^{\epsilon}(H+c)^{\epsilon'},\ R(H+c)^{\epsilon'}\langle x^{\mathrm{cm}}\rangle^{\epsilon},$$

*is bounded for some* $c > -\inf\sigma(H) + 1$ *and* $\epsilon, \epsilon' > 0$. *Then* $R \in \mathcal{R}$.

PROOF. Suppose that $(H+c)^{\epsilon'}\langle x^{\mathrm{cm}}\rangle^{\epsilon}R$ is bounded, and let $F \in C_0^\infty(\mathbb{R}^2)$. Then

$$F(K_{\mathrm{cm}})R = \Big(F(K_{\mathrm{cm}})\langle x^{\mathrm{cm}}\rangle^{-\epsilon}(H+c)^{-\epsilon'}\Big)\Big((H+c)^{\epsilon'}\langle x^{\mathrm{cm}}\rangle^{\epsilon}R\Big);$$

the first term is compact by (1.3.23) and pseudodifferential calculus, and the second one is assumed to be bounded. Hence $R \in \mathcal{R}$. The remaining cases are similar. □

The next lemma will be used to estimate the errors due to the cutoffs $F_a$ in (3.7.5).

LEMMA 3.7.19. *Let* $a \in \mathcal{A}^{\mathrm{n}}$, *and let* $F_{a,1} = F(\frac{|y_a^{\mathrm{c}}|}{\langle x_a^{\mathrm{cm}}\rangle} \geq c_1)$ *for some* $c_1 > 0$. *Then* $F(K_{\mathrm{cm}})(H+c)^{-1}F_{a,1}q_a$ *is compact for any* $F \in C_0^\infty(\mathbb{R}^2)$.

PROOF. Let $F \in C_0^\infty(\mathbb{R}^2)$. The operators

$$F(K_{\mathrm{cm}})(D_y + \mathbf{A}y)_{\mathrm{cm}},$$

$$(H + c)^{-\frac{1}{2}}(D_y - \mathbf{A}y)_{\mathrm{cm}},$$

are bounded. We multiply the first one by $(H + c)^{-1/2}$, the second one by $F(K_{\mathrm{cm}})$, and take the difference. Using also that $H$ and $K$ commute, we obtain that

$$\|F(K_{\mathrm{cm}})(H + c)^{-\frac{1}{2}}(\mathbf{A}y)_{\mathrm{cm}}\| < \infty. \tag{3.7.72}$$

From Lemma 3.7.20 below we have

$$\langle(\mathbf{A}y)_{\mathrm{cm}}\rangle \geq c\langle x^{\mathrm{cm}}\rangle \text{ on } \operatorname{supp} q_a F_{a,1}.$$

Hence $F(K_{\mathrm{cm}})(H+c)^{-\frac{1}{2}}F_{a,1}q_a\langle x^{\mathrm{cm}}\rangle$ is bounded, and $F(K_{\mathrm{cm}})(H+c)^{-1}F_{a,1}q_a$ is compact. □

LEMMA 3.7.20. *For any $c_1 > 0$, the constants $\{\delta_a, \epsilon_a\}$ in the definition of $q_a$ may be chosen sufficiently small so that for $a \in \mathcal{A}^{\mathrm{n}}$:*

$$\langle x^{\mathrm{cm}}\rangle \leq C\langle(\mathbf{A}y)_{\mathrm{cm}}\rangle \text{ on } \operatorname{supp} q_a F(\frac{|y_a^{\mathrm{c}}|}{\langle x_a^{\mathrm{cm}}\rangle} \geq c_1).$$

PROOF. Let $a \in \mathcal{A}^{\mathrm{n}}$. We may assume that $a = (\{1, 2\}, \{3\})$, so that:

$$y_a^{\mathrm{c}} = (0, 0, y_3),\ y_a^{\mathrm{n}} = (\frac{m_1y_1 + m_2y_2}{m_1 + m_2}, \frac{m_1y_1 + m_2y_2}{m_1 + m_2}, 0).$$

Using that $q_1 + q_2 = 0, q_3 \neq 0$, we obtain

$$(\mathbf{A}y)_{\mathrm{cm}} = \frac{1}{m_1 + m_2 + m_3}(m_1v, m_2v, m_3v),$$

where $v = J(q_1(y_1 - y_2) + q_3y_3)$. Hence on $\operatorname{supp} q_a$ we have

$$\begin{aligned} |y_3| &\leq c_2\langle(\mathbf{A}y)_{\mathrm{cm}}\rangle + c_2(y_1 - y_2) \\ &\leq c_2\langle(\mathbf{A}y)_{\mathrm{cm}}\rangle + c_3(y^a) \\ &\leq c_2\langle(\mathbf{A}y)_{\mathrm{cm}}\rangle + c_4\delta_a\langle x^{\mathrm{cm}}\rangle, \end{aligned} \tag{3.7.73}$$

where $c_i$ denote constants depending only on the masses and charges of the particles. On the other hand, on $\operatorname{supp} q_a$ we have $|x^a| \leq \delta_a|x^{\mathrm{cm}}|$, hence $\langle x^{\mathrm{cm}}\rangle \leq c_6\langle x_a^{\mathrm{cm}}\rangle$. Hence

$$\langle x^{\mathrm{cm}}\rangle \leq c_6\langle x_a^{\mathrm{cm}}\rangle \leq c_1^{-1}c_6|y_a^{\mathrm{c}}| = c_1^{-1}c_5|y_3| \tag{3.7.74}$$

on the support of $q_aF(\frac{|y_a^{\mathrm{c}}|}{\langle x_a^{\mathrm{cm}}\rangle} \geq c_1)$. Combining (3.7.73) and (3.7.74), we obtain that

$$\langle x^{\mathrm{cm}}\rangle \leq C\langle(\mathbf{A}y)_{\mathrm{cm}}\rangle \text{ on } \operatorname{supp} q_aF(\frac{|y_a^{\mathrm{c}}|}{\langle x_a^{\mathrm{cm}}\rangle} \geq c_1),$$

provided that $c_1^{-1}c_4c_5\delta_a < 1$. □

PROOF OF THEOREM 3.7.17. To prove that $H \in C^1(M)$, we check that the conditions of Lemma 3.2.2 are satisfied. Indeed by Proposition 3.7.1 the assumptions of Lemma 1.2.5 are satisfied for $M$; by Remark 1.2.6 they are also satisfied for $H$. Finally (3.2.4) follows from Lemma 3.7.2.

We now prove *(ii)*. Fix $\lambda \notin \tau$, and let $f \in C_0^\infty(\mathbb{R})$ be supported in $(\lambda - \delta, \lambda + \delta)$, where $\delta > 0$ will eventually be taken small enough. We also fix a $c > -\inf \sigma(H) + 1$. We denote by the generic letter $R$ an operator in $\mathcal{R}$, which may change from line to line. All commutators below will be considered as quadratic forms on $\mathcal{D}(N)$; by the now familiar argument, we may compute them using the Leibniz rule.

Let

$$\mathbf{1} = \sum_a j_a^2(x^{\mathrm{cm}})$$

be an $N$-body partition of unity as in Proposition 2.3.4. Moreover, we may choose $j_a$ so that for all $a \in \mathcal{A}$, $\sharp a = 2$:

$$\begin{aligned} &q_a = 1 \text{ on a neighbourhood of } \operatorname{supp} j_a, \\ &\operatorname{supp} j_a \cap \operatorname{supp} q_b = \emptyset \text{ for all } b \in \mathcal{A},\ b \neq a. \end{aligned} \tag{3.7.75}$$

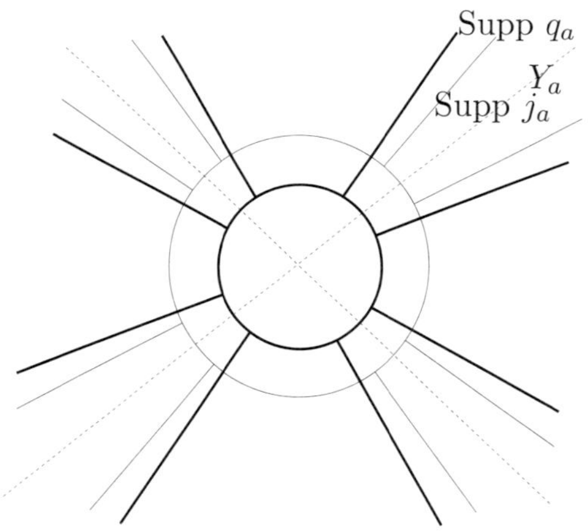


**Fig. 3.2.** Supports of $q_a$ and $j_a$

We first claim that

$$\begin{aligned} f(H)[H, \mathrm{i}M]f(H) &= \textstyle\sum_{a \neq a_{\max}} j_a^2 f(H)[H, \mathrm{i}M]f(H) \\ &= \textstyle\sum_{a \neq a_{\max}} j_a f(H) j_a [H, \mathrm{i}M] f(H) + R \\ &= \textstyle\sum_{a \neq a_{\max}} j_a f(H_a) j_a [H, \mathrm{i}M] f(H) + R. \end{aligned} \tag{3.7.76}$$

Indeed, it suffices to prove that:

$$\begin{aligned} &j_{a_{\max}}^2 f(H)[H, \mathrm{i}M]f(H) \in \mathcal{R}, \\ &j_a[f(H), j_a][H, \mathrm{i}M]f(H) \in \mathcal{R}, \\ &j_a(f(H) - f(H_a))j_a[H, \mathrm{i}M]f(H) \in \mathcal{R}. \end{aligned} \tag{3.7.77}$$

By Lemma 3.7.2, $(H+c)^{-1/2}[H, \mathrm{i}M]f(H)$ is bounded. Hence by Lemma 3.7.18 we only need to prove that

$$\begin{aligned} &(H+c)^{\epsilon}\langle x^{\mathrm{cm}}\rangle^{\epsilon'} j_{a_{\max}}^2 f(H)(H+c)^{1/2}, \\ &(H+c)^{\epsilon}\langle x^{\mathrm{cm}}\rangle^{\epsilon'} j_a[f(H), j_a](H+c)^{1/2}, \\ &(H+c)^{\epsilon}\langle x^{\mathrm{cm}}\rangle^{\epsilon'} j_a(f(H) - f(H_a))j_a(H+c)^{1/2} \end{aligned} \tag{3.7.78}$$

are bounded. The first term in (3.7.78) is bounded using Lemma 2.3.9*(ii)* and that $\langle x^{\mathrm{cm}}\rangle^{\epsilon'} j_{a_{\max}}^2 \in S_{\mathrm{cl}}^0(X^{\mathrm{cm}})$. Using Lemma 2.3.9*(ii)* again, we find that

$$\langle x^{\mathrm{cm}}\rangle j_a[f(H), j_a](H+c),\ (H+c)\langle x^{\mathrm{cm}}\rangle j_a[f(H), j_a] \text{ are bounded,}$$

which implies by interpolation that the second term in (3.7.78) is bounded. The last term is similar, except that Lemma 2.3.9*(i)* is used instead. Hence we have proved (3.7.76).

Next, we will prove that for $a \in \mathcal{A}$, $a \neq a_{\max}$:

$$j_a f(H_a) j_a[I_a, \mathrm{i}M]f(H) \in \mathcal{R}. \tag{3.7.79}$$

By Lemma 3.7.7*(i)–(ii)*, we have

$$j_a[I_a, \mathrm{i}M] = j_a[I_{a,1}, \mathrm{i}M] + R_a,$$

where $I_{a,1} = j_a' I_a$ for some $j_a' \in S_{\mathrm{cl}}^0(X^{\mathrm{cm}})$ satisfying *(iii)* of Proposition 2.3.4, and

$$\langle x^{\mathrm{cm}}\rangle R_a f(H) \text{ is bounded.} \tag{3.7.80}$$

Hence (3.7.79) will follow if we prove that:

$$j_a f(H_a) j_a[I_{a,1}, \mathrm{i}M]f(H) \in \mathcal{R}, \tag{3.7.81}$$

$$j_a f(H_a) R_a f(H) \in \mathcal{R}. \tag{3.7.82}$$

We begin with (3.7.82). By Lemma 3.7.18 it suffices to prove that

$$(H+c)\langle x^{\mathrm{cm}}\rangle j_a f(H_a) R_a f(H) \text{ is bounded.}$$

By Lemma 1.2.4, this will follow from:

$$(H+c) f(H_a)\langle x^{\mathrm{cm}}\rangle j_a R_a f(H) \text{ is bounded.}$$

But that follows from (3.7.80).

We now prove (3.7.81). By *(V4)*, $I_{a,1}$ satisfies (3.7.20) with $\epsilon = \mu_1$. Hence by (3.7.21),

$$(H+c)^{-1/2}[I_{a,1}, \mathrm{i}M](H+c)^{-1/2}\langle x^{\mathrm{cm}}\rangle^{\epsilon} \text{ is bounded,} \tag{3.7.83}$$

for $0 < \epsilon \leq \min(1, \mu_l)$. From Lemma 2.3.9*(ii)* we obtain by commuting $j_a$ through $f(H_a)$ that

$$j_a f(H_a) j_a (H+c)^{1/2} \text{ is bounded,}$$

and, by Lemma 1.2.4:

$$\langle x^{\text{cm}}\rangle^{-\epsilon}(H+c)^{1/2} f(H)(H+c)\langle x^{\text{cm}}\rangle^{\epsilon} \text{ is bounded.}$$

Hence

$$j_a f(H_a) j_a [I_{a,1}, \mathrm{i}M] f(H)(H+c)\langle x^{\text{cm}}\rangle^{\epsilon} \text{ is bounded.}$$

By Lemma 3.7.18, this implies (3.7.81).

Since

$$[H, \mathrm{i}M] = [H_a, \mathrm{i}M] + [I_a, \mathrm{i}M],$$

we obtain from (3.7.76) and (3.7.79) that

$$f(H)[H, \mathrm{i}M] f(H) = \sum_{a \neq a_{\max}} j_a f(H_a) j_a [H_a, \mathrm{i}M] f(H) + R. \tag{3.7.84}$$

We will now prove that for $a \neq a_{\max}$

$$j_a f(H_a) j_a [H_a, \mathrm{i}M] f(H) = j_a f(H_a)[H_a, \mathrm{i}G_a] f(H_a) j_a + R, \tag{3.7.85}$$

where $G_a$ was defined in (3.7.8) for $a \in \mathcal{A}^{\text{n}}$ and $G_a = A$ for $a \in \mathcal{A}^{\text{c}}$.

*Case 1:* $a \in \mathcal{A}^{\text{n}}$. Using (3.7.75), we write

$$\begin{aligned} j_a[H_a, \mathrm{i}M] &= j_a[H_a, \mathrm{i}q_a M_a q_a] \\ &= j_a[H_a, \mathrm{i}q_a] M_a q_a + j_a M_a [H_a, \mathrm{i}q_a] + j_a q_a [H_a, \mathrm{i}M_a] q_a. \end{aligned} \tag{3.7.86}$$

By (3.7.75) again, $\operatorname{supp} \nabla q_a \cap \operatorname{supp} j_a = \emptyset$. Hence the first term on the right-hand side of (3.7.86) is 0, and the second term has the form $B_1 + \langle B_2, D - \mathbf{A}x\rangle$, where $\langle x^{\text{cm}}\rangle B_i$, $i = 1, 2$ are bounded by Lemma 3.7.7*(ii)*. It follows that

$$\begin{aligned} &\langle x^{\text{cm}}\rangle (H+c) j_a f(H_a) j_a M_a [H_a, \mathrm{i}q_a] f(H) \\ &= \langle x^{\text{cm}}\rangle (H+c) j_a f(H_a) j_a (B_1 + \langle B_2, D - \mathbf{A}x\rangle) f(H) \end{aligned}$$

is bounded, where we also used Lemma 1.2.4. Thus the left-hand side of (3.7.85) equals

$$j_a f(H_a) j_a [H_a, \mathrm{i}M_a] q_a f(H) + R. \tag{3.7.87}$$

Next, we claim that (3.7.87) equals

$$j_a f(H_a) j_a [H_a, \mathrm{i}G_a] q_a f(H) + R. \tag{3.7.88}$$

To verify this, by Lemma 3.7.8*(i)* it suffices to prove that

$$j_a f(H_a) j_a R_a q_a f(H) \in \mathcal{R},$$

where $R_a$ is as in (3.7.24). But this follows from Lemma 3.7.19.

By (3.7.26) and Lemma 3.7.18, (3.7.88) is equal to

$$j_a f(H_a)[H_a, \mathrm{i}G_a] j_a q_a f(H) + R. \tag{3.7.89}$$

We claim that (3.7.89) equals

$$j_a f(H_a)[H_a, \mathrm{i}G_a] j_a f(H_a) + R. \tag{3.7.90}$$

Since $j_a q_a = j_a$, it suffices to prove that

$$j_a f(H_a)[H_a, \mathrm{i}G_a] j_a (f(H) - f(H_a)) \in \mathcal{R}. \tag{3.7.91}$$

By Lemma 2.3.9*(i)*,

$$j_a(f(H) - f(H_a))(H + c)\langle x^{\mathrm{cm}}\rangle^{\epsilon}, \ (H + c) j_a (f(H) - f(H_a))\langle x^{\mathrm{cm}}\rangle^{\epsilon}$$

are bounded; the first of these estimates implies also that

$$j_a(f(H) - f(H_a))\langle x^{\mathrm{cm}}\rangle^{\epsilon}(H + c)$$

is bounded. Hence by interpolation:

$$(H + c)^{1/2} j_a (f(H) - f(H_a))\langle x^{\mathrm{cm}}\rangle^{\epsilon}(H + c)^{1/2} \text{ is bounded.} \tag{3.7.92}$$

We have

$$\begin{aligned} & j_a f(H_a)[H_a, \mathrm{i}G_a] j_a (f(H) - f(H_a))\langle x^{\mathrm{cm}}\rangle^{\epsilon}(H + c)^{1/2} \\ = & \left(j_a f(H_a)[H_a, \mathrm{i}G_a](H + c)^{-1/2}\right) \\ & \times\left((H + c)^{1/2} j_a (f(H) - f(H_a))\langle x^{\mathrm{cm}}\rangle^{\epsilon}(H + c)^{1/2}\right), \end{aligned}$$

which is bounded by (3.7.92) and (3.7.25). By Lemma 3.7.18, this proves (3.7.91).

To complete the proof of (3.7.85) for $a \in \mathcal{A}^{\mathrm{n}}$, it suffices to verify that (3.7.90) equals

$$j_a f(H_a)[H_a, \mathrm{i}G_a] f(H_a) j_a + R. \tag{3.7.93}$$

But this follows by exactly the same argument as in the proof of (3.7.91), except that Lemma 2.3.9*(ii)* is used instead of *(i)*.

*Case 2:* $a \in \mathcal{A}^{\mathrm{c}}$, $\sharp a = 2$. This case is similar to Case 1 but simpler; the details are left to the reader.

*Case 3:* $a = a_{\min}$. We have $H_{a_{\min}} = H_0$, $M_{a_{\min}} = G_{a_{\min}} = A$. By Lemma 2.3.7,

$$A = \sum_a q_a A q_a,$$

hence

$$\begin{aligned} [H_0, \mathrm{i}M] = & \ [H_0, \mathrm{i}\textstyle\sum_{a \neq a_{\max}} q_a M_a q_a] \\ = & \ [H_0, \mathrm{i}A] + [H_0, \mathrm{i}\textstyle\sum_{\sharp a = 2} q_a (M_a - A) q_a] \\ & -[H_0, \mathrm{i}q_{a_{\max}} A q_{a_{\max}}]. \end{aligned} \tag{3.7.94}$$

It is easily checked that

$$j_{a_{\min}} f(H_0) j_{a_{\min}} [H_0, \mathrm{i} q_{a_{\max}} A q_{a_{\max}}] f(H) \in \mathcal{R}.$$

This takes care of the last term in (3.7.94). To estimate the second last term, we use Lemma 3.7.9. We have

$$\begin{aligned} & [H_0, \mathrm{i} q_a (M_a - A) q_a] \\ = \; & [H_0, \mathrm{i} q_a](M_a - A) q_a + q_a (M_a - A)[H_0, \mathrm{i} q_a] \\ & + \mathrm{i} q_a (H_0 (M_a - A) - (M_a - A) H_0) q_a. \end{aligned} \tag{3.7.95}$$

Observe that $\nabla q_a$ and $j_{a_{\min}} q_a$ are supported in a region as in (3.7.27). Using also (3.7.28) and Lemma 3.7.19, we conclude that

$$j_{a_{\min}} f(H_0) j_{a_{\min}} [H_0, \mathrm{i} q_a (M_a - A) q_a] f(H) \in \mathcal{R}.$$

Thus, by (3.7.94):

$$\begin{aligned} & j_{a_{\min}} f(H_0) j_{a_{\min}} [H_0, \mathrm{i} M] f(H) \\ & = j_{a_{\min}} f(H_0) j_{a_{\min}} [H_0, \mathrm{i} A] f(H) + R. \end{aligned}$$

The rest of the proof of (3.7.85) is as in Case 1.

We are now ready to finish the proof of the theorem by combining the channel Mourre estimates for $H_a$ into one for $H$. From (3.7.84) and (3.7.85) we have

$$f(H)[H, \mathrm{i} M] f(H) = \sum_{a \neq a_{\max}} j_a f(H_a)[H_a, \mathrm{i} G_a] f(H_a) j_a + R. \tag{3.7.96}$$

Since $\lambda \notin \tau$, we may apply to each term on the right-hand side of (3.7.96) Theorem 3.3.4 if $a \in \mathcal{A}^{\mathrm{c}}$ and Proposition 3.6.8 if $a \in \mathcal{A}^{\mathrm{n}}$. We obtain that if $f$ is supported in $[\lambda - \delta, \lambda + \delta]$ for $\delta > 0$ small enough, there is a $c_0 > 0$ such that

$$\begin{aligned} & f(H)[H, \mathrm{i} M] f(H) \\ \geq \; & c_0 \textstyle\sum_{a \neq a_{\max}} j_a f^2(H_a) j_a + R \\ = \; & c_0 \textstyle\sum_a j_a f^2(H) j_a + R \\ = \; & c_0 f^2(H) + R, \end{aligned}$$

where we again used Lemmas 2.3.9 and 3.7.18. This concludes the proof of the theorem. □

PROOF OF THEOREM 3.7.3. *(i)* is the same as in Theorem 3.7.17. To prove *(ii)*, it suffices to consider the eigenvalues of $H$ belonging to the interval $I$ (which was fixed at the beginning of this section).

Let $U_{\mathrm{cm}}$ be the reducing unitary transformation constructed in Section 1.3.5. We recall that $\tilde{H} = U_{\mathrm{cm}} H U_{\mathrm{cm}}^*$, and define

$$\tilde{M} := U_{\mathrm{cm}} M U_{\mathrm{cm}}^*.$$

Note that by *(i)*, $\tilde{H} \in C^1(\tilde{M})$.

Let $\lambda \in I$, and suppose that there is a sequence of eigenvalues $\lambda_n \in \sigma_{\rm pp}(H) = \sigma_{\rm pp}(\tilde{H})$ such that $\lambda_n \to \lambda$. Since, by (1.3.23),

$$\tilde{H} = \mathbf{1}_{L^2(Y_{\rm cm}^{(1)})} \otimes \tilde{H}^{\rm cm},$$

we may find eigenvectors $\psi_n$ of $\tilde{H}^{\rm cm}$ and a vector $v \in L^2(Y_{\rm cm}^{(1)})$ such that $u_n = v \otimes \psi_n$ is a normalized eigenvector of $\tilde{H}$ corresponding to the eigenvalue $\lambda_n$.

Pick $F \in C_0^\infty(\mathbb{R}^2)$ such that $F(y_{\rm cm}^{(1)}, D_{y_{\rm cm}^{(1)}})v \neq 0$, and let

$$B := F(y_{\rm cm}^{(1)}, D_{y_{\rm cm}^{(1)}}).$$

From Theorem 3.7.17 we have

$$[\tilde{H}, \mathrm{i}\tilde{M}] \geq c(\lambda) + \tilde{R} \text{ at } \tilde{H} = \lambda,$$

where $c(\lambda) > 0$ and $B\tilde{R}B$ is compact. Hence

$$B[\tilde{H}, \mathrm{i}\tilde{M}]B \geq c(\lambda)B^2 + K \text{ at } \tilde{H} = \lambda, \tag{3.7.97}$$

where $K = B\tilde{R}B$ is compact. Since $[B, \tilde{H}] = 0$, $Bu_n$ is an eigenvector of $\tilde{H}$ with eigenvalue $\lambda_n$. Moreover, since $Bv \neq 0$, we have $\|Bu_n\| \geq c_0$, $\mathrm{w-}\lim_{n\to\infty} Bu_n = 0$. Applying (3.7.97) and the virial relation to the sequence $Bu_n$ as in the proof of Lemma 3.2.6, we arrive at a contradiction.

Finally, *(iii)* follows from (3.7.97) and Lemma 3.1.4*(i)*. □

**3.7.4. Estimating the second commutator.** In this subsection we prove Proposition 3.7.5.

We must prove that $H \in C^{1+\epsilon}(M)$ for some $\epsilon > 0$, i.e., that

$$[(H+c)^{-1}, \mathrm{i}M] \in C^\epsilon(M),$$

where $c > -\inf\sigma(H) + 1$ is fixed throughout this subsection. By Theorem 3.7.3*(i)*, $H \in C^1(M)$, hence by Proposition 3.2.3*(iii)* we may write

$$[(H+c)^{-1}, \mathrm{i}M] = -(H+c)^{-1}[H, \mathrm{i}M](H+c)^{-1}. \tag{3.7.98}$$

Let $q_a'$, $q_a''$ be functions in $S_{\rm cl}^0(X^{\rm cm})$ satisfying the conditions *(i)*–*(v)* of Proposition 2.3.4 (with constants different from those for $q_a$, but still small enough so that the results of Subsections 3.7.1–3.7.3 apply) and such that:

$$\begin{aligned} q_a'' &\equiv 1 \text{ on a neighbourhood of } \operatorname{supp} q_a, \\ q_a' &\equiv 1 \text{ on a neighbourhood of } \operatorname{supp} q_a''. \end{aligned} \tag{3.7.99}$$

We then have

$$\begin{aligned} [H, \mathrm{i}M] &= \textstyle\sum_{\#a\geq 2}[H, \mathrm{i}q_aM_aq_a] \\ &= \textstyle\sum_{\#a\geq 2}\Big(v_aM_aq_a + q_aM_av_a + q_a[H_a, \mathrm{i}M_a]q_a + q_a[I_a', \mathrm{i}M_a]q_a\Big) \\ &=: \textstyle\sum_{\#a\geq 2}(R_1^a + R_2^a + R_3^a + R_4^a(I_a')), \end{aligned} \tag{3.7.100}$$

where

$$v_a = [H, \mathrm{i}q_a] = \langle D - \mathbf{A}x, \nabla q_a\rangle + \text{h.c.},$$

and by *(V4)*, $I'_a = q'_a I_a$ satisfies:

$$
\begin{aligned}
&I'_a = I^{\mathrm{s}} + I^{\mathrm{l}},\\
&I^{\mathrm{s}}\langle x^{\mathrm{cm}}\rangle^2 (H+c)^{-\frac{1}{2}} \text{ is bounded,} \qquad (3.7.101)\\
&|\partial^\alpha_{x^{\mathrm{cm}}} I^{\mathrm{l}}| \le C_\alpha \langle x^{\mathrm{cm}}\rangle^{-\mu_{\mathrm{l}}-|\alpha|}, |\alpha| \ge 0, \mu_{\mathrm{l}} > 0.
\end{aligned}
$$

Let $S^{-\mu}_m$ be the Banach space

$$S^{-\mu}_m = \{f \in C^m(X^{\mathrm{cm}}) : \ |\partial^\alpha_{x^{\mathrm{cm}}} f| \le C_\alpha \langle x^{\mathrm{cm}}\rangle^{-\mu-\alpha}, \ |\alpha| \le m\}.$$

We will prove below the following proposition.

PROPOSITION 3.7.21. *Let $R^a_i$, $i = 1, 2, 3$, be defined by (3.7.100), and*

$$R^a_4(J) = q_a[J, iM_a]q_a,$$

*where $J$ is a real-valued function on $X$. Then:*

*(i) $(H+c)^{-1}R^a_i(H+c)^{-1} \in C^1(M)$, $i = 1, 2, 3$, if (3.7.101) holds for some $\mu_{\mathrm{l}} > 0$;*

*(ii) $(H+c)^{-1}R^a_4(J)(H+c)^{-1} \in C^1(M)$ if we assume that*

$$
\begin{aligned}
&J = J^{\mathrm{s}} + J^{\mathrm{l}},\\
&J^{\mathrm{s}}\langle x^{\mathrm{cm}}\rangle^2 (H+c)^{-\frac{1}{2}} \textit{ is bounded,} \qquad (3.7.102)\\
&J^{\mathrm{l}} \in S^{-1}_m \textit{ for } m \textit{ large enough.}
\end{aligned}
$$

Proposition 3.7.5 is now obtained from Proposition 3.7.21 and the following interpolation argument. Consider the linear mapping

$$T : \ J \to (H+c)^{-1}q_a[J, \mathrm{i}M_a]q_a(H+c)^{-1}.$$

We proved in Subsection 3.7.2 that $T(J) \in C^0(M)$ if $J$ satisfies (3.7.101) for some $\mu_{\mathrm{l}} > 0$. In fact the same result holds (and the same proof applies) if the second equation in (3.7.101) is replaced by:

$$J^{\mathrm{l}} \in S^{-\mu}_m \text{ for } m \text{ large enough;}$$

this is because the results from pseudodifferential calculus used in Subsection 3.7.2 (Lemmas A.4, A.5) continue to hold if the symbol estimates are satisfied for only a finite (but sufficiently large) number of derivatives.

By a result of [**ABG**] ([**ABG**], equation (5.2.22)), $C^\epsilon(M)$ for $0 < \epsilon < 1$ are the real interpolation spaces between $C^0(M)$ and $C^1(M)$:

$$C^\epsilon(M) = (C^0(M), C^1(M))_{1-\epsilon,\infty}$$

(see [**ABG**], Definition 2.2.3). On the other hand, by Lemma A.3 in the Appendix we have

$$S_m^{-\mu} \subset (S_m^{-1}, S_m^{-\mu_0})_{\theta,\infty},\ \theta = \frac{1-\mu}{1-\mu_0}.$$

From Proposition 3.7.21*(ii)* and from the proof of Theorem 3.7.3*(i)* we have for any $\mu_0 > 0$ and for $m$ sufficiently large:

$$T : S_m^{-1} \to C^1(M),$$

$$T : S_m^{-\mu_0} \to C^0(M).$$

Hence by the real interpolation theorem (see e.g., [**ABG**], Theorem 2.6.1) we have for any $\mu > 0$:

$$T : S_m^{-\mu} \to C^\epsilon(M),\ \epsilon > 0.$$

This together with Proposition 3.7.21*(i)* implies Proposition 3.7.5.

It remains to prove Proposition 3.7.21. Fix a function $J$ satisfying (3.7.102), and denote $R_4^a(J) = R_4^a$. By Proposition 3.2.1 it suffices to prove that

$$\begin{aligned}&((H+c)^{-1}R_j^a(H+c)^{-1}u, Mu) - (Mu, (H+c)^{-1}R_j^a(H+c)^{-1}u)\\ &=: Q_j^a(u,u),\end{aligned}$$

defined as a quadratic form on $\mathcal{D}(M)$, is bounded in the topology of $H$.

By Theorem 3.7.3 and Proposition 3.2.3*(iii)*, *(v)*, $(H+c)^{-1}$ preserves $\mathcal{D}(M)$ and

$$M(H+c)^{-1} = (H+c)^{-1}M + (H+c)^{-1}[H,M](H+c)^{-1}$$

holds as an identity between operators in $B(\mathcal{D}(M), \mathcal{H})$. By Lemma 3.7.2, it also holds on $B(\mathcal{D}(M), \mathcal{D}((H+c)^{1/2}))$; in particular,

$$M(H+c)^{-1} \in B(\mathcal{D}(M), \mathcal{D}((H+c)^{1/2})).$$

Thus we may rewrite $Q_j^a(u,u)$ as

$$\begin{aligned}&Q_j^a(u,u)\\ = \quad & (R_j^a(H+c)^{-1}u, M(H+c)^{-1}u) - (M(H+c)^{-1}u, R_j^a(H+c)^{-1}u)\\ & -(R_j^a(H+c)^{-1}u, (H+c)^{-1}[H,M](H+c)^{-1}u)\\ & -((H+c)^{-1}[H,M](H+c)^{-1}u, R_j^a(H+c)^{-1}u).\end{aligned}$$

The last two terms are bounded by Lemma 3.7.2. Hence it suffices to prove that

$$(H+c)^{-1}[R_j^a, M](H+c)^{-1} \text{ is bounded, } j = 1,2,3,4. \tag{3.7.103}$$

Morally (though not formally), (3.7.103) is equivalent to the following proposition.

PROPOSITION 3.7.22. *Assume (V1), (V4) with* $\mu = 1$*, (M1), (S), and fix* $c > -\inf \sigma(H) + 1$*. Then*

$$(H+c)^{-1}[[H, M], M](H+c)^{-1} \text{ is bounded}, i = 1, 2, 3, 4.$$

PROOF OF (3.7.103). We first introduce some notation. We will write

$$B = O((H+c)^2),$$

if $B$ is an operator such that $(H+c)^{-1}B(H+c)^{-1}$ is bounded. We will also write

$$M'_a = q'_a M_a q'_a,$$

where $q'_a$ were defined after (3.7.98). In particular, we have $q_a M_a q_a = q_a M'_a q_a$ and $R^a_j = q'_a R^a_j q'_a$. We furthermore recall that $M_{a_{\min}} = A$, and observe that

$$[R^a_j, q_b M_b q_b] = 0 \text{ for } a \neq b, \ \sharp a = \sharp b = 2,$$

since for such $a, b$ the supports of $q_a$ and $q_b$ are disjoint. Hence it suffices to prove that for all $a \in \mathcal{A}$, $\sharp a \geq 2$:

$$[R^a_j, q_a M_a q_a] = O((H+c)^2), \ i = 1, 2, 3, 4, \tag{3.7.104}$$

and that for all $a \in \mathcal{A}$, $\sharp a = 2$:

$$[[H_a + J, q_a M_a q_a], q_{a_{\min}} A q_{a_{\min}}] = O((H+c)^2), \tag{3.7.105}$$

$$[[H_0 + J, q_{a_{\min}} A q_{a_{\min}}], q_a M_a q_a] = O((H+c)^2). \tag{3.7.106}$$

Instead of (3.7.104) we will actually prove a slightly stronger statement:

$$[R^a_j, j_a M_a j_a] = O((H+c)^2), \ i = 1, 2, 3, 4, \tag{3.7.107}$$

where $j_a$ is any function in $S^0_{\mathrm{cm}}(X)$ supported in

$$\{x \in X : \ |x^a| \leq \delta'_a, \ |x^b| \geq \epsilon'_a \ \forall b \not\leq a\} \cap \mathrm{Int}\,\{x \in X : \ q'_a(x) = 1\},$$

$q'_a$ were defined at the beginning of this subsection, and $\delta'_a, \epsilon'_a > 0$ are small enough. (Recall that the space $S^0_{\mathrm{cm}}(X)$ was defined at the beginning of Subsection 3.7.2.) (3.7.107) with $j_a = q_a$ is (3.7.104); the statement with more general $j_a$ will be used in the proof of (3.7.105)–(3.7.106).

We rewrite (3.7.107) as:

$$[v_a M_a q_a, j_a M_a j_a] = O((H+c)^2), \tag{3.7.108}$$

$$[q_a M_a v_a, j_a M_a j_a] = O((H+c)^2), \tag{3.7.109}$$

$$[q_a [H_a, iM_a] q_a, j_a M_a j_a] = O((H+c)^2), \tag{3.7.110}$$

$$[q_a [J, iM_a] q_a, j_a M_a j_a] = O((H+c)^2). \tag{3.7.111}$$

We now proceed to prove these estimates. We will henceforth assume that $a \in \mathcal{A}^{\mathrm{n}}$; the case $a \in \mathcal{A}^{\mathrm{c}}$ is considerably simpler, since then $M_a = A$.

In addition to the lemmas proved earlier in this section, we will need the following technical results. The proofs are only sketched, since the arguments used here are almost the same as in Subsection 3.7.2.

LEMMA 3.7.23. *Let $q \in S^0_{\mathrm{cm}}(X)$. Then:*

$$(H+c)^{1/2}\langle x^{\mathrm{cm}}\rangle^{-1} j_a M_a j_a (H+c)^{-1} \text{ is bounded;} \tag{3.7.112}$$

$$(H+c)^{1/2}[q, j_a M_a j_a](H+c)^{-1/2} \text{ is bounded;} \tag{3.7.113}$$

$$(H+c)^{-1/2}\langle x^{\mathrm{cm}}\rangle^{-2} j_a M_a j_a \langle x^{\mathrm{cm}}\rangle \text{ is bounded;} \tag{3.7.114}$$

$$\langle x^{\mathrm{cm}}\rangle [q, j_a M_a j_a]\langle x^{\mathrm{cm}}\rangle^{-1} \text{ is bounded;} \tag{3.7.115}$$

$$(H+c)^{-1/2} v_a \langle x^{\mathrm{cm}}\rangle \text{ is bounded.} \tag{3.7.116}$$

PROOF. (3.7.116) follows easily by direct calculation.By Lemma 3.7.7 *(i)*, *(iv)*, to prove (3.7.112) and (3.7.113) it suffices to verify that

$$(H+c)^{1/2}[\langle x^{\mathrm{cm}}\rangle^{-1} j_a M_a j_a (H+c)^{-1/2}, (H+c)^{-1/2}],$$

$$(H+c)^{1/2}[q j_a M_a j_a, (H+c)^{-1/2}],\ (H+c)^{1/2}[j_a M_a j_a q, (H+c)^{-1/2}]$$

are bounded. This is done as in the proof of (3.7.30), using (3.7.34), Lemma 3.7.7, and Lemma 3.7.7*(iii)*,*(v)*; the details are left to the reader.
To prove (3.7.114) and (3.7.115), it suffices to estimate $[\langle x^{\mathrm{cm}}\rangle^{-1}, j_a M_a j_a]$ using Lemma 3.7.7*(iv)*. □

LEMMA 3.7.24. *Let $a \in \mathcal{A}^{\mathrm{n}}$. Assume that $c > -\inf \sigma(H^{a,\mathrm{n}}) + 1$, and let $B^a = [H^{a,\mathrm{n}}, iA^a]$. Then:*

$$\langle x^a\rangle[\chi_a(H^{a,\mathrm{n}}), j_a](H^{a,\mathrm{n}}+c),\ [\chi_a(H^{a,\mathrm{n}}), j_a]\langle x^a\rangle(H^{a,\mathrm{n}}+c) \text{ are bounded;} \tag{3.7.117}$$

$$\langle x^{\mathrm{cm}}\rangle[\chi_a(H^{a,\mathrm{n}}), j_a](H^{a,\mathrm{n}}+c),\ [\chi_a(H^{a,\mathrm{n}}), j_a]\langle x^{\mathrm{cm}}\rangle(H^{a,\mathrm{n}}+c) \text{ are bounded;} \tag{3.7.118}$$

$$[\chi_a(H^{a,\mathrm{n}}), F_a]\langle x^{\mathrm{cm}}\rangle j_a \text{ is bounded;} \tag{3.7.119}$$

$$(H^{a,\mathrm{n}}+c)[A^a, \chi_a(H^{a,\mathrm{n}})],\ (H+c)^{-1/2}[B^a, \chi_a(H^{a,\mathrm{n}})]\langle x^a\rangle \text{ are bounded.} \tag{3.7.120}$$

PROOF. By Remark 3.7.14 we have $[\chi_a(H^{a,\mathrm{n}}), B] = -[\chi_{a,1}(H^{a,\mathrm{n}}), B]$ for $B = j_a, A^a, B^a$, where $\chi_{a,1} \in C_0^\infty(\mathbb{R})$. (3.7.117) follows now from Lemma 2.3.9*(ii)*. (3.7.118) and (3.7.119) are proved by the same arguments as in the proof of Lemma 2.3.9*(ii)*, using also (3.7.3) and that $\langle x^a\rangle \sim \langle x^{\mathrm{cm}}\rangle$ on $\operatorname{supp}\nabla j_a$. Finally, (3.7.120) is proved using (1.2.21) and *(V4)*, *(M1)*. □

**Proof of (3.7.108)–(3.7.109).** We only prove (3.7.108), as (3.7.109) is its transpose. We write

$$[v_a M_a q_a, j_a M_a j_a] = [v_a M'_a q_a, j_a M'_a j_a] = I_1 + I_2 + I_3 + I_4,$$

where:

$$I_1 = [v_a, j_a M_a j_a] M'_a q_a, \quad I_2 = v_a [M'_a, j_a] M'_a q_a j_a,$$

$$I_3 = v_a j_a M'_a [M'_a, j_a] q_a, \quad I_4 = v_a M'_a [q_a, j_a M_a j_a].$$

We first prove that $I_4 = O((H+c)^2)$. Indeed, we have

$$\begin{aligned} &(H+c)^{-1} I_4 (H+c)^{-1} \\ = \ & \Big((H+c)^{-1} v_a \langle x^{\mathrm{cm}}\rangle\Big)\Big(\langle x^{\mathrm{cm}}\rangle^{-1} M'_a (H+c)^{-1/2}\Big) \\ & \times\Big((H+c)^{1/2}[q_a, j_a M_a j_a](H+c)^{-1}\Big); \end{aligned}$$

the right-hand side is bounded by (3.7.116), (3.7.113), and Lemma 3.7.7*(i)*. By the same argument plus (3.7.32), $I_3 = O((H+c)^2)$. Next, we write

$$\begin{aligned} &(H+c)^{-1} I_2 (H+c)^{-1} \\ = \ & \Big((H+c)^{-1} v_a \langle x^{\mathrm{cm}}\rangle\Big)\Big(\langle x^{\mathrm{cm}}\rangle^{-1} [M'_a, j_a] \langle x^{\mathrm{cm}}\rangle\Big) \\ & \times\Big(\langle x^{\mathrm{cm}}\rangle^{-1} M'_a q_a j_a (H+c)^{-1}\Big); \end{aligned}$$

this is bounded by (3.7.116), (3.7.115), and Lemma 3.7.7*(i)*. Finally, we have

$$I_1 = [v_a, j_a A j_a] M'_a q_a + [v_a, T_a] M'_a q_a,$$

where $T_a$ is as in Lemma 3.7.9. By Lemma 3.7.7*(i)* and (3.7.112), it suffices to prove that:

$$(H+c)^{-1}[v_a, j_a A j_a]\langle x^{\mathrm{cm}}\rangle \text{ is bounded,}$$

$$(H+c)^{-1}[v_a, T_a]\langle x^{\mathrm{cm}}\rangle (H+c)^{-1/2} \text{ is bounded.}$$

The first estimate may be verified by a straightforward (but cumbersome) computation; the second one follows from (3.7.28).

**Proof of (3.7.110).** We write the left-hand side of (3.7.110) as

$$\Big([q_a, j_a M_a j_a][H_a, \mathrm{i} M_a] q_a + \text{h.c.}\Big) + q_a[[H_a, iM_a], j_a M_a j_a] q_a.$$

Hence it suffices to prove that:

$$[q_a, j_a M_a j_a][H_a, \mathrm{i} M_a] q_a = O((H+c)^2), \tag{3.7.121}$$

$$q_a[[H_a, \mathrm{i} M_a], j_a M_a j_a] q_a = O((H+c)^2). \tag{3.7.122}$$

We have

$$\begin{aligned}&(H+c)^{-1}[q_a, j_aM_aj_a][H_a, \mathrm{i}M_a]q_a(H+c)^{-1}\\ =\ &\Big((H+c)^{-1}[q_a, j_aM_aj_a](H+c)^{1/2}\Big)\times\\ &\Big((H+c)^{-1/2}q_a'[H_a, \mathrm{i}M_a]q_a(H+c)^{-1}\Big)\end{aligned}$$

which is bounded by (3.7.113) and (3.7.59). It remains to prove (3.7.122).

We have

$$\begin{aligned}M_a &= A_a + \chi_a(H^{a,\mathrm{n}})A^a\chi_a(H^{a,\mathrm{n}}) + F_aG^aF_a,\\ [H_a, \mathrm{i}M_a] &= D_{z_a}^2 + \chi_a(H^{a,\mathrm{n}})B^a\chi_a(H^{a,\mathrm{n}})\\ &+[H_a, \mathrm{i}F_a]G^aF_a + F_aB_2F_a + F_aG^a[H_a, iF_a],\end{aligned}$$

where we recall from Lemmas 3.7.12 and 3.7.24 that

$$\begin{aligned}B^a &= [H^{a,\mathrm{n}}, \mathrm{i}A^a] = D_{z^a}^2 + z^a\nabla_{z^a}V^a,\\ B_2 &= [H^{a,\mathrm{n}}, \mathrm{i}G^a].\end{aligned}$$

We need to prove that:

$$q_a[D_{z_a}^2, j_aM_aj_a]q_a = O((H+c)^2), \tag{3.7.123}$$

$$q_a[\chi_a(H^{a,\mathrm{n}})B^a\chi_a(H^{a,\mathrm{n}}), j_aM_aj_a]q_a = O((H+c)^2), \tag{3.7.124}$$

$$q_a[F_aB_2F_a, j_aM_aj_a]q_a = O((H+c)^2), \tag{3.7.125}$$

$$q_a[v_a'G_aF_a, j_aM_aj_a]q_a = O((H+c)^2), \tag{3.7.126}$$

where $v_a' = [H_a, \mathrm{i}F_a]$.

- We begin with (3.7.123). For $q \in S_{\mathrm{cm}}^0(X)$, we have by direct computation:

$$(H+c)^{-1}q_a[D_{z_a}^2, q]\langle x^{\mathrm{cm}}\rangle \text{ is bounded.} \tag{3.7.127}$$

By (3.7.127) and Lemma 3.7.7*(i)*,

$$q_a[D_{z_a}^2, j_a]M_aq_aj_a = O((H+c)^2).$$

Hence the left-hand side of (3.7.123) equals

$$\begin{aligned}&q_aj_a[D_{z_a}^2, \mathrm{i}M_a]q_aj_a + O((H+c)^2)\\ =\ &q_aj_a[D_{z_a}^2, \mathrm{i}(A_a + F_aG^aF_a)]q_aj_a + O((H+c)^2)\\ =\ &2q_aj_aD_{z_a}^2q_aj_a + \Big(q_aj_a[D_{z_a}^2, \mathrm{i}F_a]G^aF_aq_aj_a + \text{h.c.}\Big) + O((H+c)^2).\end{aligned}$$

Clearly, $2q_aj_aD_{z_a}^2q_aj_a = O((H+c)^2)$. Hence it suffices to prove that

$$(H+c)^{-1}q_aj_a[D_{z_a}^2, F_a]G^aF_aq_aj_a(H+c)^{-1} \text{ is bounded.}$$

But this follows from (3.7.127) and Lemma 3.7.11*(iii)*.

• Next, we prove (3.7.124). We have

$$\begin{aligned}&[\chi_a(H^{a,\mathrm{n}})B^a\chi_a(H^{a,\mathrm{n}}), j_aM_aj_a]\\ =\ &[\chi_a(H^{a,\mathrm{n}})B^a\chi_a(H^{a,\mathrm{n}}), j_a]M_aj_a + \text{h.c.}\\ &+j_a[\chi_a(H^{a,\mathrm{n}})B^a\chi_a(H^{a,\mathrm{n}}), M_a]j_a.\end{aligned}\tag{3.7.128}$$

We first verify that

$$q_a[\chi_a(H^{a,\mathrm{n}})B^a\chi_a(H^{a,\mathrm{n}}), j_a]M_aq_aj_a = O((H+c)^2).\tag{3.7.129}$$

By Lemma 3.7.7*(i)*, it suffices to prove that

$$\begin{aligned}&(H+c)^{-1}q_a[\chi_a(H^{a,\mathrm{n}})B^a\chi_a(H^{a,\mathrm{n}}), j_a]\langle x^{\mathrm{cm}}\rangle\\ =\ &(H+c)^{-1}q_a[\chi_a(H^{a,\mathrm{n}}), j_a]B^a\chi_a(H^{a,\mathrm{n}})\langle x^{\mathrm{cm}}\rangle\\ &+(H+c)^{-1}q_a\chi_a(H^{a,\mathrm{n}})[B^a, j_a]\chi_a(H^{a,\mathrm{n}})\langle x^{\mathrm{cm}}\rangle\\ &+(H+c)^{-1}q_a\chi_a(H^{a,\mathrm{n}})B^a[\chi_a(H^{a,\mathrm{n}}), j_a]\langle x^{\mathrm{cm}}\rangle\end{aligned}\tag{3.7.130}$$

is bounded. We begin with the first term. By (3.7.118), we need only prove that

$$\langle x^{\mathrm{cm}}\rangle^{-1}(H^{a,\mathrm{n}}+c)^{-1}B^a\chi_a(H^{a,\mathrm{n}})\langle x^{\mathrm{cm}}\rangle \text{ is bounded.}$$

This, however, follows by direct computation, using *(V4)* and that $(H^{a,\mathrm{n}}+c)^{-1}$ and $\chi_a(H^{a,\mathrm{n}})$ preserve $\mathcal{D}(\langle x^{\mathrm{cm}}\rangle)$. The last term in (3.7.130) is bounded by the same argument. Finally, we have to check that

$$(H+c)^{-1}\chi_a(H^{a,\mathrm{n}})[D^2_{z_a}, j_a]\langle x^{\mathrm{cm}}\rangle \text{ is bounded.}$$

But this again follows by direct computation.

By (3.7.128) and (3.7.129), (3.7.124) will follow from

$$\begin{aligned}&q_aj_a[\chi_a(H^{a,\mathrm{n}})B^a\chi_a(H^{a,\mathrm{n}}), M_a]q_aj_a\\ =\ &q_aj_a[\chi_a(H^{a,\mathrm{n}})B^a\chi_a(H^{a,\mathrm{n}}),\\ &\chi_a(H^{a,\mathrm{n}})A^a\chi_a(H^{a,\mathrm{n}}) + F_aG^aF_a]q_aj_a\\ =\ &O((H+c)^2).\end{aligned}\tag{3.7.131}$$

We have

$$\begin{aligned}&q_aj_a[\chi_a(H^{a,\mathrm{n}})B^a\chi_a(H^{a,\mathrm{n}}), \chi_a(H^{a,\mathrm{n}})A^a\chi_a(H^{a,\mathrm{n}})]q_aj_a\\ =\ &q_aj_a\chi_a(H^{a,\mathrm{n}})[B^a, \chi_a(H^{a,\mathrm{n}})]\chi_a(H^{a,\mathrm{n}})A^a\chi_a(H^{a,\mathrm{n}})q_aj_a + \text{h.c.}\\ &+q_aj_a\chi_a(H^{a,\mathrm{n}})[\chi_a(H^{a,\mathrm{n}}), A^a]B^a\chi_a(H^{a,\mathrm{n}})^2q_aj_a + \text{h.c.}\\ &+q_aj_a\chi_a(H^{a,\mathrm{n}})^2[B^a, A^a]\chi_a(H^{a,\mathrm{n}})^2q_aj_a.\end{aligned}\tag{3.7.132}$$

Since $\chi_a(H^{a,\mathrm{n}}), q_a, j_a$ preserve the domains of $\langle x\rangle$, $\langle D\rangle$, $H$, to prove that (3.7.132) is $O((H+c)^2)$ it suffices to verify that

$$(H+c)^{-1}[B^a, \chi_a(H^{a,\mathrm{n}})]\langle x^a\rangle,$$

$$(H+c)^{-1}[\chi_a(H^{a,\mathrm{n}}), A^a],$$

$$B^a(H+c)^{-1},$$

$$(H+c)^{-1}[B^a, A^a]\langle x^a\rangle$$

are bounded. The first two estimates follow from (3.7.120); the remaining ones are verified directly from *(M1)*.

It remains to prove that

$$q_a j_a[\chi_a(H^{a,\mathrm{n}})B^a\chi_a(H^{a,\mathrm{n}}), F_a G^a F_a]q_a j_a = O((H+c)^2). \tag{3.7.133}$$

Using Lemma 3.7.6*(ii)*, we rewrite the left-hand side of (3.7.133) as

$$q_a j_a \chi_a(H^{a,\mathrm{n}})B^a[\chi_a(H^{a,\mathrm{n}}), F_a]G^a F_a q_a j_a + \text{h.c.}.$$

Since $q_a, j_a, \chi_a(H^{a,\mathrm{n}})$ preserve $\mathcal{D}(H)$, $(H+c)^{-1}q_a j_a \chi_a(H^{a,\mathrm{n}})B^a$ is bounded by *(M1)*. It remains to check that

$$[\chi_a(H^{a,\mathrm{n}}), F_a]G^a F_a q_a j_a (H+c)^{-1} \text{ is bounded.}$$

We rewrite this as

$$\begin{aligned} & [\chi_a(H^{a,\mathrm{n}}), F_a]\langle x^{\mathrm{cm}}\rangle q_a'\langle x^{\mathrm{cm}}\rangle^{-1}G^a F_a q_a j_a (H+c)^{-1} \\ + \quad & [\chi_a(H^{a,\mathrm{n}}), F_a](1-q_a')G^a F_a q_a j_a (H+c)^{-1}; \end{aligned}$$

the first term is bounded by (3.7.119) and Lemma 3.7.11*(iii)*, and the second term is bounded by Lemma 3.7.13*(ii)*. This ends the proof of (3.7.124).

- We now prove (3.7.125). We may rewrite the left-hand side as $I_1 + I_2 + I_3 + I_4$, where:

$$I_1 = q_a F_a[B_2, j_a]F_a M_a q_a j_a + \text{h.c.},$$

$$I_2 = q_a j_a[F_a B_2 F_a, A_a]q_a j_a,$$

$$I_3 = q_a j_a[F_a B_2 F_a, \chi_a(H^{a,\mathrm{n}})A^a\chi_a(H^{a,\mathrm{n}})]q_a j_a,$$

$$I_4 = q_a j_a[F_a B_2 F_a, F_a G^a F_a]q_a j_a.$$

By Lemma 3.7.12*(ii)* and Lemma 3.7.7*(i)*, we have $I_1 = O((H+c)^2)$. Since $B_2$ commutes with $A_a$, we have

$$I_2 = q_a j_a[F_a, A_a]B_2 F_a q_a j_a + \text{h.c.} = O((H+c)^2),$$

by Lemma 3.7.6 and direct computation.

Next, using Lemma 3.7.6*(ii)* we write

$$I_3 = q_a j_a\Big(F_a B_2[F_a, \chi_a(H^{a,\mathrm{n}})]A^a\chi_a(H^{a,\mathrm{n}}) + \text{h.c.}\Big)q_a j_a.$$

By Remark 3.7.14 we may apply Lemma 2.3.9*(iii)* to $\chi_a(H^{a,\mathrm{n}})$, hence

$$I_3 = q_a j_a\Big(F_a B_2[F_a, \chi_a(H^{a,\mathrm{n}})]q_a' A^a \chi_a(H^{a,\mathrm{n}}) + \text{h.c.}\Big)q_a j_a + \text{ bounded;}$$

using (3.7.119) and that $\langle x^{\mathrm{cm}}\rangle^{-1}A^a\chi_a(H^{a,\mathrm{n}})(H+c)^{-1}$ is bounded, we obtain that $I_3 = O((H+c)^2)$. Finally, we rewrite $I_4$ as

$$\begin{aligned} I_4 = \ & q_a j_a\Big(F_a[F_a, G^a]F_a B_2 F_a + \text{h.c.}) \\ & +([B_2, F_a]G^a F_a + \text{h.c.}) + F_a[B_2, G^a]F_a\Big)q_a j_a. \end{aligned}$$

By Lemma 3.7.6 and 3.7.12*(ii)*,

$$\langle x^{\mathrm{cm}}\rangle\langle y_a^{\mathrm{n}}\rangle^{-1}[F_a, G^a],\ \langle x^{\mathrm{cm}}\rangle\langle y_a^{\mathrm{n}}\rangle^{-1}[B_2, F_a],\ [B_2, G_a]$$

are bounded; using also Lemma 3.7.10, we find that $I_4$ is bounded. This ends the proof of (3.7.125).

- Finally, we prove (3.7.126). We may write the left-hand side of (3.7.126) as $I_1 + I_2 + I_3$, where:

$$I_1 = q_a[v_a', j_a M_a j_a]G^a F_a q_a,$$

$$I_2 = q_a v_a'[G^a, j_a M_a j_a]F_a q_a,$$

$$I_3 = q_a v_a' G^a[F_a, j_a M_a j_a]q_a.$$

We first estimate $I_1$. We have

$$\begin{aligned} I_1 = \ & q_a[v_a', j_a]M_a' j_a G^a F_a q_a + q_a j_a[v_a', M_a]j_a G^a F_a q_a \\ & + q_a j_a M_a'[v_a', j_a]G^a F_a q_a. \end{aligned}$$

Recall that

$$\begin{aligned} v_a' &= [H_a, iF_a] = \langle D - \mathbf{A}x, \nabla F_a\rangle + \text{h.c.} \\ &= 2\langle D - \mathbf{A}x, \nabla F_a\rangle - i\nabla^2 F_a, \end{aligned} \tag{3.7.134}$$

so that $[v_a', j_a] = -2i\nabla j_a \cdot \nabla F_a = O(\langle x^{\mathrm{cm}}\rangle^{-2})$. Hence the first and third terms in $I_1$ are $O((H+c)^2)$ by Lemmas 3.7.7*(i)*, 3.7.11*(iii)*, and (3.7.114). It remains to estimate the second term. We write $[v_a', M_a]$ as

$$\begin{aligned} & [v_a', A_a] + [v_a', \chi_a(H^{a,\mathrm{n}})A^a\chi_a(H^{a,\mathrm{n}})] + [v_a', F_a]G^a F_a \\ & + F_a G^a[v_a', F_a] + F_a[v_a', G^a]F_a. \end{aligned}$$

It is easily verified that $q_a j_a[v_a', A_a]j_a G^a F_a q_a = O((H+c)^2)$, by explicit computation and by (3.7.3) and Lemma 3.7.11*(i)*. The same argument applies to the next three terms, except that we also use Remark 3.7.14 and Lemma 3.7.12*(i)*. Thus the difficult term is

$$q_a j_a F_a[v_a', G^a]F_a j_a G^a F_a q_a.$$

By Lemma 3.7.11*(iii)*, it suffices to prove that

$$(H+c)^{-1}q_a j_a F_a[v'_a, G_a]F_a j_a\langle x^{\mathrm{cm}}\rangle \text{ is bounded.} \tag{3.7.135}$$

Recall from Lemma 3.7.6 that

$$G_a = B_0 + \langle y^n_a, B_1\rangle,\ B_i = \phi_i(D_{y^{\mathrm{n}}_a} + (\mathbf{A}x)^{\mathrm{n}}_a),\ i=0,1,$$

and plug in (3.7.134) for $v'_a$. The term $[\nabla^2 F_a, G_a]$ is easily estimated using Lemmas 3.7.12*(ii)*, 3.7.10, and (3.7.3). Next, since $F_a$ depends only on $x_a$, we have $\nabla_{x^a} F_a = 0$, hence

$$\langle D-\mathbf{A}x, \nabla F_a\rangle = \langle (D-\mathbf{A}x)^{\mathrm{n}}_a, \nabla_{x^{\mathrm{n}}_a} F_a\rangle + \langle (D-\mathbf{A}x)^{\mathrm{c}}_a, \nabla_{x^{\mathrm{c}}_a} F_a\rangle.$$

Clearly $B_i$ commute with $(D-\mathbf{A}x)^{\mathrm{c}}_a$. Furthermore, since all components of $D_a$ commute with all components of $(\mathbf{A}x)^{\mathrm{n}}_a$, $B_i$ commute also with $(D-\mathbf{A}x)^{\mathrm{n}}_a$. Using moreover that

$$(H+c)^{-1}q_a j_a F_a (D-\mathbf{A}x)^{\mathrm{n}}_a = C q'_a F'_a,$$

where $C$ is bounded and $q'_a, F'_a$ are functions similar to $q_a, F_a$ with slightly larger supports, we reduce the proof of (3.7.135) to verifying that

$$q'_a F'_a[\nabla F_a, B_0]F_a j_a\langle x^{\mathrm{cm}}\rangle,$$

$$q'_a F'_a\langle \nabla F_a, B_1\rangle F_a j_a\langle x^{\mathrm{cm}}\rangle,$$

$$q'_a F'_a\langle y^{\mathrm{n}}_a\rangle[\nabla F_a, B_1]F_a j_a\langle x^{\mathrm{cm}}\rangle,$$

are bounded. But this follows again from (3.7.3)*(ii)*, *(iii)*, and Lemmas 3.7.12, 3.7.10.

This concludes the proof of (3.7.126) and of (3.7.110).

**Proof of (3.7.111).** This will be done by undoing the commutator. It suffices to prove that

$$(H+c)^{-1}q_a[J, \mathrm{i}M_a]q_a j_a M_a j_a (H+c)^{-1} \tag{3.7.136}$$

is bounded. We rewrite (3.7.136) as

$$\Big((H+c)^{-1}q_a[J,\mathrm{i}M'_a]\langle x^{\mathrm{cm}}\rangle (H+c)^{-1/2}\Big)\times$$
$$\Big((H+c)^{1/2}\langle x^{\mathrm{cm}}\rangle^{-1} q_a j_a M_a j_a (H+c)^{-1}\Big).$$

The first term is bounded by Lemma 3.7.7*(iii)*, since $J$ satisfies (3.7.102). The second term is bounded by (3.7.112).

**Proof of (3.7.105)–(3.7.106).** We will only prove (3.7.105), since the proof of (3.7.106) is exactly the same except that $a$ and $a_{\max}$ are interchanged. Denote $\mathbf{I} = [H_a + J, iq_a M_a q_a]$. We first claim that

$$[\mathbf{I}, q_{a_{\min}} A q_{a_{\min}}] = [\mathbf{I}, q''_a q_{a_{\min}} A q_{a_{\min}} q''_a] + O((H+c)^2). \tag{3.7.137}$$

Indeed, it suffices to prove that

$$\mathbf{I} q_{a_{\min}} A q_{a_{\min}} = \mathbf{I} q''_a q_{a_{\min}} A q_{a_{\min}} q''_a + O((H+c)^2). \tag{3.7.138}$$

By (3.7.100) and (3.7.99) we have $\mathbf{I} = \mathbf{I}(q_a'')^2$, hence

$$\begin{aligned} \mathbf{I} q_{a_{\min}} A q_{a_{\min}} &= \mathbf{I}(q_a'')^2 q_{a_{\min}} A q_{a_{\min}} \\ &= \mathbf{I} q_a'' q_{a_{\min}} A q_{a_{\min}} q_a'' + \mathbf{I} q_a''', \end{aligned}$$

where $q_a''' = q_a'' q_{a_{\min}} [q_a'', A] q_{a_{\min}} \in S^0_{\mathrm{cl}}(X^{\mathrm{cm}})$. It suffices to prove that

$$\mathbf{I} q_a''' = O((H+c)^2).$$

We write:

$$\begin{aligned} &(H+c)^{-1} \mathbf{I} q_a''' (H+c)^{-1} \\ &= \Big((H+c)^{-1} \mathbf{I} (H+c)^{-1}\Big) + \Big((H+c) q_a''' (H+c)^{-1}\Big). \end{aligned}$$

The first term is bounded by Lemma 3.7.2. The second term is equal to

$$q_a''' + [(H+c), q_a'''](H+c)^{-1} = q_a''' + (\langle D - \mathbf{A}x, q_a''' \rangle + \text{h.c.})(H+c)^{-1},$$

which is clearly bounded. This proves (3.7.138).

By (3.7.137) it suffices to prove that

$$[\mathbf{I}, j_a A j_a] = O((H+c)^2), \tag{3.7.139}$$

where $j_a = q_a'' q_{a_{\min}}$. Let $T_a = j_a M_a j_a - j_a A j_a$. Then (3.7.139) will follow from:

$$[\mathbf{I}, j_a M_a j_a] = O((H+c)^2), \tag{3.7.140}$$

$$[\mathbf{I}, T_a] = O((H+c)^2). \tag{3.7.141}$$

(3.7.140) follows from (3.7.107); it remains to prove (3.7.141). This will be done by undoing the commutator. We have

$$\begin{aligned} &(H+c)^{-1} \mathbf{I} T_a (H+c)^{-1} \\ = \;& \Big((H+c)^{-1} \mathbf{I} (H+c)^{-1/2}\Big)\Big((H+c)^{-1/2}(H+c) T_a (H+c)^{-1}\Big). \end{aligned}$$

The first term on the right-hand side is bounded by Lemma 3.7.7*(iii), (v)*. The second term may be rewritten as

$$(H+c)^{-1/2} T_a + (H+c)^{-1/2} [H, T_a] (H+c)^{-1}.$$

By Lemma 3.7.9, $T_a$ is bounded. To estimate the second term, it suffices to prove that

$$(H+c)^{-1/2}[H, j_a A j_a](H+c)^{-1}, \; (H+c)^{-1/2}[H, j_a M_a j_a](H+c)^{-1}$$

are bounded. But this again follows from Lemma 3.7.7*(iii), (v)*. Similarly, $(H+c)^{-1} T_a \mathbf{I} (H+c)^{-1}$ is bounded, which proves (3.7.141) and hence (3.7.105). This (finally!) concludes the proof of Proposition 3.7.5. □

## 3.8. Two-dimensional dispersive systems

In this section we present the 2–dimensional analogues of the results of Section 3.7. In the 2–dimensional case, the only possible scattering channels are those corresponding to cluster decompositions in $\mathcal{A}^{\mathrm{n}}$; accordingly, our conjugate operator $M$ will be a sum of auxiliary conjugate operators $M_a$ for $a \in \mathcal{A}^{\mathrm{n}}$ only.

The threshold set is

$$\tau := \cup_{a<a_{\max}} \tau_a, \tag{3.8.1}$$

where:

$$\tau_{a_{\min}} := \sigma(\tilde{H}^{a_{\min}}),$$

$$\tau_a = \tau_{a_{\min}} \cup \sigma(\tilde{H}^a),\ a \in \mathcal{A}^{\mathrm{c}}, \sharp a = 2,$$

and $\tau_a$ for $a \in \mathcal{A}^{\mathrm{n}}$, $\sharp a = 2$, was defined in Section 3.6. The set $\tau$ is closed and discrete.

Fix an interval $I \subset \mathbb{R} \setminus \tau$, and let

$$\mathbf{1} = \sum_{a\in\mathcal{A}} q_a^2(y^{\mathrm{cm}})$$

be an $N$-body partition of unity as in Proposition 2.3.4, with the constants $\delta_a, \epsilon_a$ sufficiently small. We also define for $a \in \mathcal{A}^{\mathrm{n}}$:

$$F_a = F(\frac{|y_a^{\mathrm{c}}|}{\langle y_a^{\mathrm{cm}}\rangle} \leq c_0), \tag{3.8.2}$$

where $c_0 > 0$ is small enough. As in Section 3.7, we have

$$|D_y^\alpha F_a| \leq C_\alpha \langle y^{\mathrm{cm}}\rangle^{-|\alpha|},\ |\alpha| \geq 0,\ \text{on } \operatorname{supp} q_a. \tag{3.8.3}$$

Let $U_a$, $a \in \mathcal{A}$, be the 2–dimensional modifications of the reducing transformations defined in Section 2.2. For each $a \in \mathcal{A}^{\mathrm{n}}$, let $\tilde{G}_I^a$ be the conjugate operator constructed in Section 3.6 for the interval $I$; recall that $\tilde{G}_I^a$ is a finite sum of conjugate operators as in Proposition 3.6.4. Since the interval $I$ will be fixed throughout this section, we may drop the subscripts $I$ and write $\tilde{G}^a$ instead of $\tilde{G}_I^a$.

We define for $a \in \mathcal{A}^{\mathrm{n}}$:

$$G^a = U_a^* \tilde{G}^a U_a,\ M_a = F_a G^a F_a,$$

on $\mathcal{D}(N)$, where $N = D_y^2 + y^2$. The conjugate operator $M$ is

$$M := \sum_{a\in\mathcal{A}^{\mathrm{n}}} q_a M_a q_a. \tag{3.8.4}$$

The main results in the 2–dimensional case are the following.

Proposition 3.8.1. *Assume (V1), (V4). Then $M$ satisfies the hypotheses of Nelson's commutator theorem (Lemma 1.2.5) with the comparison operator $N$; in particular, $M$ is essentially self-adjoint on $\mathcal{D}(N)$.*

We continue to use $M$ to denote the unique self-adjoint extension of the operator $M$ defined above.

LEMMA 3.8.2. *Assume (V1), (V4), and let* $c > -\inf \sigma(H)+1$. *Then the quadratic form* $(H+c)^{-\frac{1}{2}}[H, \mathrm{i}M](H+c)^{-\frac{1}{2}}$ *extends to a bounded operator on* $L^2(X)$.

THEOREM 3.8.3. *Assume (V1), (V4). Then:*

*(i)* $H \in C^1(M)$;

*(ii) the eigenvalues of* $H$ *may accumulate only at* $\tau$. *Consequently,* $\tau \cup \sigma_{\mathrm{pp}}(H)$ *is a closed countable set;*

*(iii) for any* $\lambda \in I \backslash \tau \cup \sigma_{\mathrm{pp}}(H)$ *and* $F \in C_0^\infty(\mathbb{R}^2)$, *there is a* $c(\lambda) > 0$ *such that*

$$F(K_{\mathrm{cm}})[H, \mathrm{i}M]F(K_{\mathrm{cm}}) \geq c(\lambda) \text{ at } H = \lambda.$$

PROPOSITION 3.8.4. *Assume (V1), (V4). Then* $H \in C^{1+\epsilon}(M)$ *for some* $\epsilon > 0$.

LEMMA 3.8.5. *Assume (V1), (V4).*

*(i) We have:*

$$H \in C^1(\langle y^{\mathrm{cm}} \rangle); \tag{3.8.5}$$

$$\langle y^{\mathrm{cm}} \rangle \in C^1(M); \tag{3.8.6}$$

$$[M, \langle y^{\mathrm{cm}} \rangle]\langle y^{\mathrm{cm}} \rangle^{-1} \text{ is bounded;} \tag{3.8.7}$$

$$\langle y^{\mathrm{cm}} \rangle^{-1} M(H+\mathrm{i})^{-1} \text{ is bounded;} \tag{3.8.8}$$

$$-C\langle y^{\mathrm{cm}} \rangle \leq \chi(H)M\chi(H) \leq C\langle y^{\mathrm{cm}} \rangle \text{ for } \chi \in C_0^\infty(\mathbb{R}). \tag{3.8.9}$$

*(ii) Let* $I(t, y)$ *be a time-dependent potential such that*

$$|\partial_y^\alpha I(t, y)| \leq C_\alpha(\langle y \rangle + t)^{-1-\epsilon}, \; |\alpha| \geq 0. \tag{3.8.10}$$

*Then the commutator* $[M, I(t, y)]$, *defined as a quadratic form on* $\mathcal{D}(M)$, *extends to a bounded operator with norm bounded by* $Ct^{-\epsilon}$.

The proofs are identical to those for 3−dimensional systems given in Section 3.7, with the considerable simplification that the $z$ variables (hence also $A_a$, $G^0$, etc.) are omitted. The only additional argument needed at the end of the proof of the Mourre estimate is that $F(K_{\mathrm{cm}})f(H_a)$ is compact for all $F \in C_0^\infty(\mathbb{R}^2)$ and for all $f \in C_0^\infty(\mathbb{R})$ supported away from $\bigcup_{\sharp a \geq 2} \sigma(H^a)$. This follows from the fact that for $f$ as above $f(U_{\mathrm{cm}} H_a U_{\mathrm{cm}}^*)$ is compact on $L^2(\hat{Y}_{\mathrm{cm}}^{\mathrm{c}} \times Y^{\mathrm{cm}})$, which will be proved in Section 6.5 (see equation (6.5.2).

CHAPTER 4

# Basic Propagation Estimates

*Propagation estimates* are estimates localizing the evolution $e^{-itH}$ (modulo small errors) in appropriately chosen regions of the phase space. For the free evolution generated by $H = -\frac{1}{2}\Delta$, such estimates are classic and may be proved e.g., by stationary phase arguments ([**RS**]; see also the proof of Proposition 1.4.6 in this book). A powerful idea of Enss [**E2**], [**E3**] was that phase-space localization of the evolution generated by $H = -\frac{1}{2}\Delta + V$ could be used to prove asymptotic completeness. $N$-body propagation estimates similar to those considered in this chapter were first proved by Sigal and Soffer [**SS1**]–[**SS3**]; subsequent extensions, modifications and improvements are due to Graf [**Gr**], Dereziński [**De3**], Skibsted [**Sk1**], Gérard [**G2**], and many others. Propagation estimates play a key role in the proof of $N$-particle asymptotic completeness [**SS1**], [**Gr**], [**De1**], and provide meaningful information about the behaviour of the system even in cases when asymptotic completeness might not hold. (In this book, as in many other papers, propagation estimates are proved under weaker assumptions on the pair interactions than those needed for asymptotic completeness.) We refer the interested reader to [**DG1**] for a thorough and complete presentation of the subject.

The present chapter will be devoted to "standard" propagation estimates, which do not involve the magnetic field in any nontrivial way. The propagation estimates of the next chapter will be specific to the case of magnetic Hamiltonians.

We first prove the existence of the *asymptotic energy* and the *maximal velocity estimates*; these are relatively simple estimates based on conservation of energy. We then proceed to the existence of the *asymptotic velocity* in the direction of the field. The heuristic is that, as $t \to \infty$, the system either remains in a bound state (as discussed in Chapter 1) or else breaks up into two or more disjoint clusters. We will prove that in the latter case:

- the distances between different clusters increase proportionally to $t$;
- the size of each cluster is bounded by $o(t)$;
- the bound on the size of clusters can be improved to $O(t^{\alpha})$ for some $0 < \alpha < 1$; under more restrictive assumptions on the pair potentials (the decay rate must be at least $\mu_l > \sqrt{3} - 1$) it can be further improved to $O(1)$.

The estimates of Section 4.3 are partial results in that direction, concerning the dynamics of the system in the $z$ variables. (The $y$ variables will be treated by rather different methods in the next chapter.)

For $N$-body Hamiltonians without a magnetic field, the results of Section 4.2 are essentially due to Sigal and Soffer [**SS2**], [**SS3**], and the construction of asymptotic velocity (Section 4.3) – to Dereziński [**De3**]. Below we state the magnetic analogues of these results; the proofs for the non-magnetic case (see [**DG1**]) carry over verbatim to our setup with only notational changes, therefore we either only sketch them here or omit them altogether.

The last two sections will concern the "dispersive" case – three particles with at least one neutral pair. We will combine the Mourre estimate proved in Section 3.7 with an abstract result of Section 4.4 to obtain a *minimal velocity estimate* for such systems. Similar estimates were first proved by Sigal and Soffer [**SS2**], [**SS3**]; here we prove them in a slightly more general setting.

To avoid unnecessary clutter and repetition, we state all results of this chapter only for the 3–dimensional case ($n = 3$). The results of Section 4.2 and Subsection 4.4.2, including all proofs, remain valid in the 2–dimensional case (with the variables $x$ replaced by $y$); the results of Subsection 4.4.1 are of an abstract nature; and the results of Section 4.3 are specific to the 3–dimensional case.

## 4.1. Preliminaries

Throughout this chapter we will denote by $U(t,s)$ the unitary evolution generated by a time-dependent Hamiltonian

$$H(t) := H + I(t,x), \tag{4.1.1}$$

where $H$ is an $N$-particle magnetic Hamiltonian, and $I(t,x)$ is a time-dependent real-valued potential. (The precise definition of $U(t,s)$ is given in Subsection 4.1.1.) In our proof of asymptotic completeness, the results obtained here will be applied with $H$ replaced by a channel Hamiltonian $H_a$ and with $I$ equal to the "long-range tail" of the interaction between different clusters in $a$.

We will always assume that $V_{ij}$ satisfy *(V1)*, *(SC)*, and that

$$|\partial_x^\alpha I(t,x)| \le Ct^{-|\alpha|-\mu_1}, \ |\alpha| \le 2. \tag{4.1.2}$$

We will denote $U(t,0)$ simply by $U(t)$.

### 4.1.1. Defining the dynamics.

A *unitary dynamics* will be a strongly continuous two-parameter family of unitary operators $U(t,s)$, $t, s \in \mathbb{R}$, such that

$$U(s,s) = \mathbf{1}, \ U(t,t')\,U(t',s) = U(t,s).$$

We wish to define $U(t,s)$ so that

$$\frac{\partial}{\partial t}U(t,s) = -\mathrm{i}H(t)U(t,s),$$
$$\frac{\partial}{\partial s}U(t,s) = \mathrm{i}U(t,s)H(s), \tag{4.1.3}$$

if the derivatives on the left-hand side are appropriately understood. We will rely on an abstract framework introduced in [**DG1**, Section B.3], which we now recall briefly. We will denote by $L^1_{\mathrm{loc}}([T_1,T_2],B(\mathcal{H}))$ the space of all Bochner integrable functions $[T_1,T_2] \ni t \to R(t) \in B(\mathcal{H})$. (For the precise definition and a discussion of Bochner integrability, see [**DS**].) We will also say that a function $[T_1,T_2] \ni t \mapsto B(t) \in B(\mathcal{H})$ is in $W^{1,1}([T_1,T_2],B(\mathcal{H}))$ if there is a Bochner integrable function $[T_1,T_2] \ni t \mapsto R(t) \in B(\mathcal{H})$ such that

$$B(t_2) - B(t_1) = \int_{t_1}^{t_2} R(t)\mathrm{d}t,\ t_1,t_2 \in [T_1,T_2].$$

We write $R(t) = \frac{\mathrm{d}}{\mathrm{d}t}B(t)$ if the above equality holds.

DEFINITION 4.1.1. [**DG1**] Let $B$ be a densely defined, positive, and invertible operator, and let $\mathbb{R} \ni t \mapsto H(t)$ be a function with values in self-adjoint operators. We say that the unitary dynamics $U(t,s)$ is *$B$-regularly generated* by the time-dependent Hamiltonian $H(t)$ if:

*(i)* the mapping $s \mapsto U(t,s)B^{-1}$ is in $W^{1,1}(\mathbb{R},B(\mathcal{H}))$, $\mathcal{D}(B) \subset \mathcal{D}(H(s))$ for almost all $s$, and $H(s)B^{-1} \in L^1_{\mathrm{loc}}(\mathbb{R},B(\mathcal{H}))$;

*(ii)* $\partial_s U(t,s)B^{-1} = \mathrm{i}U(t,s)H(s)B^{-1}$;

*(iii)* $B^{-1/2}[H(t),B]B^{-1/2}$, defined for almost all $t$ as a quadratic form on $\mathcal{D}(B^{1/2})$, extends to an operator-valued function in $L^1_{\mathrm{loc}}(\mathbb{R},B(\mathcal{H}))$.

The condition *(ii)* states that the second equation in (4.1.3) holds on the domain of $B$; by conjugation, it automatically implies that

$$\partial_t B^{-1}U(t,s) = -\mathrm{i}B^{-1}H(t)U(t,s).$$

Assuming (4.1.2), it is straightforward to construct a $B$-regularly generated unitary dynamics $U(t,s)$ for $H(t)$ given by (4.1.1), where $B = H + c$ for some $c$ sufficiently large. Indeed, let

$$U(t,s) := \mathrm{e}^{-\mathrm{i}tH}T(\mathrm{e}^{-\int_s^t V(u)\mathrm{d}u})\mathrm{e}^{\mathrm{i}sH}, \tag{4.1.4}$$

where $V(t) := \mathrm{e}^{\mathrm{i}tH}I(t,x)\mathrm{e}^{-\mathrm{i}tH}$, and

$$T\left(\mathrm{e}^{\int_s^t V(u)\mathrm{d}u}\right) := \sum_{n=0}^{\infty} \int\cdots\int_{t\geq u_n \geq \cdots \geq u_1 \geq s} V(u_n)\cdots V(u_1)\mathrm{d}u_n \ldots \mathrm{d}u_1$$

is the time-ordered exponential.

LEMMA 4.1.2. *Define $H(t)$ by (4.1.1), and assume that (V1), (4.1.2) hold. Then the dynamics $U(t,s)$ defined by (4.1.4) is $(H+c)$-regularly generated, provided that $c > 0$ is large enough.*

PROOF. By Proposition B.3.6 of [**DG1**], it suffices to check *(iii)*. We have

$$[H, \mathrm{i}H(t)] = [H, \mathrm{i}I(t,x)] = \frac{1}{2}(\langle D - \mathbf{A}x, \nabla_x I\rangle + \mathrm{h.c.}), \tag{4.1.5}$$

as quadratic forms on $\mathcal{D}(H)$; hence *(iii)* follows from Proposition 1.2.1 and (4.1.2). □

**4.1.2. Wave operators.** We will often have to compare two unitary dynamics $U_i(t)$, $i = 1, 2$, generated (in the sense of Definition 4.1.1) by two time-dependent Hamiltonians $H_i(t)$. We will usually formulate results of this type in terms of *wave operators*

$$W_{12}^{+} = \mathrm{s-}\lim_{t\to\infty} U_2(t)^* U_1(t) P_1, \tag{4.1.6}$$

where $P_1$ is a suitable projection. For $u \in \operatorname{Ran} P_1$, the equality $W_{12}^{+}u = u^{+}$ means that

$$\|U_1(t)u - U_2(t)u^{+}\| \to 0 \text{ as } t \to +\infty.$$

Hence, if $W_{12}^{+}$ exist, $U_1(t)$ (with initial conditions in $\operatorname{Ran} P_1$) has the same large-time asymptotics as $U_2(t)$. Typical applications involve approximating a complicated evolution (such as $\mathrm{e}^{-\mathrm{i}tH}$) by a simpler one (such as $\mathrm{e}^{-\mathrm{i}t\Delta/2}$). We will say that $W_{12}^{+}$ *intertwine* $U_1(t)$ and $U_2(t)$ on appropriate subspaces.

The notion of wave operators goes back to the 1940's [**Mø**], and has since become one of the main concepts of scattering theory. Indeed, asymptotic completeness for many physical systems – including those considered here – is usually stated in terms of wave operators. We defer a more detailed discussion of that subject to Chapter 6; here we only review some elementary properties of wave operators which will be needed in the sequel. All of the results here are well known in scattering theory, see e.g., [**RS**] for proofs and further discussion.

The wave operators $W_{12}^{+}$ defined by (4.1.6), whenever they exist, are isometries from $\operatorname{Ran} P_1$ to $\operatorname{Ran} W_{12}^{+}$, and:

$$W_{12}^{+*} W_{12}^{+} = P_1,$$

$$W_{12}^{+} W_{12}^{+*} \text{ is the projection on } \operatorname{Ran} W_{12}^{+}.$$

Let $P_2$ be a projection operator, then $\operatorname{Ran} P_2 \subset \operatorname{Ran} W_{12}^{+}$ if and only if the "inverse" wave operators

$$W_{21}^{+} = \mathrm{s-}\lim_{t\to\infty} U_1(t)^* U_2(t) P_2$$

exist. In particular, if $P_2$ is the projection on $\operatorname{Ran} W_{12}^{+}$, then $W_{21}^{+} = W_{12}^{+*}$.

The existence of wave operators is often proved using (sometimes quite sophisticated) variations on the following general theme, known as *Cook's method* [**Co**]. In order to prove that the limit

$$\lim_{t\to\infty} U_2(t)^* U_1(t) u$$

exists for $u$ in a suitable subspace of $\mathcal{H}$, it suffices to verify that the derivative

$$\frac{\mathrm{d}}{\mathrm{d}t}(U_2(t)^*U_1(t)u) = \mathrm{i}U_2(t)^*(H_2(t) - H_1(t))U_1(t)u$$

is integrable in norm as $t \to \infty$, i.e., that

$$\int_{t_0}^{\infty} \|(H_2(t) - H_1(t))U_1(t)u\|\mathrm{d}t < \infty.$$

One then proceeds to prove (via propagation estimates) that the trajectory $U_1(t)u$ is mostly localized in the region of phase space where $H_2(t) - H_1(t)$ vanishes sufficiently fast as $t \to \infty$, which yields the desired conclusion.

Traditionally, Cook's method was used to prove the existence of wave operators with $U_1(t)$ equal to the free evolution $\mathrm{e}^{\frac{1}{2}\mathrm{i}t\Delta}$. Enss [**E2**] was the first to apply it to proving asymptotic completeness. Arguments of this type have been used rather widely in scattering theory ever since, and will often appear throughout this book.

In our proof of asymptotic completeness we will produce a finite sequence of comparison dynamics $U_{a,0}, U_{a,1}, \dots,$ for $\mathrm{e}^{-\mathrm{i}tH}$, and prove the existence of the wave operators intertwining each two consecutive dynamics in the sequence; at each step a different propagation estimate will be used. To conclude that the first and last dynamics are intertwined by wave operators, we will rely on the *chain rule*, which we now explain. Define $W_{12}^+$ by (4.1.6), and let

$$W_{23}^+ = \mathrm{s-}\lim_{t\to\infty} U_3(t)^*U_2(t)P_2,$$

where $U_3(t)$ is the unitary evolution generated by a third time-dependent Hamiltonian $H_3(t)$, and $P_2$ is the projection on $\mathrm{Ran}\, W_{12}^+$. Assume that $W_{12}^+$ and $W_{23}^+$ exist. Then the wave operators

$$W_{13}^+ = \mathrm{s-}\lim_{t\to\infty} U_3(t)^*U_1(t)P_1$$

exist and are isometries from $\mathrm{Ran}\, P_1$ to $\mathrm{Ran}\, W_{13}^+$. This procedure may of course be iterated any number of times.

**4.1.3. Time-dependent observables.** A *time-dependent observable* is an operator-valued function

$$\Phi : [T_1, T_2] \ni t \to \Phi(t) \in B(\mathcal{H}).$$

We will usually require some regularity of $\Phi$, see e.g., the conditions of Proposition 4.1.3 (which will always be satisfied for the observables we will work with). We denote by $\mathbf{D}\Phi(t)$ the Heisenberg derivative of $\Phi(t)$ associated with $U(t)$:

$$\mathbf{D}\Phi(t) := \frac{\mathrm{d}}{\mathrm{d}t}\Phi(t) + [H(t), \mathrm{i}\Phi(t)].$$

We will also consider *asymptotic observables* of the form

$$\Phi^+ = \mathrm{s-}\lim_{t\to\infty} U(t)^*\Phi(t)U(t).$$

Such observables appeared implicitly in [**SS2**]–[**SS4**], and were first studied systematically in [**De3**]. Of particular importance is the *asymptotic velocity* observable [**De3**], which was crucial in the proof of asymptotic completeness for $N$-body long-range systems [**De1**].

For future reference, we observe the following relation between asymptotic observables and wave operators. Define $U_i(t)$, $i = 1, 2$, as in Subsection 4.1.2, and let $\Phi(t)$ be a time-dependent observable. Assume that the asymptotic observables

$$\Phi_i^+ = \text{s-}\lim_{t\to\infty} U_i(t)^*\Phi(t)U_i(t), \; i = 1, 2,$$

exist. In the sequel we will often use wave operators $W_{12}^+$ of the form (4.1.6), where $P_1$ is a spectral projection of $\Phi_1^+$. Then it is easy to verify that $W_{12}^+$ intertwines $\Phi_1^+$ and $\Phi_2^+$, i.e.,

$$W_{12}^+\Phi_1^+ = \Phi_2^+W_{12}^+. \tag{4.1.7}$$

We also record the trivial observation that

$$\text{s-}\lim_{t\to\infty} \mathrm{e}^{\mathrm{i}tH}\Phi(t)\mathrm{e}^{-\mathrm{i}tH} = \text{s-}\lim_{t\to\infty} \mathrm{e}^{\mathrm{i}(t+s)H}\Phi(t+s)\mathrm{e}^{-\mathrm{i}(t+s)H};$$

hence, if the asymptotic observable $\Phi^+$ defined by the above limit exists and if $\|\Phi(t+s) - \Phi(t)\| \to 0$ as $t \to +\infty$, then $\Phi^+$ commutes with $\mathrm{e}^{\mathrm{i}sH}$ and therefore with $H$.

We end this section by stating several technical results which will be useful in the sequel.

In order to prove the existence of the required wave operators and asymptotic observables, we will often have to differentiate expressions of the form $U_2(t)^*\Phi(t)U_1(t)$, where $U_1, U_2$ are unitary evolutions. The following proposition, which we borrow from [**DG1**] (Proposition B.3.4), will provide a rigorous justification for such calculations.

PROPOSITION 4.1.3. *Let $U_i(t, s)$, $i = 1, 2$, be $B$-regularly generated by two time-dependent Hamiltonians $H_i(t)$. Assume that*

$$\Phi(t) \in W^{1,1}([T_1, T_2], B(\mathcal{H})), \; \|B^{\frac{1}{2}}\Phi(t)B^{-\frac{1}{2}}\| \le C \textit{ for almost all } t,$$

*and that*

$$H_2(t)\Phi(t) - \Phi(t)H_1(t),$$

*originally defined for almost all $t$ as a quadratic form on $\mathcal{D}(B^{1/2})$, extends to an operator-valued function in $L^1_{\mathrm{loc}}([T_1, T_2], B(\mathcal{H}))$. Then*

$$\begin{aligned} &U_2(s, t_2)\Phi(t_2)U_1(t_2, s) - U_2(s, t_1)\Phi(t_1)U_1(t_1, s) \\ &= \int_{t_1}^{t_2} U_2(s, u)\left(\mathrm{D}_{21}\Phi(t)\right)U_1(u, s)\mathrm{d}u, \end{aligned} \tag{4.1.8}$$

*where*

$$\mathrm{D}_{21}\Phi(t) := \frac{\mathrm{d}}{\mathrm{d}t}\Phi(t) + \mathrm{i}H_2(t)\Phi(t) - \mathrm{i}\Phi(t)H_1(t). \tag{4.1.9}$$

Note that $\mathrm{D}_{21} = \mathbf{D}$ if $U_1(t) = U_2(t) = U(t)$.

Finally, we review some standard techniques used to prove propagation estimates. The first one is a version of the Putnam-Kato theorem developed in [**SS1**] (see also [**DG1**, Lemma B.4.1]).

LEMMA 4.1.4. *Let $\Phi(t)$ be a family of self-adjoint operators satisfying the conditions of Proposition 4.1.3. Assume that $\Phi(t)$ is uniformly bounded, and that there are operator valued functions $B(t)$ and $B_i(t)$, $i = 1, \dots, n$, such that:*

$$\mathbf{D}\Phi(t) \geq C_0 B^*(t)B(t) - \sum_{i=1}^{n} B_i^*(t)B_i(t),$$
$$\int_1^\infty \|B_i(t)U(t)\phi\|^2 \mathrm{d}t \leq C\|\phi\|^2, \quad i = 1, \dots, n,$$

*for some $C_0 > 0$. Then there is a constant $C_1 > 0$ such that*

$$\int_1^\infty \|B(t)U(t)\phi\|^2 \mathrm{d}t \leq C_1\|\phi\|^2. \tag{4.1.10}$$

PROOF. We have

$$\begin{aligned} C_2\|u\|^2 \geq \; & \langle U(T)\phi, \Phi(T)U(T)\phi\rangle - \langle U(1)\phi, \Phi(1)U(1)\phi\rangle \\ = \; & \textstyle\int_1^\infty \langle U(t)\phi, D\Phi(t)U(t)\phi\rangle \mathrm{d}t \\ \geq \; & C_0 \textstyle\int_1^\infty \|B(t)U(t)\phi\|^2 \mathrm{d}t - \sum \int_1^\infty \|B_i(t)U(t)\phi\|^2 \mathrm{d}t \\ \geq \; & C_0 \textstyle\int_1^\infty \|B(t)U(t)\phi\|^2 \mathrm{d}t + nC\|\phi\|^2; \end{aligned}$$

the lemma follows. □

Many other variations on the same theme are possible, see e.g., [**SS1**]–[**SS4**], [**Gr**], [**Sk1**]. We will rely on a related lemma proved in [**DG1**, Lemma B.4.2].

LEMMA 4.1.5. *Let $U_i(t), i = 1, 2$, be the unitary evolutions generated (in the sense of Definition 4.1.1) by the time-dependent Hamiltonians $H_i(t)$. Let $\mathcal{D}$ be a dense subspace of $\mathcal{H}$. Let $\Phi(t)$ is a uniformly bounded function with values in self-adjoint operators satisfying the conditions of Proposition 4.1.3. Define $\mathrm{D}_{21}$ by (4.1.9).*

*(i) Assume that for $\phi \in \mathcal{D}$*

$$\int_1^\infty \|(\mathrm{D}_{21}\Phi(t))U_1(t)\phi\| \,\mathrm{d}t < \infty.$$

*Then the limit*

$$\mathrm{s-}\lim_{t\to\infty} U_2^*(t)\Phi(t)U_1(t) \tag{4.1.11}$$

*exists.*

*(ii) Assume that:*

$$\begin{aligned}
&|(\psi_2|\mathrm{D}_{21}\Phi(t)\psi_1)| \leq \sum_{i=1}^{n} \|B_{2i}(t)\psi_2\|\|B_{1i}(t)\psi_1\|,\\
&\int_1^\infty \|B_{2i}(t)U_2(t)\phi\|^2 \mathrm{d}t \leq C\|\phi\|^2, \quad \phi \in \mathcal{H}, \quad i = 1, \dots, n, \\
&\int_1^\infty \|B_{1i}(t)U_1(t)\phi\|^2 \mathrm{d}t \leq C\|\phi\|^2, \quad \phi \in \mathcal{D}_1, \quad i = 1, \dots, n.
\end{aligned} \tag{4.1.12}$$

*Then the limit (4.1.11) exists.*

## 4.2. Asymptotic energy and maximal velocity estimates

We first construct the *asymptotic energy* observable.

PROPOSITION 4.2.1. *Let $\chi \in C_0^\infty(\mathbb{R})$. Then the norm limit*

$$\lim_{t\to+\infty} U(t)^*\chi(H)U(t) =: \gamma^+(\chi) \tag{4.2.1}$$

*exists. Furthermore, there is a unique densely defined self-adjoint operator $H^+$, which we will call* asymptotic energy, *such that $\gamma^+(\chi) = \chi(H^+)$.*

PROOF. It is easily verified from (1.2.21), (4.1.5), (4.1.2) that

$$\mathbf{D}\chi(H) \in L^1(\mathrm{d}t); \tag{4.2.2}$$

by Lemma 4.1.5, this proves that the limit (4.2.1) exists. By Proposition B.2.1 of [**DG1**], to prove the existence of $H^+$ as above it suffices to verify that for any $\chi \in C_0^\infty(\mathbb{R})$, $\chi(0) = 1$:

$$\mathrm{s-}\lim_{R\to+\infty} \gamma^+(\chi_R) = \mathbf{1},$$

where $\chi_R(\lambda) = \chi(\frac{\lambda}{R})$. This in turn will follow if we prove that

$$\mathbf{D}\chi_R(H) \in L^1(\mathrm{d}t) \text{ uniformly in } R,$$

which is done by the same argument as (4.2.2). The details are left to the reader. □

*Maximal velocity estimates* are also based on the conservation of energy. They state, informally, that bounds on the total energy of the system imply bounds on its kinetic energy and therefore on its velocity.

PROPOSITION 4.2.2. *Let $\chi \in C_0^\infty(\mathbb{R})$. Then:*

*(i) if $\theta_1 > 0$ is sufficiently large, for any $0 < \theta_1 < \theta_2$ we have*

$$\int_1^{+\infty} \|\mathbf{1}_{[\theta_1,\theta_2]}(\frac{|x|}{t})\chi(H)U(t)u\|^2 \frac{\mathrm{d}t}{t} \leq C\|u\|^2;$$

*(ii) if $\theta > 0$ is sufficiently large, and if $F \in C^\infty(\mathbb{R})$ satisfies $F' \in C_0^\infty(\mathbb{R})$, $\operatorname{supp} F \subset (\theta, +\infty)$, then*

$$\mathrm{s-}\lim_{t\to+\infty} U(t)^*F(\frac{|x|}{t})\chi(H)U(t) = 0.$$

PROOF. The proof is similar to that of maximal velocity estimates for standard $N-$particle Hamiltonians. We will present in detail only the proof of *(i)*; *(ii)* can be obtained by repeating verbatim the proof of Proposition 4.2.1 of [**DG1**].

Pick $f \in C_0^\infty(\mathbb{R})$ such that $f = 1$ on $[\theta_1, \theta_2]$, and define

$$F(s) := \int_{-\infty}^{s} f^2(s_1)\mathrm{d}s_1.$$

Our propagation observable will be

$$\Phi(t) := \chi(H) F\left(\frac{|x|}{t}\right)\chi(H).$$

We compute:

$$\begin{aligned} -\mathbf{D}\Phi(t) = \;& t^{-1}\chi(H) f^2\left(\tfrac{|x|}{t}\right)\tfrac{|x|}{t}\chi(H) \\ & -\tfrac{1}{2}t^{-1}\chi(H)\left(\langle D - \mathbf{A}x, \tfrac{x}{|x|}\rangle f^2\left(\tfrac{|x|}{t}\right) + \text{h.c.}\right)\chi(H) \\ & +[I(t,x), \mathrm{i}\chi(H)]F(\tfrac{|x|}{t})\chi(H) + \text{h.c.} \end{aligned} \tag{4.2.3}$$

By (1.2.21) and (4.1.2), the last term on the right-hand side of (4.2.3) is in $L^1(\mathrm{d}t)$. Pick $\tilde{\chi} \in C_0^\infty(\mathbb{R})$, $\tilde{f} \in C_0^\infty(\mathbb{R})$ such that $\chi\tilde{\chi} = \chi$, $f\tilde{f} = f$, $0 \le \tilde{\chi} \le 1$, $0 \le \tilde{f} \le 1$, and $0 \notin \operatorname{supp}\tilde{f}$. It is easy to verify using Lemma 2.3.8 that the second term on the right-hand side of (4.2.3) is equal to:

$$\begin{aligned} & -\tfrac{1}{2t}\chi(H) f\left(\tfrac{|x|}{t}\right)\tilde{\chi}(H)\left(\langle D - \mathbf{A}x, \tfrac{x}{|x|}\rangle \tilde{f}\left(\tfrac{|x|}{t}\right) + \text{h.c.}\right) \\ & \times\tilde{\chi}(H) f\left(\tfrac{|x|}{t}\right)\chi(H) + O(t^{-2}). \end{aligned} \tag{4.2.4}$$

Using also that $(D - \mathbf{A}x)\tilde{\chi}(H)$ is bounded, we estimate (4.2.4) from below by

$$-\frac{C}{t}\chi(H) f^2(\frac{|x|}{t})\chi(H) + O(t^{-2}).$$

We now observe that $\frac{|x|}{t} \ge \theta_1$ on $\operatorname{supp} f$. Hence, if $\theta_1$ is large enough, we obtain that (4.2.3) is bounded from below by

$$C_0 t^{-1}\chi(H) f^2\left(\frac{|x|}{t}\right)\chi(H), \tag{4.2.5}$$

where $C_0 > 0$. We may therefore apply Lemma 4.1.4 and conclude that

$$\int_1^\infty \left\| f\left(\frac{|x|}{t}\right)\chi(H)\phi_t\right\|^2 \frac{\mathrm{d}t}{t} \le C\|\phi\|^2, \tag{4.2.6}$$

which yields *(i)*. □

## 4.3. Asymptotic velocity along the field

We now turn our attention to the propagation along the field in the 3−dimensional case. Since the magnetic field does not act in the $z$ direction, we expect that the propagation along the $z$ axis should be qualitatively similar to the propagation of "non-magnetic" $N$-body systems, and that suitable analogues of the results of e.g., [**De3**], [**De1**] should hold. Thus for instance we expect from the form of the equations of motion in the $z$ variables for the evolution $\mathrm{e}^{-\mathrm{i}tH}$:

$$\mathbf{D}z = [H, \mathrm{i}z] = D_z, \ \mathbf{D}D_z = [H, \mathrm{i}D_z] = -\nabla_z V(x),$$

that $D_z$ is approximately constant along the trajectories $\mathrm{e}^{-\mathrm{i}tH}u$ and that $\langle z\rangle \sim tD_z$; the appropriate rigorous statements will be given shortly. The proofs of all results in this section are exactly as in the references indicated below, except that the $x$ variables are replaced by $z$ (for example, all propagation observables of the form $\chi(H)r(t,x,D_x)\chi(H)$, where $\chi \in C_0^\infty(\mathbb{R})$ and $r(t,x,D_x)$ is a differential operator, should be replaced by $\chi(H)r(t,z,D_z)\chi(H)$.)

We begin with the existence of the *asymptotic velocity* observable and an auxiliary propagation estimate.

PROPOSITION 4.3.1. *Assume that (V1), (SC) hold. Let* $\chi \in C_0^\infty(\mathbb{R})$, $q_a \in C_0^\infty(Z)$, $\operatorname{supp} q_a \subset Z\backslash \bigcup_{b\not\leq a} Z_b$. *Then*

$$\int_1^{+\infty} \| |D_{z_a} - \frac{z_a}{t}|^{\frac{1}{2}} q_a(\frac{z}{t})\chi(H)U(t)u\|^2 \frac{\mathrm{d}t}{t} \leq C\|u\|^2.$$

THEOREM 4.3.2. *Assume (V1), (SC). Then for any* $f \in C_\infty(Z)$ *the limit*

$$\text{s-}\lim_{t\to+\infty} U(t)^* f(\frac{z}{t})U(t) \tag{4.3.1}$$

*exists. There is a unique vector of densely defined commuting self-adjoint operators* $\zeta^+$ *such that (4.3.1) is equal to* $f(\zeta^+)$. *The observable* $\zeta^+$ *will be called the* asymptotic velocity along the field *(associated to the evolution* $U(t)$*); we will usually abbreviate this to* asymptotic velocity. *Moreover, we have*

$$[\zeta^+, H^+] = 0.$$

Proposition 4.3.1 and Theorem 4.3.2 (in the non-magnetic case of course) are due to [**Gr**] and [**De3**] respectively. The proofs are also given in [**De1**] (Proposition 4.6 and Theorem 4.3), and in Section 6.6 of [**DG1**].

We now apply Theorem 4.3.2 to the evolution $\mathrm{e}^{-\mathrm{i}tH}$. As in [**De1**], [**DG1**], we may use the spectral projections of $\zeta^+$ to split up any initial condition $u$

at time $t = 0$ into parts corresponding to different asymptotic configurations along the $z$ axis. Indeed, let

$$T_a := Z_a \backslash \bigcup_{b \not\leq a} Z_b, \text{ for } a \in \mathcal{A}.$$

The family $\{T_a\}_{a \in \mathcal{A}}$ forms a partition of $Z$:

$$\bigcup_{a \in \mathcal{A}} T_a = Z, \ T_a \cap T_b = \emptyset \text{ for } a \neq b,$$

hence

$$\sum_{a \in \mathcal{A}} \mathbf{1}_{T_a}(\zeta^+) = \mathbf{1}.$$

By the definition of $\zeta^+$, $\operatorname{Ran} \mathbf{1}_{T_a}(\zeta^+)$ is the closure of the set of $u \in \mathcal{H}$ for which $\mathrm{e}^{-\mathrm{i}tH}u$ is localized, modulo errors $o(1)$ as $t \to \infty$, in a region of the configuration space where:

- $|z_i - z_j| \geq ct$, $c > 0$, if $i, j$ belong to different clusters of $a$;
- $|z_i - z_j| = o(t)$ if $i, j$ belong to the same cluster.

The first step in our proof of asymptotic completeness is to reduce the problem of studying the asymptotics of the evolution $\mathrm{e}^{-\mathrm{i}tH}$ to a similar problem for a family of time-dependent evolutions $U_{a,0}(t)$ with the above characteristics explicitly built in. The advantage of this reduction is that it removes most of the interactions between different clusters in $a$: the short-range interactions are eliminated altogether, and the long-range interactions are replaced by a time-dependent potential with good decay properties as $t \to \infty$.

For $a \in \mathcal{A}, a \neq a_{\max}$, we define

$$I_a(t,x) := \sum_{(ij) \not\leq a} F(|x_i - x_j| \geq \frac{t}{\log t}) V_{ij}^{\mathrm{l}}(x); \tag{4.3.2}$$

by *(SC)*, $I_a(t,x)$ satisfies (4.1.2). Let $U_{a,0}(t)$ be the unitary dynamics generated by the time-dependent Hamiltonian

$$H_{a,0}(t) = H_a + I_a(t,x),$$

Let $\zeta^+, \zeta^+_{a,0}$ be the asymptotic velocity observables associated to the dynamics $\mathrm{e}^{-\mathrm{i}tH}, U_{a,0}(t)$ respectively.

By Theorem 4.3.2, we may write $\mathrm{e}^{-\mathrm{i}tH}u$ for $u$ in a dense subspace of $\operatorname{Ran} \mathbf{1}_{T_a}(\zeta^+)\mathcal{H}$ as

$$\mathrm{e}^{-\mathrm{i}tH}u = J_a \mathrm{e}^{-\mathrm{i}tH}u + o(1),$$

where $J_a = F(\min_{(ij) \not\leq a} |x_i - x_j| > \epsilon t)$ for some $\epsilon > 0$. Now the heuristic is that

$$(H - H_{a,0}(t))J_a = I_a^{\mathrm{s}} J_a + (I_a^{\mathrm{l}} - I_a(t,x))J_a$$

is integrable in time, since the first term is bounded by $O(t^{-\mu_{\mathrm{s}}})$, $\mu_{\mathrm{s}} > 1$, and the second term vanishes for sufficiently large $t$. By Cook's argument, this implies the existence of the wave operators in Theorem 4.3.3 below. (Since $V_{ij}^{\mathrm{s}}$ are in fact only relatively bounded, to make this heuristic rigorous we must insert additional cutoffs in asymptotic energy. See e.g., the proof of Proposition 6.9.5 in [**DG1**]. Similar arguments will also appear in the sequel, where they will be explained in more detail.)

THEOREM 4.3.3. *Assume (V1), (SC). Then the limits*

$$\mathrm{s-}\lim_{t\to+\infty} U_{a,0}(t)^*\mathrm{e}^{-\mathrm{i}tH}\mathbf{1}_{T_a}(\zeta^+), \tag{4.3.3}$$

$$\mathrm{s-}\lim_{t\to+\infty} \mathrm{e}^{\mathrm{i}tH}U_{a,0}(t)\mathbf{1}_{T_a}(\zeta_{a,0}^+) \tag{4.3.4}$$

*exist. Denote (4.3.3) by $\Gamma_{a,0}^+$. Then $\Gamma_{a,0}^+$ is a partial isometry satisfying*

$$\Gamma_{a,0}^{+*}\Gamma_{a,0}^+ = \mathbf{1}_{T_a}(\zeta^+),\ \Gamma_{a,0}^+\Gamma_{a,0}^{+*} = \mathbf{1}_{T_a}(\zeta_{a,0}^+),$$

$$\Gamma_{a,0}^+\zeta^+ = \zeta_0^+\Gamma_{a,0}^+.$$

*Moreover, (4.3.4) equals $\Gamma_{a,0}^{+*}$.*

In the short-range case ($V_{ij}^{\mathrm{l}} \equiv 0$ in *(SC)*) we have $H_{a,0}(t) = H_a$ and $U_{a,0}(t) = \mathrm{e}^{-\mathrm{i}tH_a}$, and the reader may skip directly to Section 6.3.2, where we explain how to combine this result with the Mourre estimate of Chapter 3 to obtain asymptotic completeness. For long-range interactions, we still have to deal with the time-dependent part $I_a(t,x)$ of the effective Hamiltonians. We will not be able to eliminate that part altogether; we will, however, replace it by $I_a(t,0,z_a)$. The point of this simplification is that since $H^a$ commutes with $I_a(t,0,z_a)$, the "external" motion of clusters is decoupled from the internal motion within them. Theorem 4.3.5 below is a partial result in this direction which takes care of the $z$ variables; the propagation estimates of the next chapter, and the assumption *(C)* in the 3-particle "dispersive" case, will allow us to complete this reduction and replace the $y$ in $I_a$ by 0.

We first state a general result of [**De1**]; recall that $U(t)$ denotes the unitary evolution generated by (4.1.1).

PROPOSITION 4.3.4. *Assume that (V1), (SC), (4.1.2) hold with $\mu_{\mathrm{l}} > \sqrt{3}-1$. For $a \in \mathcal{A}$, let $U_a(t)$ be the unitary evolution generated by $H + I(t,y,z_a)$, and let $\zeta_a^+$ be the asymptotic velocity associated with $U_a(t)$. Then the strong limits*

$$\mathrm{s-}\lim_{t\to+\infty} U_a(t)^*U(t)\mathbf{1}_{T_a}(\zeta^+), \tag{4.3.5}$$

$$\mathrm{s-}\lim_{t\to+\infty} U(t)^*U_a(t)\mathbf{1}_{T_a}(\zeta_a^+) \tag{4.3.6}$$

*exist. Denote (4.3.5) by $\Gamma_a^+$. Then $\Gamma_a^+$ is a partial isometry satisfying*

$$\Gamma_a^{+*}\Gamma_a^+ = \mathbf{1}_{T_a}(\zeta^+),\ \Gamma_a^+\Gamma_a^{+*} = \mathbf{1}_{T_a}(\zeta_a^+),$$

$$\Gamma_a^+\zeta^+ = \zeta_a^+\Gamma_a^+.$$

*Moreover, (4.3.6) is equal to* $\Gamma_a^{+*}$.

The proof may be found in [**De1**] or in [**DG1**], Sections 6.13 and 6.14. It relies on a propagation estimate showing that for states in $\mathrm{Ran}\,\mathbf{1}_{T_a}(\zeta^+)$, the size of each cluster in $a$ is bounded by $C\langle t\rangle^\alpha$ for some $0 \le \alpha < 1$ (depending on the rate of decay of pair potentials), which improves on the bound $o(t)$ we mentioned earlier.

We now apply Proposition 4.3.4 to the dynamics $U_{a,0}(t)$ defined above. Let $U_{a,1}(t)$ be the dynamics generated by the time-dependent Hamiltonian

$$H_{a,1}(t) = H_a + I_a(t, y, z_a).$$

THEOREM 4.3.5. *Assume that (V1), (SC) hold with* $\mu_l > \sqrt{3} - 1$, *and let* $\zeta_{a,1}^+$ *be the asymptotic velocity observable associated to* $U_{a,1}(t)$. *Then the limits*

$$\text{s-}\lim_{t\to+\infty} U_{a,1}(t)^* \mathrm{e}^{-\mathrm{i}tH} \mathbf{1}_{T_a}(\zeta^+), \tag{4.3.7}$$

$$\text{s-}\lim_{t\to+\infty} \mathrm{e}^{\mathrm{i}tH} U_{a,1}(t) \mathbf{1}_{T_a}(\zeta_{a,1}^+) \tag{4.3.8}$$

*exist. Denote (4.3.7) by* $\Gamma_{a,1}^+$. *Then* $\Gamma_{a,1}^+$ *is a partial isometry satisfying*

$$\Gamma_{a,1}^{+*}\Gamma_{a,1}^{+} = \mathbf{1}_{T_a}(\zeta^+),\ \Gamma_{a,1}^{+}\Gamma_{a,1}^{+*} = \mathbf{1}_{T_a}(\zeta_{a,1}^+),$$

$$\Gamma_{a,1}^{+}\zeta^+ = \zeta_{a,1}^+\Gamma_{a,1}^{+}.$$

*Moreover, (4.3.8) equals* $\Gamma_{a,1}^{+*}$.

## 4.4. Minimal velocity estimates

In this section we will derive an important consequence of the Mourre estimate: the *minimal velocity estimate*. The Mourre estimate, we recall, asserts that for a suitably chosen conjugate operator $A$ the inequality

$$[H, \mathrm{i}A] > c(\lambda) > 0$$

holds at all energies $\lambda$ away from a certain closed and discrete set. It follows trivially that the expected value of $A$ increases at least linearly in $t$ along the evolution. With more effort, however, one can prove a considerably stronger statement, namely that almost all propagation takes place in the region $\langle A\rangle \ge \epsilon t$ for some $\epsilon > 0$:

$$\lim_{t\to\infty} \|F(\frac{\langle A\rangle}{t} \le \epsilon)\mathrm{e}^{-\mathrm{i}tH}u\| = 0$$

and

$$\int_1^\infty \|F(\frac{\langle A\rangle}{t} \le \epsilon)\mathrm{e}^{-\mathrm{i}tH}u\|^2 \frac{\mathrm{d}t}{t} \le C\|u\|^2$$

for $u$ in a dense subset of $\mathcal{H}_{\mathrm{scatt}}$. We furthermore recall that our conjugate operators were chosen so that $\langle x^{\mathrm{cm}}\rangle^{-1}A(H+\mathrm{i})^{-1}$ is bounded; this will allow us to convert estimates on $A$ from below to similar estimates for $\langle x^{\mathrm{cm}}\rangle$. Informally, we will prove that all states in $\mathcal{H}_{\mathrm{scatt}}$ describe a breakup into at least two clusters moving apart with relative velocity bounded from below by $\epsilon > 0$.

In Subsection 4.4.1 we prove an abstract minimal velocity estimate suitable for our purposes. We then apply it in Subsection 4.4.2 to 3-particle magnetic Hamiltonians with neutral pairs, both in 2 and 3 space dimensions. The charged case is identical, modulo notational changes, to the non-magnetic $N$-body case and therefore is omitted here.

**4.4.1. Abstract estimates.** We begin with an abstract minimal velocity estimate which was first proved (under somewhat stronger assumptions) by Sigal and Soffer [**SS3**]; we formulate it in a more general setting.

PROPOSITION 4.4.1. *Let $H, A, P$ be self-adjoint operators on a Hilbert space $\mathcal{H}$. Let $I(t)$, $t \in \mathbb{R}$, be a family of self-adjoint operators bounded uniformly in $t$, and let $U(t,s)$ be the unitary propagator generated by $H(t) = H + I(t)$, defined by (4.1.4). Assume that:*

*(i) $H \in C^{1+\epsilon}(A)$ for some $\epsilon > 0$;*

*(ii) the commutator $[A, I(t)]$, defined as a quadratic form on $\mathcal{D}(A)$, extends to a bounded operator on $\mathcal{H}$ of norm $O(t^{-\epsilon})$;*

*(iii) $[(H+\mathrm{i})^{-1}, I(t)] = O(t^{-1-\epsilon})$;*

*(iv) $P$ is bounded, $[P, (H+\mathrm{i})^{-1}] = 0$, and $[P, I(t)] = O(t^{-1-\epsilon})$.*

*Let $\Delta \subset \mathbb{R}$ be an interval such that*

$$P\mathbf{1}_{\Delta}(H)[H, \mathrm{i}A]\mathbf{1}_{\Delta}(H)P \geq c_0 P\mathbf{1}_{\Delta}(H)P. \tag{4.4.1}$$

*Then for any $g \in C_0^\infty(\mathbb{R})$, $\operatorname{supp} g \subset (-\infty, c_0)$ and any $\chi \in C_0^\infty(\mathrm{Int}\Delta)$ we have:*

$$\int_1^{+\infty} \|g(\frac{A}{t})\chi(H)PU(t)u\|^2 \frac{\mathrm{d}t}{t} \leq C\|u\|^2, \tag{4.4.2}$$

$$\mathrm{s-}\lim_{t\to+\infty} g(\frac{A}{t})\chi(H)PU(t) = 0. \tag{4.4.3}$$

REMARK 4.4.2. The usual abstract minimal velocity estimate (cf. [**SS3**]) is obtained by taking $P = \mathbf{1}$. The generalization stated here corresponds to the case of a Hamiltonian $H$ commuting with another observable $K$ and satisfying a strict Mourre estimate only locally with respect to $H, K$; the operator $P$ will be a spectral projection of $K$.

REMARK 4.4.3. Instead of *(i)*, one may assume that $H \in C^1(A)$ and that the function $s \mapsto \mathrm{e}^{\mathrm{i}sA}[(H+\mathrm{i})^{-1}, A]\mathrm{e}^{-\mathrm{i}sA}$ is Dini continuous in the norm topology of $B(\mathcal{H})$ [**DG4**].

PROOF. The proof will be based on Lemma 4.1.4. We first construct a suitable propagation observable. Let $g \in C_0^\infty((-\infty, c_1))$, where $c_1 < c_0$. We set

$$F(s) = \int_s^{+\infty} g^2(s_1)\mathrm{d}s_1,$$

$$\Phi(t) = P\chi(H)F\left(\frac{A}{t}\right)\chi(H)P.$$

The Heisenberg derivative of $\Phi(t)$ is

$$\begin{aligned}
&\mathbf{D}\Phi(t) \\
&= P\chi(H)\left(\left[H, \mathrm{i}F\left(\tfrac{A}{t}\right)\right] - \tfrac{A}{t^2}g^2\left(\tfrac{A}{t}\right)\right)\chi(H)P \\
&\quad + \left[I(t), \mathrm{i}P\chi(H)F\left(\tfrac{A}{t}\right)\chi(H)P\right] \\
&= P\chi(H)\left(\left[H, \mathrm{i}F\left(\tfrac{A}{t}\right)\right] + \left[I(t), \mathrm{i}F\left(\tfrac{A}{t}\right)\right] - \tfrac{A}{t^2}g^2\left(\tfrac{A}{t}\right)\right)\chi(H)P \\
&\quad + \Big(P\left[I(t), \mathrm{i}\chi(H)\right]F\left(\tfrac{A}{t}\right)\chi(H)P + \mathrm{h.c.}\Big) \\
&\quad + \Big([I(t), \mathrm{i}P]\chi(H)F\left(\tfrac{A}{t}\right)\chi(H)P + \mathrm{h.c.}\Big);
\end{aligned} \tag{4.4.4}$$

the last two terms are $O(t^{-1-\epsilon})$ by *(iii)*, (1.2.21), and *(iv)*.

Next, we claim that for any $F \in C^\infty(\mathbb{R})$ such that $F' \in C_0^\infty(\mathbb{R})$, and for any bounded operator $B$, the following identity holds:

$$[B, \mathrm{i}F(A)] = \frac{1}{2\pi}\int_{\mathbb{R}} \widehat{F'}(\sigma)\mathrm{e}^{\mathrm{i}\sigma A}\int_0^1 \mathrm{e}^{-\mathrm{i}\sigma\theta A}[B, \mathrm{i}A]\mathrm{e}^{\mathrm{i}\sigma\theta A}\mathrm{d}\theta\mathrm{d}\sigma, \tag{4.4.5}$$

in the sense of quadratic forms on $\mathcal{D}(A)$. Indeed, this may be proved by approximating $F$ by functions in $C_0^\infty(\mathbb{R})$; the details are left to the reader. From (4.4.5) we obtain that for any family of bounded operators $B(t)$

$$\begin{aligned}
&[B(t), \mathrm{i}F(\tfrac{A}{t})] = \tfrac{1}{2\pi t}\int_{\mathbb{R}} \widehat{F'}(\sigma)\mathrm{e}^{\mathrm{i}\sigma\frac{A}{t}}\int_0^1 \mathrm{e}^{-\mathrm{i}\sigma\theta\frac{A}{t}}[B(t), \mathrm{i}A]\mathrm{e}^{\mathrm{i}\sigma\theta\frac{A}{t}}\mathrm{d}\sigma\mathrm{d}\theta \\
&= \tfrac{1}{2\pi t}\int_{\mathbb{R}^2\times[0,1]} \hat{g}(\sigma - \sigma')\hat{g}(\sigma')\mathrm{e}^{\mathrm{i}\sigma(1-\theta)\frac{A}{t}}[B(t), \mathrm{i}A]\mathrm{e}^{\mathrm{i}\sigma\theta\frac{A}{t}}\mathrm{d}\sigma\mathrm{d}\sigma'\mathrm{d}\theta \\
&= \tfrac{1}{2\pi t}\int_{\mathbb{R}^2\times[0,1]} \hat{g}(\sigma)\hat{g}(\sigma')\mathrm{e}^{\mathrm{i}(\sigma+\sigma')(1-\theta)\frac{A}{t}}[B(t), \mathrm{i}A]\mathrm{e}^{\mathrm{i}(\sigma+\sigma')\theta\frac{A}{t}}\mathrm{d}\sigma\mathrm{d}\sigma'\mathrm{d}\theta.
\end{aligned} \tag{4.4.6}$$

From (4.4.6) with $B(t) = I(t)$ and from *(ii)*, we obtain

$$[I(t), \mathrm{i}F(\frac{A}{t})] = O(t^{-1-\epsilon}). \tag{4.4.7}$$

Next, we estimate $[H, \mathrm{i}F(\frac{A}{t})]$. We will again use (4.4.6). By *(i)*, the function

$$\mathbb{R} \ni s \mapsto \Psi(s) := \mathrm{e}^{\mathrm{i}sA}[(H+\mathrm{i})^{-1}, \mathrm{i}A]\mathrm{e}^{-\mathrm{i}sA} \in B(\mathcal{H})$$

is in $C^\epsilon$ for some $\epsilon > 0$. Hence

$$\begin{aligned}
&\mathrm{e}^{\mathrm{i}(\sigma+\sigma')(1-\theta)\frac{A}{t}}\Psi(0)\mathrm{e}^{\mathrm{i}(\sigma+\sigma')\theta\frac{A}{t}} = \mathrm{e}^{\mathrm{i}\sigma\frac{A}{t}}\Psi(\tfrac{-\theta\sigma+(1-\theta)\sigma'}{t})\mathrm{e}^{\mathrm{i}\sigma'\frac{A}{t}} \\
&= \mathrm{e}^{\mathrm{i}\sigma\frac{A}{t}}\Psi(0)\mathrm{e}^{\mathrm{i}\sigma'\frac{A}{t}} + O((\tfrac{-\theta\sigma+(1-\theta)\sigma'}{t})^\epsilon).
\end{aligned}$$

It then follows from (4.4.6) (with $B(t) \equiv (H+\mathrm{i})^{-1}$) that

$$[(H+\mathrm{i})^{-1}, \mathrm{i}F(\frac{A}{t})] = \frac{1}{t}g(\frac{A}{t})[(H+\mathrm{i})^{-1}, \mathrm{i}A]g(\frac{A}{t}) + O(t^{-1-\epsilon}).$$

We rewrite it as

$$(4.4.8)\qquad \begin{aligned} &(H+\mathrm{i})^{-1}[H,\mathrm{i}F(\tfrac{A}{t})](H+\mathrm{i})^{-1} \\ = \;& \tfrac{1}{t}g(\tfrac{A}{t})(H+\mathrm{i})^{-1}[H,\mathrm{i}A](H+\mathrm{i})^{-1}g(\tfrac{A}{t}) + O(t^{-1-\epsilon}). \end{aligned}$$

We will also need the estimate

$$(4.4.9)\qquad [g(\frac{A}{t}),h(H)] = O(t^{-1}),\ h\in C_0^\infty(\mathbb{R}),$$

which is easily checked using (1.2.21) and (4.4.6).

We multiply (4.4.8) from both sides by $h(H)P$, where

$$h(H) = \chi(H)(H+\mathrm{i}).$$

Pick $\chi_1 \in C_0^\infty(\Delta)$ such that $\chi_1\chi = \chi$. From (4.4.1) and (4.4.9) we have

$$(4.4.10)\qquad \begin{aligned} &\chi(H)P[H,\mathrm{i}F(\tfrac{A}{t})]P\chi(H) \\ = \;& h(H)(H+\mathrm{i})^{-1}P[H,\mathrm{i}F(\tfrac{A}{t})]P(H+\mathrm{i})^{-1}h(H) \\ = \;& \tfrac{1}{t}\chi(H)g(\tfrac{A}{t})P\chi_1(H)[H,\mathrm{i}A]\chi_1(H)Pg(\tfrac{A}{t})\chi(H) + O(t^{-1-\epsilon}) \\ \geq \;& \tfrac{c_0}{t}P\chi(H)g^2(\tfrac{A}{t})\chi(H)P + O(t^{-1-\epsilon}). \end{aligned}$$

Finally, since $g$ is supported in $(-\infty, c_1)$, we have the trivial estimate

$$(4.4.11)\qquad -P\chi(H)\frac{A}{t^2}g^2(\frac{A}{t})\chi(H)P \geq -\frac{c_1}{t}P\chi(H)g^2(\frac{A}{t})\chi(H)P.$$

Combining (4.4.4) (and the estimate on the last two terms which follows it), (4.4.7), (4.4.10), (4.4.11), we obtain that

$$\mathbf{D}\Phi(t) \geq (c_0 - c_1)\frac{1}{t}P\chi(H)g^2(\frac{A}{t})\chi(H)P + O(t^{-1-\epsilon}).$$

By Lemma 4.1.4, this proves (4.4.2).

To prove (4.4.3), we first verify that

$$\begin{aligned} &\mathbf{D}P\chi(H)g^2(\tfrac{A}{t})\chi(H)P \\ = \;& -\tfrac{1}{t}P\chi(H)\phi(\tfrac{A}{t})\chi(H)P + O(t^{-1-\epsilon}), \end{aligned}$$

where $\phi(s) = s\frac{d}{ds}g^2(s)$. This is done by the same arguments as above; the details are left to the reader. Hence by (4.4.2) and Proposition 4.1.3:

$$\int_1^\infty \frac{d}{dt}\langle U(t)u, P\chi(H)g^2(\frac{A}{t})\chi(H)PU(t)u\rangle dt < \infty,$$

and the limit in (4.4.3) exists. Finally, it follows from (4.4.2) that this limit must be 0. □

We have proved that if the Mourre estimate holds (on a given energy interval) with the conjugate operator $A$, then the propagation takes place in the region of phase space in which $A \geq c_0 t$ for some $c_0 > 0$. For the purpose of proving asymptotic completeness, however, we will need similar localization of the propagation in the $x$ variables. The next few lemmas will allow us to replace $A$ in Proposition 4.4.1 by another observable $B$ (sometimes called a *gauge*) which "dominates" $A$ in the sense that $-cB \leq A \leq cB$ for some $c > 0$. We will later apply these lemmas with $B = \langle x^{\mathrm{cm}} \rangle$.

We will need the following result of [**ABG**] (see [**ABG**, Lemma 7.2.15, Proposition 7.2.16]).

LEMMA 4.4.4. *Let $H, A$ be two self-adjoint operators on $\mathcal{H}$ such that $H \in C^1(A)$, and let $\chi \in C_0^\infty(\mathbb{R})$. Then:*

*(i) $\chi(H)$ preserves $\mathcal{D}(A)$;*

*(ii) the operator $A_\chi = \chi(H)A\chi(H)$ is essentially self-adjoint on $\mathcal{D}(A)$;*

*(iii) $H \in C^1(A_\chi)$;*

*(iv) $[H, \mathrm{i}A_\chi] = \chi(H)[H, \mathrm{i}A]\chi(H)$.*

If $H, A$ are as in Lemma 4.4.4, the closure of the operator $\chi(H)A\chi(H)$ on $\mathcal{D}(A)$ will be again denoted by $\chi(H)A\chi(H)$.

LEMMA 4.4.5. *Let $A, B$ be two self-adjoint operators on a Hilbert space $\mathcal{H}$. Assume that:*

*(i) $B \geq 1/2$ and $-cB \leq A \leq cB$,*

*(ii) $B \in C^1(A)$,*

*and that one of the following holds:*

*(iii) $[B, A]B^{-1}$ is bounded,*

*(iii') for any $f \in C_0^\infty(\mathbb{R})$, $[f(\frac{B}{t}), A] = O(1)$ as $t \to \infty$.*

*Then for any $c_0 > 0$ there is an $\epsilon_0 > 0$ such that*

$$F\left(\frac{B}{t} \leq \epsilon_0\right) F\left(\frac{|A|}{t} \geq c_0\right) = O(t^{-1}).$$

PROOF. We first claim that *(i), (ii), (iii)* imply *(iii')*. By (1.2.21), we have:

$$\begin{aligned} &[f(\tfrac{B}{t}), A] \\ &= \tfrac{\mathrm{i}}{2\pi} \textstyle\int_{\mathbb{C}} \partial_{\bar{z}} \tilde{f}(z)[(z - \tfrac{B}{t})^{-1}, A] dz \wedge d\bar{z} \\ &= \tfrac{\mathrm{i}}{2\pi} \textstyle\int_{\mathbb{C}} \partial_{\bar{z}} \tilde{f}(z)(z - \tfrac{B}{t})^{-1}[\tfrac{B}{t}, A](z - \tfrac{B}{t})^{-1} dz \wedge d\bar{z} \\ &= \tfrac{\mathrm{i}}{2\pi} \textstyle\int_{\mathbb{C}} \partial_{\bar{z}} \tilde{f}(z)(z - \tfrac{B}{t})^{-1}[\tfrac{B}{t}, A]B^{-1}B(z - \tfrac{B}{t})^{-1} dz \wedge d\bar{z}. \end{aligned} \tag{4.4.12}$$

where $\tilde{f}$ is an almost analytic extension of $f$ supported in in $|\mathrm{Re}\, z| \leq C$, $|\mathrm{Im}\, z| \leq 1$. (Expanding the commutator $[(z - \frac{B}{t})^{-1}, A]$ in the above equation is justified by *(ii)*.) Hence *(iii')* will follow if we verify that

$$(z - \frac{B}{t})^{-1} = O(|\mathrm{Im}\, z|^{-1}), \ B(z - \frac{B}{t})^{-1} = O(\langle t \rangle \langle z \rangle |\mathrm{Im}\, z|^{-1}). \tag{4.4.13}$$

The first of the above estimates is trivial. To prove the second one, it suffices to check that for $\lambda \in \mathbb{R}$:

$$|\lambda(z - t^{-1}\lambda)^{-1}| \leq C\langle t \rangle \langle z \rangle |\mathrm{Im}\, z|^{-1},$$

or equivalently:

$$|\lambda^{-1}(xt+\mathrm{i}yt)-1|\geq C^{-1}\langle z\rangle^{-1}|y|,$$

where $z=x+\mathrm{i}y$, $x,y\in\mathbb{R}$. If $\lambda\geq 2\langle z\rangle t$, this follows from

$$\left|\frac{xt-\lambda}{\lambda}\right|\geq\left|\frac{\lambda/2}{\lambda}\right|\geq\frac{1}{2}\geq\frac{1}{2}\frac{|y|}{\langle z\rangle},$$

and if $\lambda\leq 2\langle z\rangle t$ – from

$$\left|\frac{yt}{\lambda}\right|\geq\left|\frac{yt}{2\langle z\rangle t}\right|=\left|\frac{y}{2\langle z\rangle}\right|.$$

This ends the proof of (4.4.13) and of *(iii')*.

It remains to prove the lemma under the assumptions *(i)*, *(ii)*, *(iii')*. We will prove that

$$F(\frac{B}{t}\leq\epsilon_0)F(\frac{A}{t}\geq c_0)=O(t^{-1})$$

for $\epsilon_0$ small enough; applying this to $A$ and $-A$, we obtain the lemma. We will denote:

$$R(t)=F(\frac{B}{t}\leq\epsilon_0),\ R_1(t)=F(\frac{B}{t}\leq 2\epsilon_0).$$

Observe that since $B\geq 1/2$, $R(t)$ and $R_1(t)$ are of the form $f(\frac{B}{t})$ for $f\in C_0^\infty(\mathbb{R})$. By *(ii)* and Lemma 4.4.4, $R_1(t)$ preserves $\mathcal{D}(A)$ and the operator

$$A_1(t):=R_1(t)AR_1(t)$$

is essentially self-adjoint on $\mathcal{D}(A)$.

Fix $c_0>0$. For $\epsilon_0$ small enough, we obtain from *(i)* that

$$\pm A_1(t)\leq cR_1(t)BR_1(t)\leq\frac{1}{2}c_0t, \tag{4.4.14}$$

hence

$$F(\frac{A_1(t)}{t}\geq c_0)=0.$$

Thus to prove the lemma it suffices to show that

$$R(t)\left(F(\frac{A}{t}\geq c_0)-F(\frac{A_1(t)}{t}\geq c_0)\right)=O(t^{-1}). \tag{4.4.15}$$

We may write $F(s)=(s+\mathrm{i})F_{-1}(s)$, where $F_{-1}\in S_{\mathrm{cl}}^{-1}(\mathbb{R})$. Then

$$\begin{aligned}&R(t)\left(F(\tfrac{A}{t}\geq c_0)-F(\tfrac{A_1(t)}{t}\geq c_0)\right)\\ =\ &R(t)\left(\tfrac{A}{t}-\tfrac{A_1(t)}{t}\right)F_{-1}(\tfrac{A}{t})\\ &+R(t)(\tfrac{A_1(t)}{t}+\mathrm{i})\left(F_{-1}(\tfrac{A}{t})-F_{-1}(\tfrac{A_1(t)}{t})\right).\end{aligned} \tag{4.4.16}$$

Using *(iii')* with $f(\frac{B}{t}) = R_1(t)$, we estimate

$$
\begin{aligned}
& R(t)(\tfrac{A}{t} - \tfrac{A_1(t)}{t}) \\
= \;& R(t)R_1(t)\tfrac{A}{t} - R(t)R_1(t)\tfrac{A}{t}R_1(t) \\
= \;& \tfrac{1}{t}R(t)[R_1(t), A] = O(t^{-1}).
\end{aligned} \tag{4.4.17}
$$

Hence the first term on the right-hand side of (4.4.16) is $O(t^{-1})$. Using again *(iii')*, we rewrite the second term as

$$(\frac{A_1(t)}{t} + \mathrm{i})R(t)\left(F_{-1}(\frac{A}{t}) - F_{-1}(\frac{A_1(t)}{t})\right) + O(t^{-1}).$$

By Proposition 1.2.9, we may find an almost-analytic extension $\tilde{F}_{-1}$ of $F_{-1}$ satisfying

$$|\partial^{\alpha}_{\bar{z},z}\partial_{\bar{z}}\tilde{F}_{-1}(z)| \leq C_{N,\alpha}|\mathrm{Im}\, z|^N \langle z\rangle^{-2-|\alpha|-N}, \quad \forall N \in \mathbb{N}, \alpha \in \mathbb{N}^2. \tag{4.4.18}$$

By (4.4.14), it suffices to prove that

$$
\begin{aligned}
& R(t)\left(F_{-1}(\tfrac{A}{t}) - F_{-1}(\tfrac{A_1(t)}{t})\right) \\
= \;& \tfrac{\mathrm{i}}{2\pi}\int_{\mathbb{C}} \partial_{\bar{z}}\tilde{F}_{-1}(z)R(t)\left((z - \tfrac{A}{t})^{-1} - (z - \tfrac{A_1(t)}{t})^{-1}\right) dz \wedge d\bar{z} \\
= \;& \tfrac{\mathrm{i}}{2\pi}\int_{\mathbb{C}} \partial_{\bar{z}}\tilde{F}_{-1}(z)R(t)(z - \tfrac{A}{t})^{-1}(\tfrac{A}{t} - \tfrac{A_1(t)}{t})(z - \tfrac{A_1(t)}{t})^{-1} dz \wedge d\bar{z} \\
= \;& O(t^{-1}).
\end{aligned}
$$

But this follows from (4.4.18), the identity

$$R(t)(z - \frac{A}{t})^{-1} = (z - \frac{A}{t})^{-1}R(t) + (z - \frac{A}{t})^{-1}[R(t), \frac{A}{t}](z - \frac{A}{t})^{-1},$$

and *(iii')*. This completes the proof of the lemma. □

We will also need an extension of Lemma 4.4.5, stated in Lemma 4.4.8, to the case when the gauge observable $B$ dominates only a localized version of $A$, namely $\chi_1(H)A\chi_1(H)$ for some $\chi_1 \in C_0^\infty(\mathbb{R})$. We first prove two auxiliary technical lemmas.

LEMMA 4.4.6. *Let $H, A$ be self-adjoint operators on $\mathcal{H}$ such that $H \in C^1(A)$. Let $\chi, \chi_1 \in C_0^\infty(\mathbb{R})$ satisfy $\chi_1\chi = \chi$, and let $F \in S^0(\mathbb{R})$. Then*

$$\chi(H)\left(F(\frac{A}{t}) - F(\frac{\chi_1(H)A\chi_1(H)}{t})\right) = O(t^{-1}).$$

PROOF. By Proposition 3.2.3*(iv)* we have

$$[A, f(H)] \text{ is bounded for } f \in C_0^\infty(\mathbb{R}). \tag{4.4.19}$$

As in the proof of Lemma 4.4.5, we write $F(s) = (s+\mathrm{i})F_{-1}(s)$ for $F_{-1} \in S^{-1}_{\mathrm{cl}}(\mathbb{R})$, and denote by $\tilde{F}_{-1}$ an almost analytic extension of $F_{-1}$ satisfying (4.4.18). Let $A_1 = \chi_1(H)A\chi_1(H)$. Then

$$\begin{aligned}\chi(H)(F(\tfrac{A}{t}) - F(\tfrac{A_1}{t})) &= \chi(H)(\tfrac{A}{t} - \tfrac{A_1}{t})F_{-1}(\tfrac{A}{t}) \\ &+\chi(H)(\tfrac{A_1}{t}+\mathrm{i})(F_{-1}(\tfrac{A}{t}) - F_{-1}(\tfrac{A_1}{t})).\end{aligned} \tag{4.4.20}$$

We have $\chi(H)(A - A_1) = \chi(H)\chi_1(H)[\chi_1(H), A]$ in the sense of quadratic forms on $\mathcal{D}(A)$, hence by (4.4.19)

$$\chi(H)(\frac{A}{t} - \frac{A_1}{t}) = O(t^{-1}). \tag{4.4.21}$$

It remains to estimate the second term on the right-hand side of (4.4.20). By (4.4.19),

$$\begin{aligned}&\chi(H)(\tfrac{A_1}{t}+\mathrm{i})(F_{-1}(\tfrac{A}{t}) - F_{-1}(\tfrac{A_1}{t})) \\ =\ &(\tfrac{A_1}{t}+\mathrm{i})\chi(H)(F_{-1}(\tfrac{A}{t}) - F_{-1}(\tfrac{A_1}{t})) + O(t^{-1}),\end{aligned} \tag{4.4.22}$$

Furthermore,

$$\begin{aligned}&(A+\mathrm{i})\chi(H)(F_{-1}(\tfrac{A}{t}) - F_{-1}(\tfrac{A_1}{t})) \\ =\ &\tfrac{\mathrm{i}}{2\pi}\int_{\mathbb{C}} \partial_{\bar{z}}\tilde{F}_{-1}(z)(A+\mathrm{i})\chi(H)\left((z-\tfrac{A}{t})^{-1} - (z-\tfrac{A_1}{t})^{-1}\right)dz\wedge d\bar{z} \\ =\ &\tfrac{\mathrm{i}}{2\pi}\int_{\mathbb{C}} \partial_{\bar{z}}\tilde{F}_{-1}(z)(A+\mathrm{i})\chi(H)(z-\tfrac{A}{t})^{-1}(\tfrac{A}{t} - \tfrac{A_1}{t})(z-\tfrac{A_1}{t})^{-1}dz\wedge d\bar{z} \\ =\ &\tfrac{\mathrm{i}}{2\pi}\int_{\mathbb{C}} \partial_{\bar{z}}\tilde{F}_{-1}(z)(A+\mathrm{i})(z-\tfrac{A}{t})^{-1}\chi(H)(\tfrac{A}{t} - \tfrac{A_1}{t})(z-\tfrac{A_1}{t})^{-1}dz\wedge d\bar{z} \\ &+\tfrac{\mathrm{i}}{2\pi}\int_{\mathbb{C}} \partial_{\bar{z}}\tilde{F}_{-1}(z)(A+\mathrm{i})(z-\tfrac{A}{t})^{-1}[\tfrac{A}{t}, \chi(H)] \\ &\times(z-\tfrac{A}{t})^{-1}(\tfrac{A}{t} - \tfrac{A_1}{t})(z-\tfrac{A_1}{t})^{-1}dz\wedge d\bar{z}.\end{aligned}$$

Using also (4.4.13), (4.4.19), (4.4.18), (4.4.21), we obtain that

$$(A+\mathrm{i})\chi(H)(F_{-1}(\frac{A}{t}) - F_{-1}(\frac{A_1}{t})) = O(1). \tag{4.4.23}$$

Similarly,

$$\chi(H)(F_{-1}(\frac{A}{t}) - F_{-1}(\frac{A_1}{t})) = O(t^{-1}). \tag{4.4.24}$$

Since $\chi(H)$ preserves $\mathcal{D}(A)$, $A_1(A+\mathrm{i})^{-1}$ is bounded. Hence from (4.4.23), (4.4.24) we have

$$(\frac{A_1}{t}+\mathrm{i})\chi(H)(F_{-1}(\frac{A}{t}) - F_{-1}(\frac{A_1}{t})) = O(t^{-1}).$$

This proves that (4.4.20) is bounded by $O(t^{-1})$, and concludes the proof of the lemma. □

LEMMA 4.4.7. *Let $H, A, B$ be self-adjoint operators on $\mathcal{H}$, $\chi \in C_0^\infty(\mathbb{R})$. Assume that:*

*(i) $B \geq 1/2$, $H \in C^1(A)\cap C^1(B)$, $B \in C^1(A)$,*

*(ii)* $(B+1)^{-1}A\chi(H)$ *extends from* $\mathcal{D}(A)$ *to a bounded operator on* $\mathcal{H}$.

*Let* $A_\chi = \chi(H)A\chi(H)$ *be the self-adjoint operator constructed in Lemma 4.4.4. Then* $B \in C^1(A_\chi)$.

PROOF. By Proposition 3.2.1 it suffices to prove that

$$|((B+1)^{-1}u, A_\chi u) - (A_\chi u, (B+1)^{-1}u)| \leq C\|u\|^2, \tag{4.4.25}$$

for all $u \in \mathcal{D}(A_\chi)$. Furthermore, since by Lemma 4.4.4 $\mathcal{D}(A)$ is a core for $A_\chi$, we only need to prove (4.4.25) for $\mathcal{D}(A)$. We have the operator identity:

$$[\chi(H), (B+1)^{-1}] = (B+1)^{-1}[B, \chi(H)](B+1)^{-1}. \tag{4.4.26}$$

Since $H \in C^1(B)$, by Proposition 3.2.3*(iv)* $[\chi(H), B]$ is bounded. From (4.4.26) we have

$$\begin{aligned} & ((B+1)^{-1}u, \chi(H)A\chi(H)u) - (\chi(H)A\chi(H)u, (B+1)^{-1}u) \\ = \; & ((B+1)^{-1}\chi(H)u, A\chi(H)u) - (A\chi(H)u, (B+1)^{-1}\chi(H)u) \\ & +((B+1)^{-1}[\chi(H), B](B+1)^{-1}u, A\chi(H)u) \\ & +(A\chi(H)u, (B+1)^{-1}[\chi(H), B](B+1)^{-1}u). \end{aligned}$$

Since $B \in C^1(A)$ and $\chi(H)$ preserves $\mathcal{D}(A)$, the first term on the right-hand side is bounded by $C\|\chi(H)u\|^2$. Since $(B+1)^{-1}A\chi(H)$ is bounded, the second and third term are bounded by $C\|u\|^2$. Thus (4.4.25) holds, and the lemma has been proved. □

LEMMA 4.4.8. *Let* $H, A, B$ *be self-adjoint operators on* $\mathcal{H}$, *and let* $\Delta \subset \mathbb{R}$ *be an interval. Assume that:*

*(i)* $H \in C^1(A) \cap C^1(B)$, $B \in C^1(A)$,

*(ii)* $[B, A](B+\mathrm{i})^{-1}$ *and* $(B+i)^{-1}A(H+\mathrm{i})^{-1}$ *are bounded,*

*(iii)* $B \geq 1/2$,

*(iv)* $-cB \leq \chi_1(H)A\chi_1(H) \leq cB$ *for some* $\chi_1 \in C_0^\infty(\mathbb{R})$, $\chi_1 \equiv 1$ *on* $\Delta$.

*Then for any* $c_0 > 0$ *there is an* $\epsilon_0 > 0$ *such that for all* $\chi \in C_0^\infty(\Delta)$:

$$F(\frac{B}{t} \leq \epsilon_0)\chi(H)F(\frac{|A|}{t} \geq c_0) = O(t^{-1}).$$

PROOF. Let $\chi_1$ be as in *(iv)*, and let $A_1$ be the operator $\chi_1(H)A\chi_1(H)$ constructed in Lemma 4.4.4. By Lemma 4.4.6, we have

$$\chi(H)F(\frac{|A|}{t} \geq c_0) = \chi(H)F(\frac{|A_1|}{t} \geq c_0) + O(t^{-1}). \tag{4.4.27}$$

Next, we will prove that

$$\begin{aligned} & F(\tfrac{B}{t} \leq \epsilon_0)\chi(H)F(\tfrac{|A|}{t} \geq c_0) \\ = \; & \chi(H)F(\tfrac{B}{t} \leq \epsilon_0)F(\tfrac{|A_1|}{t} \geq c_0) + O(t^{-1}). \end{aligned} \tag{4.4.28}$$

Indeed, by *(iii)* we may write $F(\frac{B}{t} \leq \epsilon_0)$ as $g(\frac{B}{t})$ for some $g \in C_0^\infty(\mathbb{R})$. Hence (4.4.28) will follow if we prove that

$$[f(H), g(\frac{B}{t})] \in O(t^{-1}),\ f, g \in C_0^\infty(\mathbb{R}). \tag{4.4.29}$$

Let $\tilde{g}$ be an almost analytic extension of $g$, then

$$\begin{aligned} & [f(H), g(\tfrac{B}{t})] \\ = & \tfrac{\mathrm{i}}{2\pi} \int_{\mathbb{C}} \partial_{\bar{z}} \tilde{g}(z) [f(H), (z - \tfrac{B}{t})^{-1}] dz \wedge d\bar{z} \\ = & \tfrac{\mathrm{i}}{2\pi} \int_{\mathbb{C}} \partial_{\bar{z}} \tilde{g}(z) (z - \tfrac{B}{t})^{-1} [f(H), \tfrac{B}{t}] (z - \tfrac{B}{t})^{-1} dz \wedge d\bar{z} \\ = & O(t^{-1}), \end{aligned}$$

by (4.4.19). Note that the expansion of the commutator above is justified since $f(H)$ is bounded.

It remains to show that there is an $\epsilon_0$ such that

$$F(\frac{B}{t} \leq \epsilon_0) F(\frac{|A_1|}{t} \geq c_0) = O(t^{-1}); \tag{4.4.30}$$

the lemma will then follow from (4.4.27), (4.4.28), and (4.4.30).

To prove (4.4.30), it suffices to verify that $B$, $A_1$ satisfy the assumptions *(i), (ii), (iii')* of Lemma 4.4.5. Indeed, *(i)* follows from the hypotheses *(iii)–(iv)* above, and by Lemma 4.4.7:

$$B \in C^1(A_1), \tag{4.4.31}$$

which is *(ii)*.

It remains to prove that *(iii')* of Lemma 4.4.5 holds. Let $f \in C_0^\infty(\mathbb{R})$. By (4.4.31), $f(\frac{B}{t})$ preserves the domain of $A_1$, hence $[f(\frac{B}{t}), A_1]$ is defined (for each $t$) as a bounded operator from $\mathcal{D}(A_1)$ to $\mathcal{H}$. We need to prove that

$$[f(\frac{B}{t}), A_1] = O(1),\ t \to \infty. \tag{4.4.32}$$

By Lemma 4.4.4, $\mathcal{D}(A)$ is a core for $A_1$, hence by density it suffices to consider $[f(\frac{B}{t}), A_1]$ on $\mathcal{D}(A)$. Observe that since $B \in C^1(A)$, $f(\frac{B}{t})$ preserves $\mathcal{D}(A)$. Hence on $\mathcal{D}(A)$ the following identities are valid: as in (4.4.12), we have

$$\begin{aligned} & [f(\tfrac{B}{t}), \chi_1(H) A \chi_1(H)] \\ = & \tfrac{\mathrm{i}}{2\pi} \int_{\mathbb{C}} \partial_{\bar{z}} \tilde{f}(z) [(z - \tfrac{B}{t})^{-1}, \chi_1(H) A \chi_1(H)] dz \wedge d\bar{z}, \end{aligned} \tag{4.4.33}$$

and, as in the proof of Lemma 4.4.7,

$$\begin{aligned} & [(z - \tfrac{B}{t})^{-1}, \chi_1(H) A \chi_1(H)] \\ = & \chi_1(H) (z - \tfrac{B}{t})^{-1} [\tfrac{B}{t}, A] (z - \tfrac{B}{t})^{-1} \chi_1(H) \\ & + (z - \tfrac{B}{t})^{-1} [\tfrac{B}{t}, \chi_1(H)] (z - \tfrac{B}{t})^{-1} A \chi_1(H) \\ & + \chi_1(H) A (z - \tfrac{B}{t})^{-1} [\tfrac{B}{t}, \chi_1(H)] (z - \tfrac{B}{t})^{-1}. \end{aligned}$$

Since $H \in C^1(B)$, by Proposition 3.2.3*(iv)* we have

$$[B, \chi_1(H)] \text{ is bounded.} \tag{4.4.34}$$

By *(ii)*,

$$(B + \mathrm{i})^{-1} A \chi_1(H) \text{ is bounded.} \tag{4.4.35}$$

From *(ii)*, (4.4.34), (4.4.35), and (4.4.13), we see that the integrand in (4.4.33) is $O(\langle z \rangle |\mathrm{Im}\, z|^{-1})$, which proves (4.4.32). □

From Proposition 4.4.1 and Lemma 4.4.8 we obtain the following.

PROPOSITION 4.4.9. *Let $A$, $H$, $I(t)$, $P$, $B$ be self-adjoint operators, let $\Delta \subset \mathbb{R}$ be an interval, and suppose that the hypotheses of Proposition 4.4.1 and Lemma 4.4.8 are satisfied. Then for any $\chi \in C_0^\infty(\Delta)$ and for $\epsilon_0 > 0$ small enough we have:*

$$\int_1^{+\infty} \| F(\frac{B}{t} \leq \epsilon_0) \chi(H) P U(t) u \|^2 \frac{\mathrm{d}t}{t} \leq C \|u\|^2, \tag{4.4.36}$$

$$\mathrm{s-}\lim_{t \to +\infty} F(\frac{B}{t} \leq \epsilon_0) \chi(H) P U(t) = 0. \tag{4.4.37}$$

PROOF. We write $F(\frac{B}{t} \leq \epsilon_0)\chi(H)$ as $F(\frac{B}{t} \leq \epsilon_0)\chi(H)F(\frac{|A|}{t} \leq c_0) + F(\frac{B}{t} \leq \epsilon_0)\chi(H)F(\frac{|A|}{t} \geq c_0)$, and apply Proposition 4.4.1 and Lemma 4.4.8. □

**4.4.2. Minimal velocity estimates for 3-particle systems.** We now apply the results of Section 4.4.1 to 3−particle Hamiltonians of the type considered in Sections 3.7, 3.8. In this section we will assume that $N = 3$, that there is at least one pair of particles $(i, j)$ such that $q_i + q_j = 0$, and that *(V1)*, *(V4)* hold.

We begin with the 3−dimensional case.

PROPOSITION 4.4.10. *Assume that $d = 3$, and let $H$ be as above. Assume that (V1), (V4), (M1), (S) hold. Let $I(t, x)$ be a real-valued function satisfying*

$$|\partial_x^\alpha I(t, x)| \leq C_\alpha (\langle x \rangle + t)^{-1-\mu_1}, \ |\alpha| \geq 0, \tag{4.4.38}$$

*and define $U(t)$ as in Section 4.1. Let $\chi \in C_0^\infty(\mathbb{R} \backslash \tau \cup \sigma_{\mathrm{pp}}(H))$ and $F_1 \in C_0^\infty(\mathbb{R}^2)$. Then there is an $\epsilon_0 > 0$ such that:*

$$\int_1^{+\infty} \| F(\frac{|x^{\mathrm{cm}}|}{t} \leq \epsilon_0) \chi(H) F_1(K_{\mathrm{cm}}^\perp) U(t) u \|^2 \frac{\mathrm{d}t}{t} \leq C \|u\|^2,$$

$$\mathrm{s-}\lim_{t \to +\infty} F(\frac{|x^{\mathrm{cm}}|}{t} \leq \epsilon_0) \chi(H) F_1(K_{\mathrm{cm}}^\perp) U(t) = 0.$$

PROOF. By a covering argument and the Cauchy-Schwarz inequality, it suffices to prove the proposition for $\chi$ supported near some $\lambda \in \mathbb{R}\backslash\tau\cup\sigma_{\mathrm{pp}}(H)$. Fix $\lambda$, let $I \subset \mathbb{R}\backslash\tau$ be an interval such that $\lambda \in I$, and let $M$ be the conjugate operator for $I$ constructed in Section 3.7.

The proposition will follow from Proposition 4.4.9 if we verify that the assumptions of Proposition 4.4.1 and Lemma 4.4.8 are satisfied with

$$A = M,\ P = F_1(K_{\mathrm{cm}}^{\perp}),\ B = \langle x^{\mathrm{cm}}\rangle.$$

We begin with Proposition 4.4.1. The assumption *(i)* is satisfied by Proposition 3.7.5; *(ii)* follows from (4.4.38) and Lemma 3.7.4; *(iii)* is obtained directly from (4.4.38); and *(iv)* follows from Definition 1.3.4, (4.4.38), and pseudodifferential calculus.

It remains to check the hypotheses of Lemma 4.4.8. Part *(i)* follows from Theorem 3.7.3*(i)*, (3.7.9), and (3.7.10); *(ii)* – from (3.7.11) and (3.7.12); *(iii)* – from the definition of $\langle x^{\mathrm{cm}}\rangle$; and *(iv)* – from (3.7.13). □

We now turn to the 2−dimensional case. The proof of Proposition 4.4.11 below follows exactly the argument in the proof of Proposition 4.4.10 and is therefore omitted.

PROPOSITION 4.4.11. *Let $d = 2$, and let $H$ be as above. Assume that (V1), (V4) hold. Let $I(t,x)$ be a real-valued function satisfying*

$$|\partial_y^\alpha I(t,y)| \leq C_\alpha(\langle y\rangle + t)^{-1-\mu_1},\ |\alpha| \geq 0. \tag{4.4.39}$$

*Define $U(t)$ as in Section 4.1. Let $\chi \in C_0^\infty(\mathbb{R}\backslash\tau\cup\sigma_{\mathrm{pp}}(H))$ and $F_1 \in C_0^\infty(\mathbb{R}^2)$. Then there is an $\epsilon_0 > 0$ such that:*

$$\int_1^{+\infty} \|F(\frac{|y^{\mathrm{cm}}|}{t} \leq \epsilon_0)\chi(H)F_1(K_{\mathrm{cm}}^{\perp})U(t)u\|^2\frac{dt}{t} \leq C\|u\|^2,$$

$$\mathrm{s-}\lim_{t\to+\infty} F(\frac{|y^{\mathrm{cm}}|}{t} \leq \epsilon_0)\chi(H)F_1(K_{\mathrm{cm}}^{\perp})U(t) = 0.$$

CHAPTER 5

# Geometrical Methods II

In this chapter we analyze the dynamics of the system in the directions transversal to the magnetic field. Our main result is that propagation across the field may take place only along the subspaces $Y_a$, $a \notin \mathcal{A}^{\mathrm{cc}}$. (Recall that $a \notin \mathcal{A}^{\mathrm{cc}}$ if and only if at least one cluster of $a$ contains a neutral subsystem.) More precisely, we prove that if $F \in S^0(Y)$ satisfies

$$\mathrm{dist}\,(\mathrm{supp}\, F, Y_a) > \epsilon > 0 \text{ for all } a \notin \mathcal{A}^{\mathrm{cc}},$$

then

$$s - \lim_{t\to\infty} F(\frac{y}{t^\alpha})\mathrm{e}^{-\mathrm{i}tH} = 0 \tag{5.0.1}$$

for some $\alpha < 1$. In particular, if the system has no neutral subsystems, there is no propagation to infinity across the field in any direction, i.e., there is an $\alpha < 1$ such that

$$s - \lim_{t\to\infty} F(\frac{|y|}{t^\alpha} \geq 1)\mathrm{e}^{-\mathrm{i}tH} = 0.$$

These results are obviously related to those of Section 1.4, where it was proved that a bound state of the system may (and usually will) escape to infinity across the field if and only if the total electric charge of the system is zero. The situation here is more complicated, though, since the clusters need not be either isolated or stable.

As in Chapter 1, we will use the conservation of external pseudomomentum. Suppose that the system breaks up into several clusters which separate at a rate of at least $O(t^\alpha)$, $\alpha > 0$, as $t \to \infty$; denote this cluster decomposition by $a$. As the distance between the clusters in $a$ increases, the intercluster interactions vanish, so that the evolution is asymptotically the same as that generated by the Hamiltonian $H_a$ of non-interacting clusters. However, the evolution generated by $H_a$ conserves the external pseudomomentum $K_a$. It turns out that, if all clusters of $a$ are charged, this together with the conservation of energy implies bounds on the growth of $|y|$. The obvious complication is that the regions of configuration space corresponding to different break-ups of the system into clusters cannot be isolated and treated separately. We thus have to accommodate the possibility of the system oscillating between different scattering channels in ways that may be very difficult to control.

Our main tool will be the *center of orbit* observable $\mathbf{C}$, defined in Section 5.1. It is similar to the pseudomomentum $K$ in that it also commutes with the free magnetic Hamiltonian $H_0$; on the other hand, it has the additional property that $\mathbf{C} - y$ is bounded by the total energy. In Section 5.2 we introduce and construct *admissible functions*. Roughly speaking, $F \in S^0(Y)$ is admissible if in each region of the configuration space $F(\mathbf{C})$ depends only on the part of $\mathbf{C}$ which is asymptotically conserved in that region. It turns out that all of the configuration space except for neighbourhoods of $Y_a$, $a \notin \mathcal{A}^{\mathrm{cc}}$, can be covered by supports of gradients of such functions. We use this to prove our propagation estimates in Section 5.3.

The methods of this chapter were first introduced in [**GŁ2**], where in particular Theorem 5.3.1 was proved. The results of Subsection 5.3.2 are new, except for the special case $N = 3$ which was considered in [**GŁ3**]. The idea of constructing observables depending only on the variables $y_a = \pi_a\, y$ in a neighbourhood of $Y_a$ is due to Graf [**Gr**]; the results of Section 5.2 extend the main geometrical construction of [**Gr**] to the case when $\pi_a$ are replaced by more general projections $P_a$, not necessarily defined by a positive definite inner product.

## 5.1. The center of orbit observable

### 5.1.1. Introduction.

DEFINITION 5.1.1. The *center of orbit* observable is

$$\mathbf{C} = \frac{1}{2}(y + \mathbf{A}^{-1} D_y),$$

where $\mathbf{A}$ is understood as an invertible mapping $Y \to Y'$.

The corresponding classical observable

$$\mathbf{c} = \frac{1}{2}(y + \mathbf{A}^{-1}\eta) \tag{5.1.1}$$

has a natural physical interpretation. The trajectory of a classical charged particle in a constant magnetic field is a helix, the result of the superposition of uniform motion in the direction of the magnetic field and circular motion with constant speed in the transversal plane. The axis of this helix is the line $y = \mathbf{c}$; in other words, $\mathbf{c}$ is the position of the center of the circle along which the particle moves in the plane transversal to the field.

The difference $y - \mathbf{c}$ (i.e., the radius of the circle) is bounded by the total energy of the particle. This can be seen from the following simple calculation:

$$\mathbf{c} - y = \frac{1}{2}(\mathbf{A}^{-1}\eta + y) - y = \frac{1}{2}\mathbf{A}^{-1}(\eta - \mathbf{A}y),$$

where the last term is bounded by $H_0 = \frac{1}{2}((\eta - \mathbf{A}y)^2 + |\zeta|^2)$.

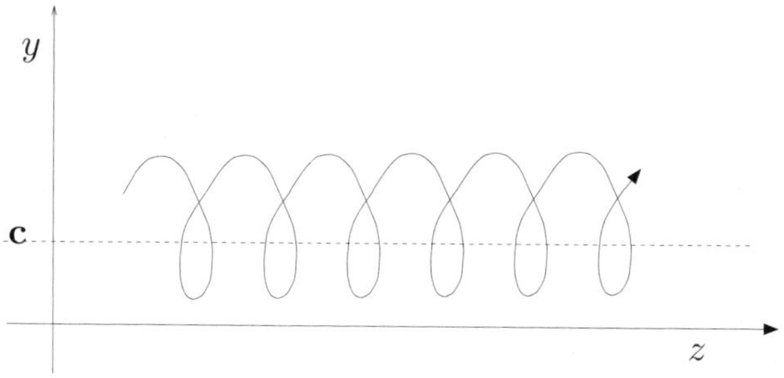


**Fig. 5.1.** The classical observable **c**

The quantum observable $\mathbf{C}$ can be interpreted in a similar way. Its basic properties are the following:

$$[H, \mathbf{C}] = 0; \tag{5.1.2}$$

$$\mathbf{C} - y \text{ is } H_0\text{-bounded.} \tag{5.1.3}$$

(In this informal discussion, we will ignore domain questions.) Observe that $\mathbf{C}$ is related to the pseudomomentum $K$ by the formula

$$\mathbf{C} = \frac{1}{2}\mathbf{A}^{-1}K^{\perp}.$$

For a single particle, $\mathbf{A}^{-1} : Y' \to Y$ is given by the matrix $\begin{pmatrix} 0 & 1/bq \\ -1/bq & 0 \end{pmatrix}$, hence $\mathbf{C}$ is essentially a multiple of $K^{\perp}$. However, for $N > 1$ particles with the $q_i$'s not all equal the observables $K$ and $\mathbf{C}$ play rather different roles in scattering theory.

Our goal in this chapter is to prove that

$$s - \lim_{t \to \infty} F(t^{-\alpha}y)\mathrm{e}^{-itH} = 0$$

for functions $F \in S^0(Y)$ supported in "forbidden" regions of configuration space, i.e., satisfying

$$\text{dist}\,(\text{supp}\, F, Y_a) > 0 \text{ for all } a \notin \mathcal{A}^{\text{cc}}.$$

The strategy of the proof is as follows. By (5.1.3), it will suffice to prove a similar estimate for $\Phi(t) = F(t^{-\alpha}\mathbf{C})$. This will be done by estimating

$\mathbf{D}\Phi(t) = \partial_t\Phi + [H, \mathrm{i}\Phi]$ and then integrating in $t$. By (5.1.2) we have

$$[H, \mathrm{i}\mathbf{C}] = [V, \mathrm{i}\mathbf{C}] = \frac{1}{2}\mathbf{A}^{-1}\nabla V.$$

For a potential $V$ such that $\nabla V$ vanishes at infinity sufficiently fast in all directions, this would readily yield the estimates we seek. However, the $N$-particle potential $V = \sum V_{ij}$ does not vanish along the subspaces $Y_a$. We therefore construct functions $F$ such that near the intersection of any $Y_a$ with $\operatorname{supp} F$ we have $F(\mathbf{C}) = F(\mathbf{C}_a^q)$, where $\mathbf{C}_a^q$ is an observable similar to $\mathbf{C}$, but with the additional property that it commutes with those pair potentials which do not vanish along $Y_a$. Then the Heisenberg derivative of $F(\mathbf{C}/t^\alpha)$ near $Y_a$ will involve only the derivatives of the remaining pair potentials, which can be easily dealt with.

The construction of such functions, though quite difficult technically, is based on a very natural idea. Namely, we introduce the *center of charge coordinates*. The center of charge of a cluster $C_j = \{i_1, \ldots, i_k\}$ is

$$Q[C_j]^{-1} \sum_{i \in C_j} q_i x_i, \tag{5.1.4}$$

where $Q[C_j] = \sum_{i \in C_j} q_i$ is the total electric charge of the cluster. Note that the center of charge of $C_j$ is well defined if and only if $Q[C_j] \neq 0$.

Let $a = \{C_1, \ldots, C_n\} \in \mathcal{A}^{\mathrm{c}}$ be a cluster decomposition consisting of only charged clusters. We define $y_a^q$ as the observable obtained from $y$ by replacing each $y_i$ by the center of charge of the cluster $C_j$ to which $i$ belongs. Observe that $y_a^q = y$ if $y \in Y_a$, hence for $y$ close to $Y_a$ the difference $y - y_a^q$ is small. Furthermore, if we replace $x$ by $\mathbf{C}$ in (5.1.4), we obtain

$$Q[C_j]^{-1} \sum_{i \in C_j} q_i \frac{1}{2q_i} J^{-1}(D_{y_i} + q_i J y_i) = (2Q[C_j]J)^{-1} \sum_{i \in C_j} (D_{y_i} + q_i J y_i),$$

which is a multiple of the total pseudomomentum of the cluster $C_j$ and therefore commutes with the pair potentials $V_{ik}$ for all $i, k \in C_j$. It therefore suffices to construct $F(y)$ so that near any $Y_a$ it depends only on the coordinates $y_a^q$ of the centers of charge of the clusters of $a$; such functions will be called *admissible* (see Section 5.2). Then $F(\mathbf{C})$ will have the properties we need.

If a cluster decomposition $a \in \mathcal{A}$ contains a neutral cluster, the center of charge of that cluster is not defined. What happens in this case is, essentially, that the conservation laws that were used in the charged case remain valid but they no longer lead to estimates on the growth of $y$. Consider, for instance, a 2-particle system with $q_1 + q_2 = 0$. Then $q_1\mathbf{C}_1 + q_2\mathbf{C}_2$ is still conserved; however, this only implies bounds on $q_1 y_1 + q_2 y_2 = q_1(y_1 - y_2)$, i.e., on the distance between the two particles.

Since it was demonstrated in Section 1.4 that bound states of neutral clusters may travel across the magnetic field with constant velocity, propagation to infinity along the subspaces $Y_a$, $a \in \mathcal{A}^{\mathrm{n}}$, is possible and usually expected. It is therefore not possible to prove an estimate of the form (5.0.1) for a function $F$ whose support has a nonempty intersection with a space $Y_a$, $a \in \mathcal{A}^{\mathrm{n}}$. Moreover, if the support of a smooth function $F$ intersects $Y_a$, it also must intersect $Y_b$ for all $b < a$, which is one reason why we actually need to assume that $a \in \mathcal{A}^{\mathrm{cc}}$ (i.e., that no cluster of $a$ contains a neutral subsystem).

One could ask whether the restriction $a \in \mathcal{A}^{\mathrm{cc}}$ in our proof might not be due only to the limitations of our methods, and whether propagation to infinity along the subspaces $Y_a \setminus \bigcup_{b \not\leq a} Y_b$, where $a \in \mathcal{A}^{\mathrm{c}} \setminus \mathcal{A}^{\mathrm{cc}}$ (i.e., all clusters of $a$ are charged but at least one of them has a neutral subsystem), can actually take place. This is a question that we are not able to answer at this time: we can neither rule out such behaviour nor give a convincing reason why it should be expected to take place.

**5.1.2. Basic properties.** The center of orbit is an observable with values in $Y$ which satisfies the following commutation relations:

$$[\langle \mathbf{C}, \eta' \rangle, i\langle \mathbf{C}, \eta'' \rangle] = -\frac{1}{2}\langle \mathbf{A}^{-1}\eta'', \eta' \rangle. \tag{5.1.5}$$

In particular, the two components of any $\mathbf{C}_i$ can never commute. We will need a functional calculus for the non-commuting observable $\mathbf{C}$, similar to the functional calculus for the external pseudomomentum $K_{\mathrm{cm}}$ defined in Subsection 1.3.6.

DEFINITION 5.1.2. For $F \in S^0(Y)$, we define

$$F(\mathbf{C}) := (F \circ \mathbf{c})^{\mathrm{w}}(y, D_y).$$

Note that since the components of $\mathbf{C}$ do not commute $F(\mathbf{C})G(\mathbf{C}) \neq FG(\mathbf{C})$.

Since $F \circ \mathbf{c}$ belongs to $S^0(T^*Y)$ for $F \in S^0(Y)$, the operator $F(\mathbf{C})$ is bounded on $L^2(X)$. Moreover, we have

$$[H_0, F(\mathbf{C})] = 0 \text{ for all } F \in S^0(Y). \tag{5.1.6}$$

We will often rely on the functional calculus formula

$$F(\mathbf{C}) = (2\pi)^{-2N} \int \hat{F}(\eta) \mathrm{e}^{\mathrm{i}\langle \mathbf{C}, \eta \rangle} d\eta, \tag{5.1.7}$$

for $F \in \mathcal{S}(Y)$; the proof of (5.1.7) is essentially the same as that of Lemma 1.3.5.

The following lemma summarizes some of the properties of $\mathbf{C}$.

LEMMA 5.1.3. *Assume that (V1) holds. Then:*

*(i) there is a constant* $c_0$ *such that for any* $F \in S^0(Y)$ *and* $\alpha > 0$ *one has*

$$\|(H + \mathrm{i})F(t^{-\alpha}\mathbf{C})(H + \mathrm{i})^{-1}\| \leq c_0 \|F(t^{-\alpha}\mathbf{C})\|.$$

*(ii) for any* $F \in S^0(Y)$,

$$(H+\mathrm{i})^{-1}\left(F(t^{-\alpha}\mathbf{C}) - F(t^{-\alpha}y)\right) = O(t^{-\alpha}).$$

*(iii) for any* $F \in S^0(Y)$ *and* $F_1 \in S^0(Y)$ *such that* $F_1 \equiv 1$ *on* $\operatorname{supp} F$,

$$(H+\mathrm{i})^{-1}F(t^{-\alpha}\mathbf{C})\,(\mathbf{1} - F_1(t^{-\alpha}y)) = O(t^{-2\alpha}),$$

$$(H+\mathrm{i})^{-1}F(t^{-\alpha}y)\,(\mathbf{1} - F_1(t^{-\alpha}\mathbf{C})) = O(t^{-2\alpha}).$$

PROOF. *(i)* Using that $[H_0, t^{-\alpha}\mathbf{C}] = 0$, we write

$$\begin{aligned} &(H+\mathrm{i})F(t^{-\alpha}\mathbf{C})(H+\mathrm{i})^{-1} \\ = \;& (H+\mathrm{i})(H_0+\mathrm{i})^{-1}F(t^{-\alpha}\mathbf{C})(H_0+\mathrm{i})(H+\mathrm{i})^{-1} \\ = \;& O(\|F(t^{-\alpha}\mathbf{C})\|), \end{aligned}$$

by Proposition 1.2.1.

*(ii)* We have

$$F(t^{-\alpha}\mathbf{c}) - F(t^{-\alpha}y) = t^{-\alpha}\langle \mathbf{c} - y, \int_0^1 \nabla F(s\mathbf{c} + (1-s)y)ds\rangle,$$

hence

$$F(t^{-\alpha}\mathbf{C}) - F(t^{-\alpha}y) = t^{-\alpha}\langle \mathbf{C} - y, R(t)\rangle,$$

where $R(t) = O(1)$. Using the identity

$$\mathbf{C} - y = \frac{1}{2}\mathbf{A}^{-1}(D_y - \mathbf{A}y),$$

and Proposition 1.2.1, we find that

$$\|(\mathbf{C} - y)^\alpha (H+\mathrm{i})^{-1}\| < \infty,\ |\alpha| \le 2. \tag{5.1.8}$$

which proves *(ii)*.

*(iii)* We have

$$\begin{aligned} &F(t^{-\alpha}\mathbf{c}) - F(t^{-\alpha}y) = t^{-\alpha}\langle \mathbf{c} - y, \nabla t^{-\alpha}y\rangle \\ &+t^{-2\alpha}\langle \mathbf{c} - y, \textstyle\int_0^1 (1-s)\nabla^2 F(s\mathbf{c} + (1-s)y)(\mathbf{c} - y)ds\rangle. \end{aligned}$$

This yields by pseudodifferential calculus

$$\begin{aligned} &F(t^{-\alpha}\mathbf{C}) - F(t^{-\alpha}y) \\ = \;& t^{-\alpha}\langle \mathbf{C} - y, \nabla F(t^{-\alpha}y)\rangle + t^{-2\alpha}\textstyle\sum_{|\beta|\le 2}(\mathbf{C} - y)^\beta R_\beta(t), \end{aligned}$$

where $R_\beta(t) = O(1)$. By (5.1.8), this proves first estimate in *(iii)*. The second one follows by interchanging $\mathbf{C}$ and $y$. □

## 5.2. Admissible functions

In this section we construct observables that will be used in the proof of our propagation estimates in Section 5.3.

**5.2.1. Center of charge coordinates.** The Lorentz forces $\vec{F}_{L,i} = q_i\vec{B} \times \vec{v}_i$ acting on the charged particles in a magnetic field are proportional to their electric charges $q_i$. This motivates us to introduce the *center of charge coordinates.*

Consider the following inner product on $Y$:

$$q(y, \overline{y}) := \sum_i q_i y_i . \overline{y}_i .$$

where $q_i$ are the electric charges of the particles. Since we do not assume that all electric charges have the same sign, the inner product $q$ need not be positive definite.

For any cluster decomposition $a \in \mathcal{A}_*$, we define the subspace $Y^{a,q}$ orthogonal to $Y_a$ in the metric $q$:

$$\begin{aligned} Y^{a,q} &:= \{y \in Y : \ q(y, \overline{y}) = 0, \forall \overline{y} \in Y_a\} \\ &= \{y \in Y : \ \textstyle\sum_{i \in C_j} q_i y_i = 0, \forall C_j \in a\}. \end{aligned}$$

Then:

$$\begin{aligned} Y_a \cap Y^{a,q} &= \{y \in Y : \ y_i = y_j \text{ if } (ij) \leq a, \text{ and} \\ &\textstyle\sum_{i \in C_j} q_i y_i = 0 \text{ for all } C_j \in a\} \\ &= \{y \in Y : \ y_i = 0 \text{ if } i \in C_j, \ Q[C_j] \neq 0\}, \end{aligned}$$

where $Q[C_j] = \sum_{i \in C_j} q_i$ is the total charge of the cluster $C_j$.

Assume that $a \in \mathcal{A}^{\mathrm{c}}$, i.e., $Q[C_j] \neq 0$ for all $C_j \in a$. Then $Y_a \cap Y^{a,q} = \{0\}$, hence we may write

$$Y = Y_a \oplus Y^{a,q},$$

and, given a vector $y \in Y$, we may define projections of $y$ on the above spaces:

$$y = y_a^q + y^{a,q}, \ y_a^q \in Y_a, \ y^{a,q} \in Y^{a,q}.$$

The projection onto $Y^a$ along $Y^{a,q}$ will be denoted by $\pi_a^q$. Explicitly, we have $\pi_a^q y = y_a^q$, where

$$(y_a^q)_j = Q[C_i]^{-1} \sum_{n \in C_i} q_n y_n \text{ if } j \in C_i \tag{5.2.1}$$

is the *center of charge* of the cluster $C_i$. Since $Y_a = \{y \in Y : \ y^{a,q} = 0\} = \{y \in Y : \ y^a = 0\}$, the following lemma holds.

LEMMA 5.2.1. *There is a constant $c_0$ such that*

$$c_0^{-1} |y^{a,q}| \leq |y^a| \leq c_0 |y^{a,q}|.$$

We will sometimes use the following alternative characterization of the space $Y^{a,q}$. Recall that the map $\mathbf{A} : Y \to Y'$ is bijective. If $Y^{a\prime} = \{\eta \in Y' : \langle y, \eta \rangle = 0, \forall y \in Y_a\}$, then

$$Y^{a,q} = \mathbf{A}^{-1}(Y^{a\prime}). \tag{5.2.2}$$

**5.2.2. Definition of admissible functions.** Recall that $\mathcal{A}^{\mathrm{c}}$ is the family of those cluster decompositions which consist only of charged clusters, and $\mathcal{A}^{\mathrm{n}}$ is the set of the remaining cluster decompositions (i.e., those which contain at least one neutral cluster).

DEFINITION 5.2.2. A function $F \in S^0(Y)$ is *admissible* if the following conditions are satisfied:

(i) for each $a \in \mathcal{A}^{\mathrm{n}}$, there is an $\epsilon > 0$ such that $F(y) = 0$ in the neighbourhood $\{|y^a| < \epsilon\}$ of $Y_a$;

(ii) for each $a \in \mathcal{A}^{\mathrm{c}}$, there is an $\epsilon > 0$ such that $F(y) = F(y_a^q)$ in the neighbourhood $\{|y^a| < \epsilon\}$ of $Y_a$.

Our main observables will be operators of the form $F(\mathbf{C})$, where $F$ is an admissible function and $\mathbf{C}$ is the center of orbit observable defined in Section 5.1.

Basic to all of our considerations here is Lemma 5.2.3 below, which follows immediately from the characterization of $Y^{a,q}$ given in (5.2.2). (It can also be checked directly, as in Subsection 5.1.1.) We will write $\mathbf{C}_a^q = \pi_a^q \mathbf{C}$, where the projection $\pi_a^q$ was defined in the previous subsection.

LEMMA 5.2.3. *Let $a \in \mathcal{A}^{\mathrm{c}}$. Then*

$$[x^a, \mathbf{C}_a^q] = 0.$$

If $F$ is admissible, Definition 5.2.2 and Lemma 5.2.3 imply that each subspace $Y_a$ has a neighbourhood $\{|y^a| < \epsilon\}$ in which $F(\mathbf{C})$ will commute with the part $V^a$ of the total $N$-body potential that does not vanish along $Y_a$. This will enable us to estimate $[H, F(\mathbf{C})]$ in Proposition 5.2.5.

We begin with a simple auxiliary lemma.

LEMMA 5.2.4. *Let $I(t,x)$ be a real time-dependent potential such that:*

$$(i)\ I(t,x) = I^{\mathrm{s}}(t,x) + I^{\mathrm{l}}(t,x),$$

$$(ii)\ \|(H+\mathrm{i})^{-1} I^{\mathrm{s}}(t,x)\| = O(t^{-\alpha\mu_{\mathrm{s}}}),$$

$$(iii)\ \partial_x^\beta I^{\mathrm{l}}(t,x) = O(t^{-\alpha(\mu_{\mathrm{l}}+|\beta|)}),\ |\beta| \le 2,$$

*where $\alpha, \mu_{\mathrm{s}}, \mu_{\mathrm{l}} > 0$ and $(2+\mu_{\mathrm{l}})\alpha > 1, \alpha\mu_{\mathrm{s}} > 1$. Then for any $F \in S^0(Y)$ we have*

$$\|(H+\mathrm{i})^{-1}[F(t^{-\alpha}\mathbf{C}), I(t,x)](H+\mathrm{i})^{-1}\| = O(t^{-1-\epsilon}),\ \epsilon > 0.$$

PROOF. We first note that by *(ii)*:

$$\begin{aligned}
&\|(H+\mathrm{i})^{-1}[F(t^{-\alpha}\mathbf{C}), I^{\mathrm{s}}(t,x)](H+\mathrm{i})^{-1}\| \\
\leq\ &\|(H+\mathrm{i})^{-1}F(t^{-\alpha}\mathbf{C})I^{\mathrm{s}}(t,x)(H+\mathrm{i})^{-1}\| \\
&+\|(H+\mathrm{i})^{-1}I^{s}(t,x)F(t^{-\alpha}\mathbf{C})(H+\mathrm{i})^{-1}\| \\
=\ &O(t^{-\alpha\mu_{\mathrm{s}}}).
\end{aligned}$$

Next, we use Lemma A.2 in the Appendix to write

$$I^{\mathrm{l}}(t,x) = I_1(t,x) + I_2(t,x),$$

where

$$I_1(t,x) = O(t^{2\delta-(2+\mu_{\mathrm{l}})\alpha}),$$

$$\partial_x^\beta I_2(t,x) = \begin{cases} O(t^{-\alpha(\mu_{\mathrm{l}}+|\beta|)}), & |\beta| \leq 2, \\ O(t^{-\alpha(\mu_{\mathrm{l}}+2)}t^{-\delta(|\beta|-2)}), & |\beta| \geq 3, \end{cases}$$

for $\delta > 0$ to be fixed later. We have

$$\|[F(t^{-\alpha}\mathbf{C}), I_1(t,x)]\| \leq C\|I_1(t,\cdot)\| = O(t^{2\delta-(2+\mu_{\mathrm{l}})\alpha});$$

by symbolic calculus, we also have $[F(t^{-\alpha}\mathbf{C}), I_2(t,x)] = O(t^{-(2+\mu_{\mathrm{l}})\alpha})$. Using that $(2+\mu_{\mathrm{l}})\alpha > 1$ and picking $\delta > 0$ small enough, we obtain the lemma. □

PROPOSITION 5.2.5. *Assume that (V1) and (SC) hold, and let $F \in S^0(Y)$ be an admissible function. Then for $\alpha > \max(\frac{1}{2}, \mu_{\mathrm{s}}^{-1})$, we have*

$$(H+\mathrm{i})^{-1}[H, F(t^{-\alpha}\mathbf{C})](H+\mathrm{i})^{-1} = O(t^{-1-\epsilon}),\ \epsilon > 0.$$

PROOF. Since $F$ is admissible, we may find cutoff functions $q_a \in S^0(Y)$ such that:

$$\begin{aligned}
&\mathbf{1} = \textstyle\sum_{a\in\mathcal{A}} q_a(y), \\
&\operatorname{supp} q_a \subset \{y \in Y \,|\, |y_b| > 0,\ \forall b \not\leq a\}, \qquad (5.2.3)\\
&\partial_{y^{a,q}} F \equiv 0, \text{ on } \operatorname{supp} q_a.
\end{aligned}$$

Similarly, there are functions $\tilde{q}_a \in S^0(Y)$ such that $\tilde{q}_a \equiv 1$ on $\operatorname{supp} q_a$, $\operatorname{supp} \tilde{q}_a \subset \{y \in Y \,|\, |y_b| > 0,\ \forall b \not\leq a\}$, and

$$\partial_{y^{a,q}} F \equiv 0 \text{ on } \operatorname{supp} \tilde{q}_a. \qquad (5.2.4)$$

We have

$$
\begin{aligned}
&(H+\mathrm{i})^{-1}[V, F(t^{-\alpha}\mathbf{C})](H+\mathrm{i})^{-1} \\
= \ & \textstyle\sum_{a\in\mathcal{A}}(H+\mathrm{i})^{-1}[q_a(t^{-\alpha}y)V, F(t^{-\alpha}\mathbf{C})](H+\mathrm{i})^{-1} \\
= \ & \textstyle\sum_{a\in\mathcal{A}}(H+\mathrm{i})^{-1}[q_a(t^{-\alpha}y)V, \tilde{q}_a(t^{-\alpha}\mathbf{C})F(t^{-\alpha}\mathbf{C})](H+\mathrm{i})^{-1} + O(t^{-2\alpha}) \\
= \ & \textstyle\sum_{a\in\mathcal{A}}(H+\mathrm{i})^{-1}[q_a(t^{-\alpha}y)V, \tilde{q}_a(t^{-\alpha}\mathbf{C})F(t^{-\alpha}\mathbf{C}_a^q)](H+\mathrm{i})^{-1} \\
& + O(t^{-2\alpha}),
\end{aligned}
$$

where we used, successively, Lemma 5.1.3*(iii)* and *(i)*, Proposition 1.2.1, and (5.2.4). Using again Lemma 5.1.3, we obtain

$$
\begin{aligned}
&(H+\mathrm{i})^{-1}q_a(t^{-\alpha}y)V(x)\tilde{q}_a(t^{-\alpha}\mathbf{C})F(t^{-\alpha}\mathbf{C}_a^q)(H+\mathrm{i})^{-1} \\
= \ & (H+\mathrm{i})^{-1}q_a(t^{-\alpha}y)V(x)F(t^{-\alpha}\mathbf{C}_a^q)(H+\mathrm{i})^{-1} + O(t^{-2\alpha}), \\
&(H+\mathrm{i})^{-1}\tilde{q}_a(t^{-\alpha}\mathbf{C})F(t^{-\alpha}\mathbf{C}_a^q)q_a(t^{-\alpha}y)V(x)(H+\mathrm{i})^{-1} \\
= \ & (H+\mathrm{i})^{-1}F(t^{-\alpha}\mathbf{C}_a^q)q_a(t^{-\alpha}y)V(x)(H+\mathrm{i})^{-1} + O(t^{-2\alpha}),
\end{aligned}
$$

so that

$$
\begin{aligned}
&(H+\mathrm{i})^{-1}[V(x), F(t^{-\alpha}\mathbf{C})](H+\mathrm{i})^{-1} \\
= \ & \textstyle\sum_{a\in\mathcal{A}}(H+\mathrm{i})^{-1}[q_a(t^{-\alpha}y)V(x), F(t^{-\alpha}\mathbf{C}_a^q)](H+\mathrm{i})^{-1} \\
& + O(t^{-2\alpha}).
\end{aligned}
\tag{5.2.5}
$$

To compute the right hand side of (5.2.5), we set

$$I_a(t,x) = q_a(t^{-\alpha}y)I_a(x),$$

then

$$
\begin{aligned}
&(H+\mathrm{i})^{-1}[q_a(t^{-\alpha}y)V(x), F(t^{-\alpha}\mathbf{C}_a^q)](H+\mathrm{i})^{-1} \\
= \ & (H+\mathrm{i})^{-1}[q_a(t^{-\alpha}y)V^a(x^a), F(t^{-\alpha}\mathbf{C}_a^q)](H+\mathrm{i})^{-1} \\
& + (H+\mathrm{i})^{-1}[I_a(t,x), F(t^{-\alpha}\mathbf{C}_a^q)](H+\mathrm{i})^{-1} \\
=: \ & R_1 + R_2.
\end{aligned}
$$

By Lemma 5.2.3, $[V^a(x^a), F(t^{-\alpha}\mathbf{C}_a^q)] = 0$. Next, by symbolic calculus we have

$$[q_a(t^{-\alpha}y), F(t^{-\alpha}\mathbf{C}_a^q)] = O(t^{-2\alpha}),$$

so that

$$R_1 = O(t^{-2\alpha}).$$

It remains to estimate $R_2$. By *(SC)*, we may write

$$I_a(t,x) = I_a^{\mathrm{s}}(t,x) + I_a^{\mathrm{l}}(t,x),$$

where

$$\|(H+\mathrm{i})^{-1}I_a^{\mathrm{s}}(t,x)\| = O(t^{-\alpha\mu_{\mathrm{s}}}),$$

$$|\partial_x^\beta I_a^{\mathrm{l}}(t,x)| \le Ct^{-\alpha(\mu_{\mathrm{l}}+|\beta|)},\ |\beta| \le 2.$$

Applying Lemma 5.2.4, we obtain that

$$R_2 = O(t^{-1-\epsilon}),\ \epsilon > 0,$$

which completes the proof of the proposition. □

**5.2.3. A geometric construction.** The geometric construction in this section will be needed to construct nontrivial admissible functions.

Consider a Euclidean space $Y$ equipped with a family of linear subspaces $\{Y_a\}_{a\in\mathcal{A}}$. As in Section 2.1, we add to this family the subspace $\{0\} =: Y_\emptyset$. Recall that $\emptyset$ is a "cluster decomposition" such that $\sharp(\emptyset) = 0$ and $\emptyset > a$ for all $a \in \mathcal{A}$. The augmented family $\mathcal{A} \cup \{\emptyset\}$ is, as before, denoted by $\mathcal{A}_*$.

For each $a \in \mathcal{A}_*$ we fix a projection $P_a : Y \to Y_a$. We assume that $\operatorname{Ran} P_a = Y_a$ for all $a \in \mathcal{A}_*$, and that the family $\{P_a\}_{a\in\mathcal{A}}$ satisfies the consistency condition

$$P_aP_b = P_bP_a = P_b \text{ for all } a \le b. \tag{5.2.6}$$

We set $Y^a = \operatorname{Ker} P_a$ and $P^a = \mathbf{1} - P_a$. We will use the shorthand $y^a = P^a y$, $y_a = P_a y$, and $y_a^b = P_a y^b = P^b y_a$ for $a \le b$.

The projections $P_a$ *need not* be the same as the projections $\pi_a$ defined in Section 2.1; consequently, the spaces $Y^a$ and the projections $y_a = \pi_a y$, $y^a = \pi^a y$, etc., may be different from those of Section 2.1. This should not cause confusion, since the latter spaces and projections will not be mentioned anywhere in this chapter. (In the next subsection, we will apply the results obtained here to $P_a = \pi_a^q$, where $\pi_a^q$ were defined in Subsection 5.2.1. Clearly, $\pi_a^q$ satisfy (5.2.6).)

By convention, $P_\emptyset = \mathbf{0}$ and $P^\emptyset = \mathbf{1}$, so that $y_\emptyset = 0$ and $y^\emptyset = y$. Note also that (5.2.6) is equivalent to $P^aP^b = P^bP^a = P^a$ for $a \le b$. Hence

$$\text{for all } a \le b,\ (y_a)_b = (y_b)_a = y_b \text{ and } (y^a)^b = (y^b)^a = y^a. \tag{5.2.7}$$

Since $Y_a = \operatorname{Ran} P_a$ and $Y^a = \operatorname{Ker} P_a$, we have

$$Y = Y_a \oplus Y^a. \tag{5.2.8}$$

Hence there is a constant $c_0 > 0$ such that for all $a \in \mathcal{A}_*$:

$$|y_a| \le c_0|y|,\ |y^a| \le c_0|y|,\ |y| \le c_0(|y^a| + |y_a|). \tag{5.2.9}$$

Recall from Section 2.1 that $a \vee b$ is the cluster decomposition for which $Y_{a\vee b} = Y_a \cap Y_b$. Then, by (5.2.8), $Y^{a\vee b} = \operatorname{Span}(Y^a, Y^b)$. We may therefore assume, increasing value of $c_0$ if necessary, that

$$|y^{a\vee b}| \le c_0(|y^a| + |y^b|) \text{ for all } a, b \in \mathcal{A}_*. \tag{5.2.10}$$

We will associate to the family $\{Y_a, P_a\}_{a\in\mathcal{A}_*}$ a family of open sets as follows. For $b \le a$, let

$$\epsilon_b^a = C_1(a)^{\sharp b-\sharp a},\ \mu_b^a = \epsilon_b^a C_2(a)^{\sharp b-\sharp a},\ D_b^a = D_0(a) D_1(a)^{\sharp b-\sharp a}, \tag{5.2.11}$$

where the constants $C_1(a) < 1$, $C_2(a) < 1$, $D_1(a) < 1$, $D_0(a)$ will be chosen later. We then define:

$$\begin{aligned} F^a_{a_{\min}} &:= \{y^a \in Y^a \mid |y^a| < D^a_{a_{\min}}\}, \\ F^a_b &:= \{y^a \in Y^a \mid |y^a_b| < D^a_b, y^b \in \epsilon^a_b \Omega^b\} \text{ for } b < a,\ b \neq a_{\min}, \\ \Omega^a &:= \cup_{b<a} F^a_b. \end{aligned} \tag{5.2.12}$$

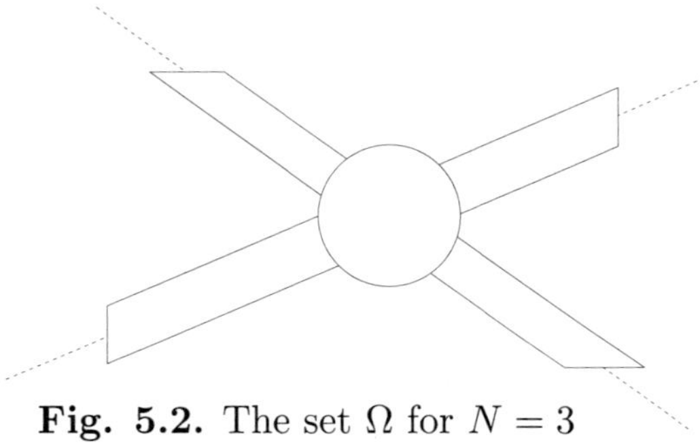

**Fig. 5.2.** The set $\Omega$ for $N = 3$

**Fig. 5.3.** The set $\Omega$ for $N = 4$

Note that all of the sets $F^a_b$ and $\Omega^a$ are open, and that $\Omega^a$ depends on the constants $D^{a'}_{b'}$ and $\epsilon^{a'}_{b'}$ for all $b' < a' \leq a$. We also define open neighbourhoods $G^a_b$ of the subspaces $Y^a_b$:

$$G^a_b := \{y^a \in Y^a \mid |y^b| < \mu^a_b\}.$$

The sets $F^{\emptyset}_b, G^{\emptyset}_b, \Omega^{\emptyset}$ will be denoted simply by $F_b, G_b, \Omega$.

Our goal in this subsection is to prove the following proposition.

PROPOSITION 5.2.6. *Fix $a \in \mathcal{A}_*$, $a \neq a_{\min}$. We may choose the constants $C_1(a) < 1$, $C_2(a) < 1$, $D_1(a) < 1$, $D_0(a)$ so that:*

*(i) for any $b \leq a$, the part of the boundary of $\Omega^a$ which is contained in $G^a_b$ depends only on $y^a_b$, i.e., it is described by equations of the form $\Phi^a_b(y^a_b) = 0$;*

*(ii) there are constants $K_1(a) > 0$ such that*

$$\Omega^a \subset \{y^a \in Y^a : \ |y^a| \leq K_1(a)\};$$

*(iii) there are constants $K_2(a) > 0$ such that*

$$\{y^a \in Y^a : \ |y^a| \leq K_2(a)\} \subset \Omega^a;$$

*(iv) $y^a \cdot \nabla_{y^a} 1_{\Omega^a}(y^a) \leq 0$ in distributional sense.*

PROOF. We will prove the proposition by decreasing induction on $\sharp a$. If $\sharp a = N - 1$, the lemma is trivial, since then the only $b \in \mathcal{A}$ such that $b < a$ is $b = a_{\min}$. Assume now that we have chosen the constants $C_1(b), C_2(b), D_1(b), D_0(b)$ for all $b \in \mathcal{A}$ with $\sharp b > k$, and consider an $a \in \mathcal{A}_*$ with $\sharp a = k$. Since $\Omega^a$ is a subset of $Y^a$ and the projections $P_a$ satisfy (5.2.6), the whole analysis takes place in the space $Y^a$. We may therefore assume without loss of generality that $a = \emptyset$ and $Y^a = Y$. To simplify the notation, we will drop the index $\emptyset$ and denote $K_1(\emptyset), K_2(\emptyset), D^{\emptyset}_b, \epsilon^{\emptyset}_b, \mu^{\emptyset}_b$ simply by $K_1, K_2, D_b, \epsilon_b, \mu_b$.

The proof will be divided in two parts. In Step 1, we will find conditions on the constants $C_1, C_2, D_1, D_0$ which guarantee that *(i)–(iv)* of Proposition 5.2.6 hold. In Step 2 we will find constants satisfying these conditions.

*Step 1).* We will first verify that *(ii)*, *(iii)*, and *(iv)* hold for any choice of the constants in (5.2.11). Since by (5.2.9),

$$|y| \leq c_0(|y^a| + |y_a|),$$

*(ii)* holds with

$$K_1 = c_0 \sup_{a < \emptyset}(D_a + K_1(a)).$$

Moreover, we have

$$F_{a_{\min}} = \{y \in Y \mid |y| < D_{a_{\min}}\} \subset \Omega,$$

hence *(iii)* is satisfied whenever

$$K_2 < D_{a_{\min}}.$$

Finally, since $0 \in \Omega$, *(iv)* is satisfied whenever $\Omega$ is star-shaped with respect to the origin. Clearly, the domain

$$\{y^a \in Y_a \mid |y^a_b| < D^a_b\}$$

is star-shaped. Since a cartesian product of star-shaped sets is star-shaped, the sets $F_a$ are star-shaped, and so is $\Omega$ as a union of star-shaped sets.

We now turn to the property *(i)*, which is the most difficult to enforce. We must prove that for all $b \in \mathcal{A}$ the set $G_b \cap \partial\Omega$ can be expressed in terms of $y_b$, where $\partial$ is used to denote the boundary of a subset of $Y$. Pick $y \in G_b \cap \partial\Omega$ so that $y \in G_b \cap \partial F_a$ for some $a \in \mathcal{A}$. Note that

$$\partial F_a = (\partial F_a)_1 \cup (\partial F_a)_2,$$

where

$$(\partial F_a)_1 = \{|y_a| \le D_a, y^a \in \partial(\epsilon_a \Omega_a)\},$$

$$(\partial F_a)_2 = \{|y_a| = D_a, y^a \in \epsilon_a \Omega_a\}.$$

We will consider the following cases separately.

*Case 1*: $a = b$, i.e., $y \in G_b \cap \partial F_b$. By the inductive hypothesis *(iii)*,

$$\{|y^b| \le K_2(b)\} \subset \Omega^b.$$

If $y \in G_b$, we have $|y^b| < \mu_b$. Hence, if

$$\mu_b < \epsilon_b K_2(b), \tag{5.2.13}$$

we obtain that $y^b \in \epsilon_b \Omega^b$ and

$$G_b \cap \partial F_b \subset (\partial F_b)_2 \subset \{|y_b| = D_b\},$$

which implies *(i)*.

*Case 2*: $a > b$. If $y \in G_b \cap (\partial F_a)_2$, then $|y_a| = D_a$. But $|y_a| = |P_a(y_b)|$ is a function of $y_b$, so that *(i)* holds.

Suppose now that $y \in G_b \cap (\partial F_a)_1$. Then $y^a \in \partial(\epsilon_a \Omega^a)$. By the inductive hypothesis *(i)*,

$$\partial(\epsilon_a \Omega_a) \cap \epsilon_a G^a_b$$

depends only on $y^a_b = P^a(y_b)$. It therefore suffices to ensure that $y^a \in \epsilon_a G^a_b$ whenever $y \in G_b$. But

$$G^a_b = \{y^a : \ |y^b| < \mu^a_b\}, \ G_b = \{y : \ |y^b| < \mu_b\},$$

hence we need only assume that

$$\mu_b < \epsilon_a \mu^a_b. \tag{5.2.14}$$

*Case 3*: $a < b$. In this case we will prove that, if (5.2.13) holds and if

$$c_0 D_a < D_b, \tag{5.2.15}$$

then $G_b \cap \partial F_a \subset \Omega$, so that $(G_b \cap \partial F_a) \cap (\partial\Omega) = \emptyset$. In fact, we will prove that $G_b \cap \partial F_a \subset F_b$.

Suppose that $y \in G_b \cap \partial F_a$. Since $y \in G_b$, we have $|y^b| < \mu_b < \epsilon_b K_2(b)$ by (5.2.13). Hence by inductive hypothesis *(iii)* we have

$$y^b \in \epsilon_b \Omega^b. \tag{5.2.16}$$

Since $y \in F_a$, we have $|y_a| \leq D_a$, so that by (5.2.9) $|y_b| = |P_b(y_a)| \leq c_0 D_a$. But from (5.2.15) we have $|y_b| < D_b$. This together with (5.2.16) implies that $y \in F_b$.

*Case 4*: $a \not\leq b, b \not\leq a$. Then $a < a \vee b, b < a \vee b$. By (5.2.10), we have

$$|y^{a\vee b}| \leq c_0(|y^a| + |y^b|).$$

Since $y^a \in \overline{\epsilon_a \Omega_a}$, by the inductive hypothesis *(ii)* we have

$$|y^a| \leq \epsilon_a K_1(a).$$

Since $y \in G_b$,

$$|y^b| < \mu_b.$$

Hence if we assume that:

$$\mu_b < \frac{1}{2c_0}\mu_{a\vee b}, \tag{5.2.17}$$

$$\epsilon_a K_1(a) < \frac{1}{2c_0}\mu_{a\vee b}, \tag{5.2.18}$$

we obtain that

$$|y^{a\vee b}| < \mu_{a\vee b}.$$

If $a \vee b < \emptyset$, then $y \in G_{a\vee b}$, which reduces this case to the case 3) with $b$ replaced by $a \vee b$. It follows that

$$(G_b \cap \partial F_a) \subset (G_{a\vee b} \cap \partial F_a) \subset F_{a\vee b} \subset \Omega.$$

If $a \vee b = \emptyset$, let

$$\mu_\emptyset < D_{a_{\min}}. \tag{5.2.19}$$

Then $|y^\emptyset| < \mu_\emptyset$ implies that $y = y^\emptyset \in F_{a_{\min}}$, so that $y \notin \partial\Omega$.

We have proved that *(i)* holds if the following conditions are satisfied: (5.2.13) for all $b \in \mathcal{A}$, (5.2.14) for all $\emptyset > a > b$, (5.2.15) for all $a < b < \emptyset$, (5.2.17) and (5.2.18) for all $a, b \in \mathcal{A}$, and (5.2.19).

*Step 2.* Relabelling the cluster decompositions involved, we see that *(i)* holds provided that for all $b < a < \emptyset$:

$$\begin{aligned} &(1)\ \mu_a < \epsilon_a K_2(a), && (4)\ \mu_b < (2c_0)^{-1}\mu_a, \\ &(2)\ \mu_b < \epsilon_a \mu_b^a, && (5)\ \epsilon_b K_1(b) < (2c_0)^{-1}\mu_a, \\ &(3)\ c_0 D_b < D_a, && (6)\ \mu_\emptyset < D_{a_{\min}}. \end{aligned} \tag{5.2.20}$$

Let

$$\alpha_1 = \tfrac{1}{2}\min_{b<\emptyset}(1, K_2(b)),\ \ \alpha_2 = \tfrac{1}{2}\min_{b<a<\emptyset}(1, \mu_b^a),$$
$$\alpha_3 = \tfrac{1}{2}\min_{b<\emptyset}\left(1, (2c_0)^{-1}, (2c_0K_1(b))^{-1}\right).$$

Then the equations (5.2.20) are satisfied if:

$$(1')\ \mu_a < \alpha_1\epsilon_a, \qquad (4')\ \mu_b < \alpha_3\mu_a,$$
$$(2')\ \mu_b < \epsilon_a\alpha_2, \qquad (5')\ \epsilon_b < \alpha_3\mu_a,$$
$$(3')\ D_b < \alpha_3 D_a, \qquad (6')\ \mu_\emptyset < D_{a_{\min}},$$

for all $b < a < \emptyset$.

Recall that $\mu_a$, $\epsilon_a$, and $D_a$ were defined in (5.2.11) with $C_1 < 1$, $C_2 < 1$, $D_1 < 1$, and that $b < a < \emptyset$ implies $\sharp b > \sharp a \geq 1$. We therefore obtain that:

$(1')$ is equivalent to $C_2^{\sharp a} < \alpha_1$, which is satisfied if $C_2 < \alpha_1$.

$(2')$ is equivalent to $\epsilon_b C_2^{\sharp b} < \epsilon_a\alpha_2$, i.e., $C_1^{\sharp b - \sharp a} C_2^{\sharp b} < \alpha_2$, which holds if $C_2 \leq \alpha_2$.

$(3')$ is equivalent to $D_1^{\sharp b - \sharp a} < \alpha_3$, which holds provided that $D_1 < \alpha_3$.

$(4')$ is equivalent to $(C_1C_2)^{\sharp b - \sharp a} < \alpha_3$, which holds if $C_1C_2 < \alpha_3$.

$(5')$ is equivalent to $C_1^{\sharp b - \sharp a} < \alpha_3 C_2^{\sharp a}$. The left hand side is maximized when $\sharp b - \sharp a = 1$, and the right hand side is minimized when $\sharp a = N$, so we only need to assume that $C_1 < \alpha_3 C_2^N$.

$(6')$ is equivalent to $1 < D_0 D_1^N$, which holds if $D_0 > D_1^{-N}$.

It thus suffices to choose $C_2, C_1, D_1, D_0$ (in this order) so that

$$(5.2.21)\qquad C_2 < \min(\alpha_1, \alpha_2, \alpha_3),\ C_1 < \alpha_3 C_2^N,\ D_1 < \alpha_3,\ D_0 > D_1^{-N}.$$

This completes the proof of the lemma. □

### 5.2.4. Construction of admissible functions.

PROPOSITION 5.2.7. *Let $a \in \mathcal{A}_*^{\mathrm{cc}}$. There is a function $\mathcal{F}^a \in C_0^\infty(Y^{a,q})$ such that:*

*(i) there is a constant $c > 0$ such that*

$$\mathcal{F}^a(y^{a,q}) \equiv 1 \textit{ for } |y^{a,q}| \leq c;$$

*(ii) for any $b \in \mathcal{A}$, $b \leq a$, there is $\epsilon_b > 0$ such that*

$$\partial_{y^{b,q}}\mathcal{F}^a(y^{a,q}) = 0 \textit{ for } |y^{b,q}| \leq \epsilon_b;$$

*(iii) $y^{a,q}\cdot\nabla\mathcal{F}^a(y^{a,q}) \leq 0$.*

PROOF. Since, as in the proof of Proposition 5.2.6, the whole analysis will take place in the subspace $Y^{a,q}$, we may assume without loss of generality that $a = \emptyset$ and $Y^{a,q} = Y$. The condition $\emptyset \in \mathcal{A}_*^{\mathrm{cc}}$ implies that for all $b \in \mathcal{A}_*$ the projections $P_b = \pi_b^q$ are well defined and $\operatorname{Ran}\pi_b^q = Y_b$; it is easy to see that they also satisfy (5.2.6).

Let $\Omega \subset Y$ be the set constructed in Proposition 5.2.6 for the family of projections $\pi_a^q$. Note that $\Omega$ depends on a set of parameters

$$\alpha = \{(C_1(b), C_2(b), D_0(b), D_1(b) : \ b \in \mathcal{A}_*\} \in \mathbb{R}^M.$$

To emphasize this dependence, we will write $\Omega = \Omega_\alpha$. From the proof of Proposition 5.2.6 (specifically from equation (5.2.21), we see that there is an open set $K$ of values of $\alpha$ for which $\Omega_\alpha$ has the required properties. Pick a function $\phi \in C_0^\infty(K)$ so that

$$0 \leq \phi(\alpha) \leq 1, \quad \int \phi(\alpha) d\alpha = 1,$$

and set

$$\mathcal{F}^a(y^{a,q}) = \int \mathbf{1}_{\Omega_\alpha^a}(y^{a,q})\phi(\alpha)d\alpha.$$

It is straightforward to verify using Proposition 5.2.6 that $\mathcal{F}^a$ has the required properties. □

PROPOSITION 5.2.8. *For $a \in \mathcal{A}^{\mathrm{cc}}$, let $\mathcal{F}^a$ be the function constructed in Proposition 5.2.7. Then:*

*(i) if $\emptyset \in \mathcal{A}_*^{\mathrm{cc}}$, $\mathcal{F}^\emptyset$ is an admissible function;*

*(ii) let $a \in \mathcal{A}^{\mathrm{cc}}$ and let $F_a \in S^0(Y_a)$ satisfy*

$$\operatorname{supp} F_a \subset \{|y^b| \geq \delta > 0\}$$

*for all $b \in \mathcal{A}$, $b \not\leq a$. Then, if $\epsilon > 0$ is small enough, the function*

$$F(y) = F_a(y_a^q)\mathcal{F}^a(\epsilon^{-1}y^{a,q})$$

*is admissible and satisfies $F(y) = 0$ in a neighbourhood $\{|y^b| \leq \sigma\}$, $\sigma > 0$, of $Y_b$ for all $b \not\leq a$.*

PROOF. *(i)* is a direct consequence of Proposition 5.2.7.

*(ii)* Let $b \in \mathcal{A}$. We will consider the cases $b \leq a$ and $b \not\leq a$ separately.

*Case 1*: $b \leq a$. Since $a \in \mathcal{A}^{\mathrm{cc}}$, we must have $b \in \mathcal{A}^{\mathrm{c}}$. $F_a(y_a^q)$ depends only on $y_b^q$, since $y_a^q = (y_b^q)_a^q$, and $\mathcal{F}^a$ depends only on $y_b^q$ in a neighbourhood of $Y_b$ by Proposition 5.2.7. Hence in this case the condition *(i)* of admissibility is satisfied.

*Case 2*: $b \not\leq a$ (note that this case includes all cluster decompositions $b \in \mathcal{A}^{\mathrm{n}}$). Let $R$ be large enough so that $\operatorname{supp} \mathcal{F}^a \subset \{|y^{a,q}| \leq R\}$. Then on $\operatorname{supp} F(y)$ we have

$$y^b = (y_a^q + y^{a,q})^b,$$

$$|y^b| \geq |(y_a^q)^b| - |(y^{a,q})^b| \geq \delta - |y^{a,q}| \geq \delta - \epsilon R.$$

If $\epsilon > 0$ is small enough, $\sigma = \delta - \epsilon R > 0$. Hence $F(y) \equiv 0$ for $|y^b| \leq \sigma$, $\sigma > 0$. This proves the proposition. □

## 5.3. Propagation estimates in the transversal direction

In this section we will prove the propagation estimates mentioned in the introduction to this chapter for the unitary evolution $U(t,0)$ generated by a time-dependent Hamiltonian

$$H(t) = H + I(t,x)$$

of the type considered in Chapter 4. We assume that *(V1)*, *(SC)* hold and that

$$|\partial_x^\alpha I(t,x)| \leq C\langle t\rangle^{-\mu_1-|\alpha|},\ |\alpha| \leq 2,\ \mu_1 > 0. \tag{5.3.1}$$

**5.3.1. The strongly charged case.** In this subsection we prove that if the system under consideration has no neutral subsystems, i.e.,

$$q_{i_1} + \cdots + q_{i_n} \neq 0 \text{ for any } \{i_1, \ldots, i_n\} \subset \{1, \ldots, N\}, \tag{5.3.2}$$

then the system cannot propagate to infinity in directions transversal to the magnetic field at a rate faster than $O(t^\alpha)$ for some $\alpha < 1$. (Note that (5.3.2) is slightly stronger than *(Q2)* in that we also assume that the total charge $Q = \sum_{i=1}^N q_i$ is non-zero.) This result will be used in the proof of asymptotic completeness in the "charged" long-range case in Subsection 6.3.3.

The main result here is the following theorem.

THEOREM 5.3.1. *Assume (V1), (SC), (5.3.1), and (5.3.2). Then*

$$s - \lim_{t\to\infty} U^*(t)F(|t^{-\alpha}\mathbf{C}| \leq 1)U(t) = \mathbf{1}$$

*for any* $\alpha > \max(1/2, \mu_{\mathrm{s}}^{-1})$.

From Theorem 5.3.1 and Lemma 5.3.3 below we obtain the following corollary.

COROLLARY 5.3.2. *Under the assumptions of Theorem 5.3.1 we have*

$$s - \lim_{t\to\infty} U^*(t)F(|t^{-\alpha}y| \leq 1)U(t) = \mathbf{1}.$$

LEMMA 5.3.3. *Assume (V1), (SC), (5.3.1), and (5.3.2). Let* $F \in S^0(Y)$, $\chi \in C_0^\infty(\mathbb{R})$, $\alpha > 0$. *Then:*

$$(i)\ (F(t^{-\alpha}y) - F(t^{-\alpha}\mathbf{C}))\chi(H) = O(t^{-\alpha});$$

$$(ii)\ \text{s-}\lim_{t\to+\infty}(F(t^{-\alpha}y) - F(t^{-\alpha}\mathbf{C}))U(t) = 0.$$

PROOF. *(i)* follows from Lemma 5.1.3*(ii)*. It remains to prove *(ii)*. Let $H^+$ be the asymptotic energy observable associated to $U(t)$, constructed in

Section 4.2. Clearly, the set of $u \in \mathcal{H}$ such that $u = \chi(H^+)u$ for some $\chi \in C_0^\infty(\mathbb{R})$ is dense in $\mathcal{H}$. For such $u$ we have

$$\begin{aligned} & (F(t^{-\alpha}y) - F(t^{-\alpha}\mathbf{C}))U(t)u \\ = \; & (F(t^{-\alpha}y) - F(t^{-\alpha}\mathbf{C}))U(t)\chi(H^+)u \\ = \; & (F(t^{-\alpha}y) - F(t^{-\alpha}\mathbf{C}))\chi(H)U(t)u + o(1) \\ = \; & o(1), \end{aligned}$$

where we used the definition of $H^+$ and part *(i)* of the lemma. By density, this proves *(ii)*. □

The proof of Theorem 5.3.1 will be based on the two propositions which follow. Both of them are stated and proved in more generality than it is necessary for the proof of Theorem 5.3.1, where they are only used with $f \equiv g \equiv 0$. The more general case will be needed in the next subsection.

PROPOSITION 5.3.4. *Assume (V1), (SC), (5.3.1), and (5.3.2). Let* $F(y)$ *be an admissible function,* $G(y) = y \cdot \nabla F(y)$, $f, g \in S^0(Y)$, $\chi \in C_0^\infty(\mathbb{R})$, *and let* $\alpha > \max(1/2, \mu_{\mathrm{s}}^{-1})$. *Assume that either*

$$G(y) \geq f^2(y) - g^2(y) \tag{5.3.3}$$

*or*

$$G(y) \leq -f^2(y) + g^2(y) \tag{5.3.4}$$

*for all* $y \in Y$, *where* $g(y)$ *satisfies*

$$\int_1^\infty \|g(t^{-\alpha}\mathbf{C})\chi(H)U(t)u\|^2 t^{-1} dt \leq C\|u\|^2. \tag{5.3.5}$$

*Then:*

*(i)*

$$\int_1^\infty \|f(t^{-\alpha}\mathbf{C})\chi(H)U(t)u\|^2 t^{-1} dt \leq C\|u\|^2;$$

*(ii) the limit*

$$\lim_{t\to\infty} (U(t)u, \chi(H)F(t^{-\alpha}\mathbf{C})\chi(H)U(t)u)$$

*exists.*

PROOF. We will only consider the case when (5.3.4) is satisfied; the case when (5.3.3) holds is similar. We denote $u_t = U(t)u$.

Let $\Phi(t) = \chi(H)F(t^{-\alpha}\mathbf{C})\chi(H)$. We have

$$\frac{d}{dt}(u_t, \Phi(t)u_t) = (u_t, D\Phi(t)u_t), \tag{5.3.6}$$

where

$$
(5.3.7)\qquad \begin{aligned} D\Phi(t) = \;& \chi(H)\Big(\tfrac{\partial}{\partial t}F(t^{-\alpha}\mathbf{C}) + i[H, F(t^{-\alpha}\mathbf{C})]\Big)\chi(H) \\ & + [I(t,x), i\chi(H)]F(t^{-\alpha}\mathbf{C})\chi(H) + \text{h.c.} \\ & + \chi(H)[I(t,x), iF(t^{-\alpha}\mathbf{C})]\chi(H). \end{aligned}
$$

By (1.2.21) and (5.3.1), $[I(t,x), \chi(H)] = O(t^{-1-\mu_1})$.

To estimate $[I(t,x), F(t^{-\alpha}\mathbf{C})]$, we write using Lemma A.2 with $\delta = \frac{1}{2}$:

$$
I(t,x) = I_1(t,x) + I_2(t,x),
$$

where:

$$
\begin{aligned} & I_1(t,x) \in O(t^{-1-\mu_1}), \\ & \partial_x^\beta I_2(t,x) \in \begin{cases} O(t^{-\mu_1-|\beta|}), \ |\beta| \le 2, \\ O(t^{-2-\mu_1-\frac{1}{2}(|\beta|-2)}), \ |\beta| \ge 2. \end{cases} \end{aligned}
$$

It follows as in the proof of Lemma 5.2.4 that

$$
[I(t,x), F(t^{-\alpha}\mathbf{C})] = O(t^{-1-\mu_1}).
$$

Since $F$ is admissible, by Proposition 5.2.5 we have

$$
(5.3.8)\qquad \chi(H)[H, F(t^{-\alpha}\mathbf{C})]\chi(H) = O(t^{-1-\epsilon})
$$

for some $\epsilon > 0$. To estimate the other term on the right-hand side of (5.3.7), we compute using symbolic calculus, (5.3.3), and the sharp Gärding inequality (Proposition 1.2.7):

$$
(5.3.9)\qquad \begin{aligned} & \tfrac{\partial}{\partial t}F(t^{-\alpha}\mathbf{C}) = -\alpha t^{-1}G(t^{-\alpha}\mathbf{C}) + O(t^{-1-\alpha}) \\ & \ge \alpha t^{-1}\big(f^2(t^{-\alpha}\mathbf{C}) - g^2(t^{-\alpha}\mathbf{C})\big) + O(t^{-1-\alpha}), \end{aligned}
$$

hence

$$
(5.3.10)\qquad \begin{aligned} & \chi(H)\tfrac{\partial}{\partial t}F(t^{-\alpha}\mathbf{C})\chi(H) \\ & \ge \alpha t^{-1}\chi(H)\big(f^2(t^{-\alpha}\mathbf{C}) - g^2(t^{-\alpha}\mathbf{C})\big)\chi(H) + O(t^{-1-\alpha}). \end{aligned}
$$

Part *(i)* of the proposition follows now from Lemma 4.1.4 and from (5.3.5), (5.3.10).

It remains to prove *(ii)*. By (5.3.7), (5.3.8), (5.3.10), that there is a constant $C > 0$ such that

$$
(5.3.11)\qquad \begin{aligned} & J(t) := (u_t, D\Phi(t)u_t) - \alpha t^{-1}(u_t, \chi(H)f^2(t^{-\alpha}\mathbf{C})\chi(H)u_t) \\ & + \alpha t^{-1}(u_t, \chi(H)g^2(t^{-\alpha}\mathbf{C})\chi(H)u_t) + Ct^{-1-\epsilon}\|u\|^2 \ge 0. \end{aligned}
$$

Hence $\int_1^T J(t)dt$ is increasing in $T$, so that

$$
\int_1^\infty J(t)dt = \lim_{T\to\infty}\int_1^T J(t)dt
$$

either converges to a finite number or is equal to $+\infty$. The same must be true for

$$\int_1^\infty (u_t, D\Phi(t)u_t)dt, \tag{5.3.12}$$

since the last three terms in $J(t)$ are integrable in $t$. However, the integral (5.3.12) cannot be infinite, since

$$\int_1^T (u_t, D\Phi(t)u_t)dt = (u_T, \Phi(T)u_T) - (u_1, \Phi(1)u_1)$$

is bounded by $C\|u\|^2$ uniformly in $T$. Hence (5.3.12) exists and is finite, so that the limit

$$\lim_{t\to\infty}(u_t, \chi(H)F(t^{-\alpha}\mathbf{C})\chi(H)u_t)$$

exists. This proves *(ii)*. □

PROPOSITION 5.3.5. *Assume (V1), (SC), (5.3.1), and (5.3.2). Let $F$ be an admissible function such that $0 \leq F \leq 1$, $G(y) = y \cdot \nabla F(y)$, and let $\alpha > \max(1/2, \mu_{\mathrm{s}}^{-1})$.*

*(i) If $F \equiv 1$ in a neighbourhood of $0$ and $G$ satisfies (5.3.4), (5.3.5), then*

$$\begin{aligned} &s-\lim_{t\to\infty} U^*(t)F(t^{-\alpha}y)U(t) = \mathbf{1}, \\ &s-\lim_{t\to\infty} U^*(t)F(t^{-\alpha}\mathbf{C})U(t) = \mathbf{1}. \end{aligned} \tag{5.3.13}$$

*(ii) If $F \equiv 0$ in a neighbourhood of $0$ and $G$ satisfies (5.3.3) and (5.3.5), then*

$$s-\lim_{t\to\infty} U^*(t)F(t^{-\alpha}y)U(t) = 0,$$

$$s-\lim_{t\to\infty} U^*(t)F(t^{-\alpha}\mathbf{C})U(t) = 0.$$

PROOF. It suffices to prove *(i)*, since *(ii)* follows by applying *(i)* to the function $1 - F$. By Lemma 5.3.3, it suffices to prove the second estimate in (5.3.13).

Pick $u$ such that $u = \chi(H^+)u$ for some $\chi \in C_0^\infty(\mathbb{R})$, and $\beta$ such that $\alpha > \beta > \max(1/2, \mu_{\mathrm{s}}^{-1})$. Let $R > 1$, $F_R = F(\cdot/R)$, $G_R = G(\cdot/R)$, and

$$\Phi_R(t) = \chi(H)F_R(t^{-\beta}\mathbf{C})\chi(H).$$

As in the proof of Proposition 5.3.4*(i)*, we have

$$\frac{\partial}{\partial t}(u_t, \Phi_R(t)u_t) = (u_t, D\Phi_R(t)u_t),$$

$$\begin{aligned} D\Phi_R(t) = \; & \chi(H)\Big(\tfrac{\partial}{\partial t}F_R(t^{-\beta}\mathbf{C}) + [H, \mathrm{i}F_R(t^{-\beta}\mathbf{C})]\Big)\chi(H) \\ & + [I(t,x), \mathrm{i}\chi(H)]F_R(t^{-\beta}\mathbf{C})\chi(H) + \mathrm{h.c.} \\ & + \chi(H)[I(t,x), \mathrm{i}F_R(t^{-\beta}\mathbf{C})]\chi(H), \end{aligned}$$

where the last two terms on the right-hand side are bounded by $O(t^{-1-\epsilon})$, $\epsilon > 0$, uniformly in $R \geq 1$. Let $s = R^{1/\beta}t$, so that $Rt^\beta = s^\beta$. By Proposition 5.2.5, with $\alpha$ and $t$ replaced by $\beta$ and $s$ respectively, we have

$$\chi(H)[H, F_R(t^{-\beta}\mathbf{C})]\chi(H) = O(s^{-1-\epsilon}) = O(R^{-\epsilon_1}t^{-1-\epsilon})$$

for some $\epsilon > 0$. Moreover, using (5.3.4), the sharp Gärding inequality (Proposition 1.2.7), and symbolic calculus, we obtain that

$$\begin{aligned}
&\chi(H)\tfrac{\partial}{\partial t}F_R(t^{-\beta}\mathbf{C})\chi(H)\\
= \;& -\beta t^{-1}\chi(H)G_R(t^{-\beta}\mathbf{C})\chi(H) + O(R^{-1}t^{-1-\beta})\\
\geq \;& \beta t^{-1}\chi(H)f_R^2(t^{-\beta}\mathbf{C})\chi(H) - \beta t^{-1}\chi(H)g_R^2(t^{-\beta}\mathbf{C})\chi(H) + O(R^{-1}t^{-1-\beta})\\
\geq \;& \beta t^{-1}\chi(H)f_R^2(t^{-\beta}\mathbf{C})\chi(H) - \beta t^{-1}\chi(H)g_R^2(t^{-\beta}\mathbf{C})\chi(H) + O(R^{-1}t^{-1-\beta}),
\end{aligned}$$

where $f_R = f(\cdot/R)$ and $g_R = g(\cdot/R)$. Hence there is a constant $C$ such that for all $R > 1$

$$\begin{aligned}
D\Phi_R(t) \geq \;& \beta t^{-1}\chi(H)f_R^2(t^{-\beta}\mathbf{C})\chi(H)\\
& -\beta t^{-1}\chi(H)g_R^2(t^{-\beta}\mathbf{C})\chi(H) - Ct^{-1-\epsilon}.
\end{aligned} \tag{5.3.14}$$

As in the proof of Proposition 5.3.4*(ii)*, we prove that the integral

$$\int_1^\infty (u_t, D\Phi_R(t)u_t)dt$$

is finite, and that the limit

$$L_R(u) := \lim_{t\to\infty}(u_t, \Phi_R(t)u(t))$$

exists. Moreover, from (5.3.14) we see that for any $t_0 \geq 1$:

$$\begin{aligned}
L_R(u) &= (u_{t_0}, \Phi_R(t_0)u_{t_0}) + \textstyle\int_{t_0}^\infty (u_t, D\Phi_R(t)u_t)dt\\
&\geq (u_{t_0}, \Phi_R(t_0)u_{t_0}) + \beta \textstyle\int_{t_0}^\infty (u_t, \chi(H)f_R^2(t^{-\beta}\mathbf{C})\chi(H)u_t)t^{-1}\,dt\\
&\quad -\beta \textstyle\int_{t_0}^\infty (u_t, \chi(H)g_R^2(t^{-\beta}\mathbf{C})\chi(H)u_t)t^{-1}\,dt - Ct_0^{-\epsilon}\|u\|^2,
\end{aligned} \tag{5.3.15}$$

where $C$ does not depend on $R$.

Let $\delta > 0$. Then for $t_0$ large enough we have

$$Ct_0^{-\epsilon} < \delta/4.$$

Next, using the same change of variables as above ($s = R^{1/\beta}t$), we obtain that

$$\begin{aligned}
I_f &:= \textstyle\int_{t_0}^\infty (u_t, \chi(H)f_R^2(t^{-\beta}\mathbf{C})\chi(H)u_t)t^{-1}\,dt\\
&= \textstyle\int_{R^{1/\beta}t_0}^\infty (u_t, \chi(H)f^2(s^{-\beta}\mathbf{C})\chi(H)u_t)s^{-1}\,ds.
\end{aligned}$$

The integral $I_f$ is convergent by Proposition 5.3.4*(i)*, and, if $t_0$ is large enough, we have

$$|\beta I_f| < \delta/4$$

for all $R > 1$. The same argument, with $f$ replaced by $g$, is used to estimate the second integral on the right hand side of (5.3.15) by $\delta/4$ for large $t_0$, uniformly in $R$. Finally, if $R$ is sufficiently large:

$$(u_{t_0}, \Phi_R(t_0)u_{t_0}) \geq (1 - \delta/4)\|u\|^2,$$

since $s - \lim_{R\to\infty} F_R(t^{-\beta}\mathbf{C}) = \mathbf{1}$. Hence by (5.3.15):

$$L_R(u) \geq (1 - \delta)\|u\|^2,$$

provided that $R$ is large enough. Thus we have proved that

$$\lim_{t\to\infty} (u_t, \chi(H)F_R(t^{-\beta}\mathbf{C})\chi(H)u_t) \geq (1 - \delta)\|u\|^2, \tag{5.3.16}$$

However, since $\alpha > \beta$, for any fixed $R$ there is a $T > 0$ such that

$$F^2(t^{-\alpha}y) \geq F_R(t^{-\beta}y), \ t > T.$$

Note that by symbolic calculus,

$$F(t^{-\alpha}\mathbf{C})^2 \geq F^2(t^{-\alpha}\mathbf{C}) + O(t^{-\alpha}). \tag{5.3.17}$$

Together with Proposition 1.2.7, this implies

$$F(t^{-\alpha}\mathbf{C})^2 \geq F_R(t^{-\beta}\mathbf{C}) - Ct^{-\beta}. \tag{5.3.18}$$

Applying Proposition 5.3.4 *(ii)* to the function $F^2$, and using (5.3.17), we obtain that

$$\lim_{t\to\infty} \|F(t^{-\alpha}\mathbf{C})\chi(H)u_t\|^2 \text{ exists.}$$

By (5.3.16), (5.3.18), this yields that

$$\lim_{t\to\infty} \|F(t^{-\alpha}\mathbf{C})\chi(H)u_t\|^2 \geq (1 - \delta)\|u\|^2.$$

Next, since $0 \leq F \leq 1$, by (5.3.17) and Proposition 1.2.7 we have

$$F(t^{-\alpha}\mathbf{C})^2 \leq \mathbf{1} + O(t^{-\alpha}),$$

hence

$$U^*(t)\chi(H)F(t^{-\alpha}\mathbf{C})^2\chi(H)U(t) \leq U^*(t)\chi^2(H)U(t) + O(t^{-\alpha}).$$

Using also that $u = \chi(H^+)u$, we find that

$$\lim_{t\to\infty} \|F(t^{-\alpha}\mathbf{C})U(t)u\|^2 = \|u\|^2,$$

hence by a density argument

$$\text{w}-\lim_{t\to\infty} U^*(t)F(t^{-\alpha}\mathbf{C})^2U(t) = \mathbf{1},$$

and

$$\text{w}-\lim_{t\to\infty} U^*(t)(\mathbf{1} - F(t^{-\alpha}\mathbf{C})^2)U(t) = 0. \tag{5.3.19}$$

Thus it remains to prove that

$$\text{s}-\lim_{t\to\infty} U^*(t)(\mathbf{1} - F(t^{-\alpha}\mathbf{C}))U(t) = 0, \tag{5.3.20}$$

Since

$$(1 - F(y))^2 \leq 1 - F^2(y),$$

we have by symbolic calculus and Proposition 1.2.7

$$(\mathbf{1} - F(t^{-\alpha}\mathbf{C}))^2 \leq \mathbf{1} - F(t^{-\alpha}\mathbf{C})^2 + O(t^{-\alpha}),$$

so that (5.3.20) follows from (5.3.19). □

PROOF OF THEOREM 5.3.1. Pick $\beta$ such that $\alpha > \beta > \max(1/2, \mu_{\mathrm{s}}^{-1})$, and let $\mathcal{F}^{\emptyset}$ be the function constructed in Proposition 5.2.7 for $a = \emptyset$. By Proposition 5.3.5 and Lemma 5.3.3 we have

$$s - \lim_{t\to\infty} \mathcal{F}^{\emptyset}(t^{-\beta}y)U(t) = \mathbf{1}.$$

However, since $\alpha > \beta$,

$$\mathcal{F}^{\emptyset}(t^{-\beta}y) \leq F(t^{-\alpha}|y| \leq 1)$$

for $t$ large enough. This proves the theorem. □

The next proposition is an important consequence of Theorem 5.3.1, which will be used in Chapter 6.

PROPOSITION 5.3.6. *Let $a \in \mathcal{A}^{\mathrm{cc}}$, and let $U_a(t), \tilde{U}_a(t)$ be the unitary evolutions generated by $H_a + I(t,x)$ and $H_a + \tilde{I}(t,x)$ respectively, where $x = (y,z)$, $I(t,x)$ is a real-valued time-dependent potential satisfying (4.1.2), and*

$$\tilde{I}(t,y,z) = I(t,0,z).$$

*Assume that $\mu_{\mathrm{l}} > \frac{1}{2}$, $\mu_{\mathrm{l}}\mu_{\mathrm{s}} > 1$. Then the limits*

$$\text{s-}\lim_{t\to+\infty} \tilde{U}_a^*(t)U_a(t) =: \Gamma_a^+, \tag{5.3.21}$$

$$\text{s-}\lim_{t\to+\infty} U_a^*(t)\tilde{U}_a(t) = \Gamma_a^{+*} \tag{5.3.22}$$

*exist and are unitary.*

PROOF. We will prove the existence of the limit (5.3.21), the proof of the existence of (5.3.22) being similar. Let $H_a^+$ be the asymptotic energy operator associated to $U_a(t)$, constructed in Proposition 4.2.1. Let $u \in \mathcal{H}$. By density, we may assume that $u = \chi^2(H_a^+)u$, $\chi \in C_0^\infty(\mathbb{R})$. Hence

$$U_a(t)u = \chi^2(H_a)U_a(t)u + o(1). \tag{5.3.23}$$

We write $H_a$ as the sum of cluster Hamiltonians:

$$H_a = \sum_{l=1}^{\sharp a} H[C_l].$$

(see (2.2.1)). Let $\mathbf{C}$ be the center of orbit observable on $Y$; we may write $\mathbf{C} = (\mathbf{C}_1, \ldots, \mathbf{C}_{\sharp a})$, where $\mathbf{C}_l$ is the projection of $\mathbf{C}$ on $Y[C_l]$. Since $a \in \mathcal{A}^{\mathrm{cc}}$, each $C_l$ is strongly charged. For $1 \leq l \leq \sharp a$, let $\mathcal{F}_l$ be the admissible function of compact support on $Y[C_l]$ constructed in Proposition 5.2.7. Define

$$F(y) := \Pi_1^{\sharp a} \mathcal{F}_l(y[C_l]), \tag{5.3.24}$$

where $y[C_l]$ be the projection of $y$ on $Y[C_l]$. Applying Propositions 5.3.4 and 5.3.5 to each $H[C_l]$, we obtain the estimates

$$U(t)u = F(t^{-\alpha}\mathbf{C})U(t)u + o(1), \tag{5.3.25}$$

$$\int_1^{+\infty} \|F_1(t^{-\alpha}C)\chi(H_a)U(t)u\|^2 \frac{\mathrm{d}t}{t} \leq C\|u\|^2, \tag{5.3.26}$$

where $U(t)$ is either $U_a(t)$ or $\tilde{U}_a(t)$, $\chi \in C_0^\infty(\mathbb{R})$, $F_1(y) = (y \cdot \nabla F(y))^{\frac{1}{2}}$, and $\alpha$ satisfies $\max(\frac{1}{2}, \mu_{\mathrm{s}}^{-1}) < \alpha < \mu_{\mathrm{l}}$. By (5.3.23) and (5.3.25), it suffices to prove that:

$$\lim_{t\to+\infty} \tilde{U}_a^*(t)F(t^{-\alpha}\mathbf{C})\chi^2(H_a)U_a(t)u \tag{5.3.27}$$

exists.

By (5.3.24) and Proposition 5.2.5, we have

$$[(H_a + \mathrm{i})^{-1}, F(t^{-\alpha}\mathbf{C})] = O(t^{-1-\epsilon}), \epsilon > 0.$$

From this and (1.2.21), we see that the existence of (5.3.27) is equivalent to the existence of

$$\lim_{t\to+\infty} \tilde{U}_a^*(t)\chi(H_a)F(t^{-\alpha}\mathbf{C})\chi(H_a)U_a(t)u. \tag{5.3.28}$$

By Lemma 4.1.5, to prove that (5.3.28) exists it suffices to prove that

$$\int_1^\infty \|(\mathrm{D}_{21}M(t))U_a(t)u\|dt < \infty, \tag{5.3.29}$$

where $M(t) = \chi(H_a)F(t^{-\alpha}\mathbf{C})\chi(H_a)$ and

$$\mathrm{D}_{21}M(t) = \partial_t M(t) + \mathrm{i}\Big((H_a + \tilde{I}(t,x))M(t) - M(t)(H_a + I(t,x))\Big).$$

We have $\mathrm{D}_{21}M(t) = I_1 + I_2 + I_3 + I_4 + I_5$, where:

$$I_1 = -\alpha t^{-1}\chi(H_a)F_1(t^{-\alpha}\mathbf{C})\chi(H_a),$$

$$I_2 = \chi(H_a)[H_a, \mathrm{i}F(t^{-\alpha}C)]\chi(H_a),$$

$$I_3 = [\tilde{I}(t,x), \mathrm{i}\chi(H_a)]F(t^{-\alpha}\mathbf{C})\chi(H_a) + \chi(H_a)F(t^{-\alpha}\mathbf{C})[\chi(H_a), \mathrm{i}I(t,x)],$$

$$I_4 = \chi(H_a)[F(t^{-\alpha}\mathbf{C}), \mathrm{i}I(t,x)]\chi(H_a),$$

$$I_5 = \chi(H_a)\mathrm{i}(\tilde{I}(t,x) - I(t,x))F(t^{-\alpha}\mathbf{C})\chi(H_a).$$

We will prove (5.3.29) for each $I_j$ separately. For $I_1$ and $I_2$, (5.3.29) follows from (5.3.26) and Proposition 5.2.5 respectively. For $I_3$, it follows from

$$[\tilde{I}(t,x), \mathrm{i}\chi(H_a)],\ [I(t,x), \mathrm{i}\chi(H_a)] = O(t^{-1-\epsilon}),$$

which is proved as in Proposition 4.2.1. To estimate $I_4$, we apply Lemma A.2 with $\delta = \frac{1}{2}$ to $I(t,x)$ and write $I(t,x) = I_1(t,x) + I_2(t,x)$, where

$$I_1(t,x) = O(t^{-1-\mu_l}),$$

$$\partial_x^\beta I_2(t,x) = \begin{cases} O(t^{-(\mu_l+|\beta|)}), & |\beta| \leq 2, \\ O(t^{-\mu_l-2-\frac{1}{2}(|\beta|-2)}), & |\beta| \geq 3. \end{cases}$$

Then $[F(t^{-\alpha}\mathbf{C}), I_1(t,x)] = O(t^{-1-\mu_l})$ by undoing the commutator, and the same holds for $[F(t^{-\alpha}\mathbf{C}), I_2(t,x)]$ by symbolic calculus. Hence (5.3.29) holds for $I_4$. Finally, by Lemma 5.1.3 we have

$$\begin{aligned}&(\tilde{I}(t,x) - I(t,x))F(t^{-\alpha}\mathbf{C})\chi(H_a) \\ &= (\tilde{I}(t,x) - I(t,x))F(|y| \leq ct^\alpha)F(t^{-\alpha}\mathbf{C})\chi(H_a) + O(t^{-2\alpha}).\end{aligned} \tag{5.3.30}$$

Since $\tilde{I}(t,x) = I(t,0,z)$, we obtain from Taylor's formula that

$$(\tilde{I}(t,x) - I(t,x))F(|y| \leq ct^\alpha) = O(t^{-1-\mu_l+\alpha}), \tag{5.3.31}$$

which is integrable since $\mu_l > \alpha > \frac{1}{2}$. □

### 5.3.2. The general case.

THEOREM 5.3.7. *Let $F \in S^0(Y)$ satisfy*

$$\operatorname{dist}(\operatorname{supp} F,\ Y_a) > \epsilon > 0 \textit{ for all } a \notin \mathcal{A}^{cc}, \tag{5.3.32}$$

*and let $\alpha > \max(1/2, \mu_s^{-1})$. Then*

$$s - \lim_{t\to\infty} F(t^{-\alpha}y)U(t) = 0. \tag{5.3.33}$$

*If, moreover, there are constants $\delta, R > 0$ such that for all $a \in \mathcal{A}^{cc}$*

$$\operatorname{supp} F \cap \{|y^a| < \delta\} \subset \bigcup_{(ij)\not\leq a} \{|y_i - y_j| < R\}, \tag{5.3.34}$$

*then*

$$\int_1^\infty \|F(t^{-\alpha}y)\chi(H)U(t)u\|^2 t^{-1}\, dt \leq C\|u\|^2 \tag{5.3.35}$$

*for any $\chi \in C_0^\infty(\mathbb{R})$.*

Theorem 5.3.7 will follow from Theorem 5.3.8 below. To state it, we first introduce some notation. Recall that for a cluster decomposition $a \in \mathcal{A}$,

$$|y|_a = \min_{(ij)\not\leq a} |y_i - y_j|.$$

We define:

$$\mathcal{Y}_{a,\delta} = \{y \in Y : \ |y|_a > \delta\},$$

$$\mathcal{Y}_{a,\delta_1,\delta_2} = \mathcal{Y}_{a,\delta_1} \setminus \overline{\mathcal{Y}_{a,\delta_2}} \text{ for } 0 < \delta_1 < \delta_2 < \infty,$$

where $\overline{\Omega}$ stands for the closure of the set $\Omega$ in $Y$. Note that $y \in \mathcal{Y}_{a,\delta_1,\delta_2}$ if and only if $|y_i - y_j| > \delta_1$ for *all* $(ij) \not\leq a$ and $|y_i - y_j| < \delta_2$ for *some* $(ij) \not\leq a$.

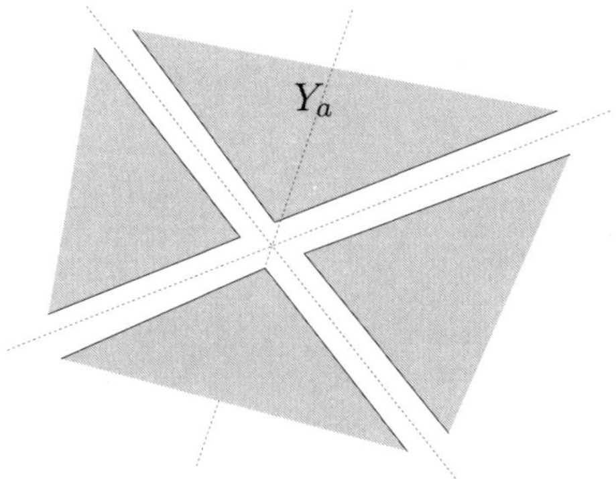


**Fig. 5.4.** The set $\mathcal{Y}_{a,\delta}$ for $N = 3$, $\sharp a = 2$

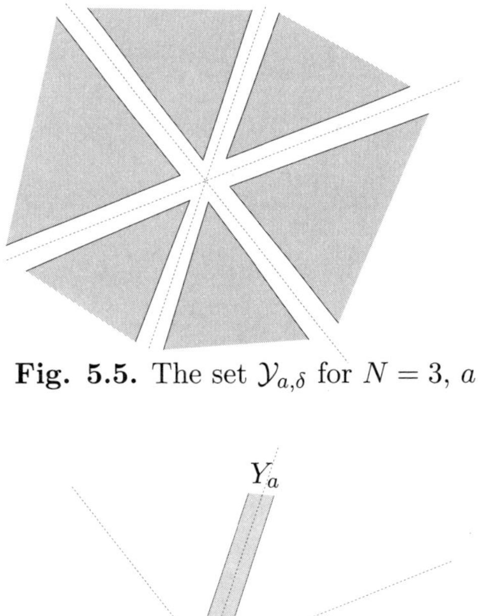

**Fig. 5.5.** The set $\mathcal{Y}_{a,\delta}$ for $N = 3$, $a = a_{\min}$

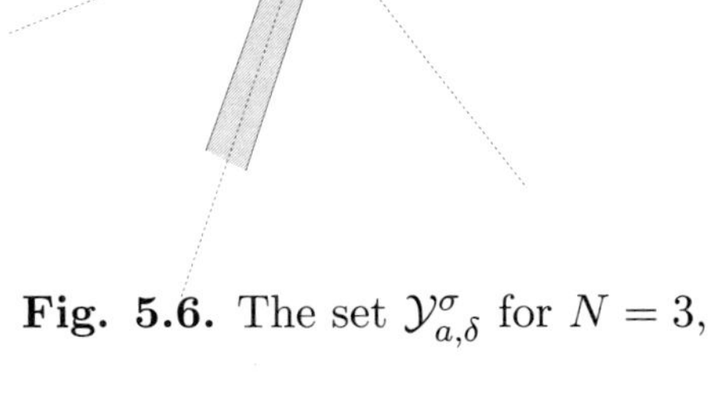


**Fig. 5.6.** The set $\mathcal{Y}^{\sigma}_{a,\delta}$ for $N = 3$, $\sharp a = 2$

THEOREM 5.3.8. *Let $a \in \mathcal{A}^{\rm cc}$ and $\alpha > \max(1/2, \mu_{\rm s}^{-1})$.*

*(i) If $F \in S^0(Y)$ is supported in $\mathcal{Y}_{a,\delta}$ for some $\delta > 0$, then*

$$s - \lim_{t\to\infty} F(t^{-\alpha}y)U(t) = 0; \tag{5.3.36}$$

*(ii) If $F \in S^0(Y)$ is supported in $\mathcal{Y}_{a,\delta_1,\delta_2}$ for some $0 < \delta_1 < \delta_2 < \infty$, then for $\chi \in C_0^\infty(\mathbb{R})$*

$$\int_1^\infty \|F(t^{-\alpha}y)\chi(H)U(t)u\|^2 t^{-1}\, dt \le C\|u\|^2. \tag{5.3.37}$$

PROOF OF THEOREM 5.3.7, GIVEN THEOREM 5.3.8. Let $F \in S^0(Y)$ satisfying (5.3.32). By Proposition 2.3.1, for any $\gamma > 0$ there is a partition of unity

$$1 = \sum_{a\in\mathcal{A}} q_a$$

such that $q_a \in S^0(Y)$ is supported in $\{|y^a| < \delta_a,\ |y|_a > \epsilon_a\}$ for some $\delta_a, \epsilon_a$ satisfying $0 < \delta_a < \gamma$, $0 < \epsilon_a < \gamma$. In particular, $q_a$ is supported in $\mathcal{Y}_{a,\epsilon_a}$, and by (5.3.32):

$$q_a F = 0 \text{ if } a \notin \mathcal{A}^{\rm cc},$$

provided that $\gamma$ is small enough. Writing

$$F = \sum_{a\in\mathcal{A}^{\rm cc}} q_a F,$$

and applying Theorem 5.3.8*(i)* to the functions $q_a F$, we obtain the first part of Theorem 5.3.7.

Assume now that $F$ also satisfies (5.3.34). We choose the partition of unity $\{q_a\}$ as above with $\gamma < \delta$, where $\delta$ is the same as in (5.3.34). Then $q_a F$ is supported in $\mathcal{Y}_{a,\delta_a,R}$, so that (5.3.35) follows by applying Theorem 5.3.8*(ii)* to functions $q_a F$. □

We first construct a family of auxiliary functions which will be used in the proof of Theorem 5.3.8. Let $\phi \in C^\infty(\mathbb{R})$ satisfy:

$$\phi(s) \equiv 0 \text{ for } s \le 1/4,$$

$$\phi(s) \equiv 1 \text{ for } s \ge 2,$$

$$\phi'(s) \ge 0 \text{ for all } s \in \mathbb{R},$$

$$s \cdot \phi'(s) \ge \theta_0 > 0 \text{ for } 1/2 \le s \le 1.$$

For a cluster decomposition $a \in \mathcal{A}$, let

$$F_{a,\delta}(y) = \prod_{(ij)\not\le a} \phi(\delta^{-1}|y_i - y_j|). \tag{5.3.38}$$

The function $F_{a,\delta}$ has the following obvious properties.

LEMMA 5.3.9. *Let $F_{a,\delta}$ be given by (5.3.38). Then:*

$$\begin{aligned} &(i) && 0 \le F_{a,\delta} \le 1, \\ &(ii) && \operatorname{supp} F_{a,\delta} \subset \mathcal{Y}_{a,\delta/4}, \\ &(iii) && F_{a,\delta} \equiv 1 \text{ on } \mathcal{Y}_{a,2\delta}, \\ &(iv) && y \cdot \nabla F_{a,\delta} \ge 0, \\ &(v) && y \cdot \nabla F_{a,\delta} \ge \theta > 0 \text{ on } \mathcal{Y}_{a,\delta/2,\delta}. \end{aligned}$$

We may now begin the proof of Theorem 5.3.8. We will proceed by induction in $a$, the first step of which will be completed in the following lemma.

LEMMA 5.3.10. *Theorem 5.3.8 holds for $a = a_{\min}$.*

PROOF. By Lemma 5.3.3, it suffices to prove the theorem with $F(t^{-\alpha}y)$ replaced by $F(t^{-\alpha}\mathbf{C})$. For any $\delta > 0$, the function $F_\delta := F_{a_{\min},\delta}$ is admissible and satisfies the hypotheses of Proposition 5.3.5*(ii)*. This implies (5.3.36). Moreover, if $f \in S^0(Y)$ is supported in $\mathcal{Y}_{a_{\min},\frac{\delta}{2},\delta}$, there is a constant $M$ such that

$$y \cdot \nabla F_\delta(y) \ge M|f(y)|$$

for all $y \in Y$. Applying Proposition 5.3.4*(i)* with $g \equiv 0$, we obtain that for such $f$

$$\int_1^\infty \|f(t^{-\alpha}\mathbf{C})\chi(H)u_t\|^2 t^{-1}\, dt \le C\|u\|^2. \tag{5.3.39}$$

Finally, let $f \in S^0(Y)$ be supported in $\mathcal{Y}_{a_{\min},\delta_1,\delta_2}$ for some $0 < \delta_1 < \delta_2 < \infty$. We can pick a finite sequence of numbers:

$$\delta_1 < \rho_1 < \rho_2 < \cdots < \rho_{n-1} < \rho_n, \ \rho_n > \delta_2,$$

so that

$$(\delta_1, \delta_2) \subset \bigcup_{i=1}^n (\rho_i/2, \rho_i). \tag{5.3.40}$$

Since

$$\mathcal{Y}_{a,\sigma_2} \subset \mathcal{Y}_{a,\sigma_1} \text{ and } \mathcal{Y}_{a,\sigma_1,\sigma_2} = \mathcal{Y}_{a,\sigma_1} \setminus \overline{\mathcal{Y}_{a,\sigma_2}}$$

for any $\sigma_1 < \sigma_2$, (5.3.40) implies that

$$\mathcal{Y}_{a_{\min},\delta_1,\delta_2} \subset \bigcup_{i=1}^n \mathcal{Y}_{a_{\min},\frac{\rho_i}{2},\rho_i}.$$

Writing $f = \sum_1^n f_i$, where $f_i \in S^0(Y)$ is supported in $\mathcal{Y}_{a_{\min},\frac{\rho_i}{2},\rho_i}$, and applying (5.3.39) to each $f_i$, we obtain part *(ii)* of the theorem for $a = a_{\min}$. □

The inductive step in the proof of Theorem 5.3.8 requires the following lemma. We define

$$\mathcal{Y}^{\sigma}_{a,\delta} = \mathcal{Y}_{a,\delta} \cap \{|y^a| < \sigma\},$$

$$\mathcal{Y}^{\sigma}_{a,\delta_1,\delta_2} = \mathcal{Y}_{a,\delta_1,\delta_2} \cap \{|y^a| < \sigma\}.$$

LEMMA 5.3.11. *Let* $0 < \delta_1 < \delta_2 < \infty$. *For any* $\sigma > 0$ *there is a* $\rho > 0$ *such that:*

$$(i)\ \mathcal{Y}_{a,\delta_1} \subset \bigcup_{b \le a} \mathcal{Y}^{\sigma}_{b,\rho_b},$$

$$(ii)\ \mathcal{Y}_{a,\delta_1,\delta_2} \subset \bigcup_{b \le a} \mathcal{Y}^{\sigma}_{b,\rho_b,\delta_2},$$

*where* $\rho_b = \rho$ *for* $b \ne a$, *and* $\rho_a = \delta_1$.

PROOF. *(i)* Let $M_0$ be a constant such that

$$|y^b| \le M_0 \cdot \max_{(ij) \le a} |y_i - y_j| \tag{5.3.41}$$

for all $b \le a$, and let $\rho = \min(\delta_1, \sigma/(2M_0N))$. Let $y \in \mathcal{Y}_{a,\delta_1}$. We consider two cases.

*Case 1.* $|y_i - y_j| > \rho$ for all $(ij) \le a$. Since for $(ij) \not\le a$ we have $|y_i - y_j| > \delta_1$ by the definition of $\mathcal{Y}_{a,\delta_1}$, we obtain that

$$y \in \mathcal{Y}_{a_{\min},\rho} = \mathcal{Y}^{\sigma}_{a_{\min},\rho},$$

where the last equality follows from the fact that $y^{a_{\min}} = 0$ for all $y \in Y$.

*Case 2.* Otherwise, let $b$ be the finest cluster decomposition such that if $(ij) \le a$ and $|y_i - y_j| < 2\rho$, then $(ij) \le b$. Namely, let

$$p_k = (i_k\, j_k),\ k = 1, 2, \dots, n,$$

be all pairs such that $p_k \le a$ and $|y_{i_k} - y_{j_k}| < 2\rho$, and let

$$b = p_1 \vee \dots \vee p_n.$$

Since $p_k \le a$ for $k = 1, \dots, n$, it follows that $b \le a$. We will prove that $y \in \mathcal{Y}^{\sigma}_{b,\rho_b}$.

If $b = a$, we have $y \in \mathcal{Y}_{a,\delta_1} = \mathcal{Y}_{a,\rho_a}$ (since $\rho_a = \delta_1$). Suppose now that $b \ne a$. If $(ij) \le a$ and $(ij) \not\le b$, then $|y_i - y_j| \ge \rho$ by the definition of $b$. If $(ij) \not\le a$, we have $|y_i - y_j| > \delta_1$ since $y \in \mathcal{Y}_{a,\delta_1}$. Hence $|y_i - y_j| > \rho$ for all $(ij) \not\le b$, so that $y \in \mathcal{Y}_{b,\rho}$.

Moreover, suppose that $(ij) \le b$. Then there is a sequence of pairs $(i_k, i_{k+1})$, $k = 1, \dots, m-1$, such that $m \le N$, $i_1 = i$, $i_m = j$, and $|y_{i_k} - y_{i_{k+1}}| < 2\rho$. Hence by the triangle inequality,

$$|y_i - y_j| < 2N\rho \le \sigma/M_0.$$

It follows that $|y^b| < \sigma$ and $y \in \mathcal{Y}^{\sigma}_{b,\rho_b}$.

*(ii)* Suppose that $y \in \mathcal{Y}_{a,\delta_1,\delta_2}$. This means that $y \in \mathcal{Y}_{a,\delta_1}$ and that $|y_i - y_j| < \delta_2$ for some $(ij) \not\le a$. By *(i)*, $y \in \mathcal{Y}^{\sigma}_{b,\rho_b}$ for some $b \le a$. Since $(ij) \not\le a$ and $b \le a$, we also have $(ij) \not\le b$. Hence $y \in \mathcal{Y}^{\sigma}_{b,\rho_b,\delta_2}$. □

PROOF OF THEOREM 5.3.8. Let $a \in \mathcal{A}$. Assume that *(i)* and *(ii)* hold for all $b < a$; we will prove that they hold for $a$.

Let $F \in S^0(Y)$ be supported in $\mathcal{Y}_{a,\delta}$ for some $\delta > 0$. Using Lemma 5.3.11, we may write for any $\sigma > 0$:

$$F = \sum_{b \leq a} F_b,$$

where $F_b \in S^0(Y)$ is supported in $\mathcal{Y}^\sigma_{b,\rho_b}$ and $\rho_a = \delta$. By the inductive assumption, (5.3.36) is satisfied for all $F_b$, $b < a$. It therefore suffices to prove that there is a $\sigma > 0$ (which may depend on $\delta$) such that (5.3.36) holds for $F$ supported in $\mathcal{Y}^\sigma_{a,\delta}$.

Similarly, to prove the part *(ii)* of the theorem it suffices to prove that for some sufficiently small $\sigma > 0$ (depending on $\delta_1$) (5.3.37) holds for functions supported in $\mathcal{Y}^\sigma_{a,\delta_1,\delta_2}$.

Let $\delta > 0$, and let

$$F = F_{a,\delta}(y^q_a)\mathcal{F}^a(\epsilon_0^{-1} y^{a,q}),$$

where $F_{a,\delta}$ was constructed in Lemma 5.3.9, and $\mathcal{F}^a$ – in Proposition 5.2.7. By Proposition 5.2.8, we may choose $\epsilon_0$ small enough so that $F$ is admissible and

$$\operatorname{supp} F \subset \bigcap_{b \not\leq a} \{|y^b| \geq \gamma\}$$

for some $\gamma > 0$. Hence

$$\operatorname{supp} F \subset \mathcal{Y}_{a,\gamma/M_0}, \tag{5.3.42}$$

where $M_0$ is the same as in (5.3.41).

We have:

$$\begin{aligned} y \cdot \nabla F(y) &= f_0 + g, \\ f_0(y) &= (y^q_a \cdot \nabla F_{a,\delta}(y^q_a))\mathcal{F}^a(\epsilon_0^{-1} y^{a,q}), \\ g(y) &= F_{a,\delta}(y^q_a) \cdot (\epsilon_0^{-1} y^{a,q} \cdot \nabla \mathcal{F}^a(\epsilon_0^{-1} y^{a,q})). \end{aligned}$$

Moreover, by Proposition 5.2.7 there is a $c > 0$ such that $\mathcal{F}^a = 1$ for $|y^{a,q}| \leq c$, i.e.,

$$\mathcal{F}^a(\epsilon_0^{-1} y^{a,q}) \equiv 1 \text{ for } |y^{a,q}| \leq c\epsilon_0.$$

By Lemma 5.2.1, there is a constant $c_0 > 0$ such that

$$c_0^{-1}|y^{a,q}| \leq |y^a| \leq c_0|y^{a,q}|. \tag{5.3.43}$$

Hence

$$\mathcal{F}^a(\epsilon_0^{-1} y^{a,q}) \equiv 1 \text{ for } |y^a| \leq c_0^{-1}|y^{a,q}| \geq c_0^{-1}c\epsilon_0. \tag{5.3.44}$$

Combining this with (5.3.42), and using Lemma 5.3.11 with $\sigma = \frac{1}{2}c_0^{-1}c\epsilon_0$, we see that for some $\rho > 0$:

$$\begin{aligned}\operatorname{supp} g \ &\subset \mathcal{Y}_{a,\gamma/M_0} \cap \{|y^a| \geq \sigma\} \\ &\subset \textstyle\bigcup_{b\leq a} \mathcal{Y}^\sigma_{b,\rho} \cap \{|y^a| \geq \sigma\} \\ &\subset \textstyle\bigcup_{b<a} \mathcal{Y}^\sigma_{b,\rho},\end{aligned}$$

since $\mathcal{Y}^\sigma_{a,\rho} \subset \{|y^a| < \sigma\}$. Applying the inductive assumption to $g$, we obtain that

$$\int_1^\infty \|g(t^{-\alpha}y)\chi(H)u_t\|^2 t^{-1}\,dt \leq C\|u\|^2.$$

Since $\mathcal{F}(y) \geq 0$ for all $y$, and, by Lemma 5.3.9*(iv)*, $y \cdot \nabla F_{a,\delta}(y) \geq 0$, we have $f_0(y) \geq 0$. By Lemma 5.3.9*(i)*:

$$y^q_a \cdot \nabla F_{a,\delta}(y^q_a) \geq \theta > 0 \text{ for } y^q_a \in \mathcal{Y}_{a,\frac{\delta}{2},\delta}.$$

Using also (5.3.44), we obtain that for $\sigma = \frac{1}{2}c_0^{-1}c\epsilon_0$:

$$f_0(y) \geq \theta \text{ for } y \in \Omega_{\sigma,\delta} := \{|y^a| \leq \sigma,\ y^q_a \in \mathcal{Y}_{a,\frac{\delta}{2},\delta}\}. \tag{5.3.45}$$

By (5.3.43), $|y^a| < \sigma$ implies that $|y^{a,q}| < c_0\epsilon$. Hence

$$|y - y^q_a| = |y^{a,q}| < c_0\sigma,$$

so that for any $\delta/2 < \delta' < \delta'' < \delta$ we have

$$\{y :\ |y^a| < \sigma,\ y \in \mathcal{Y}_{a,\delta',\delta''}\} \subset \{y :\ y^q_a \in \mathcal{Y}_{a,\frac{\delta}{2},\delta}\},$$

provided that $\sigma$ is small enough. This means that for such $\sigma$

$$\mathcal{Y}^\sigma_{a,\delta',\delta''} \subset \Omega_{\sigma,\delta}. \tag{5.3.46}$$

Let $\delta' = 6\delta/10$, $\delta'' = 9\delta/10$, and let $\sigma = \sigma(\delta)$ be chosen as above for these values of $\delta', \delta''$.

If $f \in S^0(Y)$ is supported in $\mathcal{Y}^\sigma_{a,\frac{6\delta}{10},\frac{9\delta}{10}}(\delta)$, by (5.3.45) and (5.3.46) there is a constant $M$ such that $Mf_0 \geq |f|$ for all $y$. Hence the assumptions (5.3.3), (5.3.5) of Proposition 5.3.4 are satisfied (with $f$ replaced by $M^{-1}f$). We obtain that

$$\int_1^\infty \|f(t^{-\alpha}\mathbf{C})\chi(H)u_t\|^2 t^{-1}\,dt \leq C\|u\|^2. \tag{5.3.47}$$

Moreover, by Proposition 5.3.5*(ii)* we have

$$s - \lim_{t\to\infty} F(t^{-\alpha}\mathbf{C})U(t) = 0. \tag{5.3.48}$$

We can now finish the proof of the theorem. We begin with *(i)*. It follows from (5.3.44) and Lemma 5.3.9*(iii)* that

$$F(y) \equiv 1 \text{ on } \mathcal{Y}^\sigma_{a,2\delta},$$

where $\sigma = \frac{1}{2}c_0^{-1}\epsilon_0 c$. Since $2\delta$ was arbitrary, this implies *(i)*.

We will now prove *(ii)*. Pick any $0 < \delta_1 < \delta_2 < \infty$. As in the proof of Lemma 5.3.10, we choose a finite sequence of numbers

$$0 < \rho_1 < \rho_2 < \cdots < \rho_{n-1} < \rho_n, \ \rho_1 < \delta_1, \ \rho_n > \delta_2,$$

so that

$$(\delta_1, \delta_2) \subset \bigcup_{i=1}^{n} (6\rho_i/10, \, 9\rho_i/10).$$

Let $\sigma = \min(\sigma(\rho_1), \dots, \sigma(\rho_n))$. Then

$$\mathcal{Y}^{\sigma}_{a,\delta_1,\delta_2} \subset \bigcup_{i=1}^{n} \mathcal{Y}^{\sigma}_{a,\frac{6\rho_i}{10},\frac{9\rho_i}{10}}.$$

For any function $f \in S^0(Y)$ supported in $\mathcal{Y}^{\sigma}_{a,\delta_1,\delta_2}$, we write $f = \sum_1^n f_i$, where $f_i \in S^0(Y)$ is supported in $\mathcal{Y}^{\sigma}_{a,\frac{6\rho_i}{10},\frac{9\rho_i}{10}}$. Applying (5.3.47) to each $f_i$, we obtain part *(ii)* of the theorem. □

CHAPTER 6

# Wave Operators and Scattering Theory

## 6.1. Introduction

We are now ready to formulate and prove our results on asymptotic completeness for $N$-body systems in a constant magnetic field.

Asymptotic completeness is, roughly speaking, the statement that all trajectories $\mathrm{e}^{-\mathrm{i}tH}u$ of the system follow certain prescribed asymptotics as $t \to \infty$. The asymptotics in question may be essentially classical, or it may be a combination of classical and quantum effects.

The exact statement of asymptotic completeness may be somewhat technical, and depends on the system under consideration. In our case, the dominant features of the system are:

- the geometry of the $N$-particle potential $V = \sum V_{ij}(x_i - x_j)$, which gives rise to the phenomenon of *multichannel scattering*,
- the decay of the pair potentials $V_{ij}(x_i - x_j)$ at infinity (short-range or long-range),
- the additional structure due to the presence of the magnetic field.

Traditionally, asymptotic completeness is stated in terms of the *wave operators*

$$\Omega^{\pm} = \text{s-}\lim_{t \to \pm\infty} \mathrm{e}^{\mathrm{i}tH} U_0(t), \tag{6.1.1}$$

where $U_0(t)$ is the "comparison" evolution. To explain the main concepts involved, we first discuss the (relatively simple) case when $H = -\frac{1}{2}\Delta + V$ is a one-particle Hamiltonian with $V(x)$ sufficiently regular and bounded by $O(|x|^{-1-\epsilon})$ at infinity (the *short-range* case). We may then put $U_0(t) = \mathrm{e}^{\mathrm{i}t\Delta/2}$. The existence of the wave operators (6.1.1) means that for any solution $\mathrm{e}^{\mathrm{i}t\Delta/2}u^+$ of the free Schrödinger equation there is a trajectory $\mathrm{e}^{-\mathrm{i}tH}u$ such that

$$\|\mathrm{e}^{\mathrm{i}t\Delta/2}u^+ - \mathrm{e}^{-\mathrm{i}tH}u\| \to 0 \text{ as } t \to \infty \tag{6.1.2}$$

(and similarly for $t \to -\infty$). Asymptotic completeness is the statement that the range of $\Omega^{\pm}$ is equal to the continuous spectral subspace $\mathcal{H}_{\text{cont}}(H)$ of $H$, i.e., for any $u \in \mathcal{H}_{\text{cont}}(H)$ there is a $u^+ \in \mathcal{H}$ such that (6.1.2) holds. It then follows that any $u \in \mathcal{H}$ is a superposition of a bound state and a "scattering state" with the asymptotics (6.1.2).

Asymptotic completeness has several important consequences. It implies that $H$ is unitarily equivalent to $H_0 = -\frac{1}{2}\Delta$ on $\mathcal{H}_{\rm cont}(H)$:

$$H\Omega^{\pm} = \Omega^{\pm} H_0,$$

and that $\operatorname{Ran}\Omega^- = \operatorname{Ran}\Omega^-$. In particular, if $e^{-itH}u$ was a scattering state in the past ($t \to -\infty$), it cannot be "trapped" by the potential, but rather must become a scattering state in the future. We may further define the *scattering operator*:

$$S := \Omega^{+*}\Omega^-.$$

which compares the asymptotics for $t \to -\infty$ to that for $t \to +\infty$. Asymptotic completeness implies that $S$ is unitary; this is important in physics, as it allows one to proceed with the analysis of the scattering amplitudes (the so-called optical theorem). Note that asymptotic completeness is the only currently known method of proving the unitarity of the scattering operator.

For multiparticle Hamiltonians it becomes necessary to use *channel wave operators*, corresponding to the different possible break-ups of the system into clusters. Assuming again that the pair potentials are short-range, one defines

$$\Omega_a^{\pm} = \text{s}-\lim_{t\to\pm\infty} e^{itH}e^{-itH_a}\Pi_a,\, a \in \mathcal{A}, \tag{6.1.3}$$

where $\Pi_a$ is the orthogonal projection on $\mathcal{H}_{\rm pp}(H^a)$. Asymptotic completeness means that (after the center of mass separation)

$$\mathcal{H}_{\rm c}(H) = \bigoplus_{a\in\mathcal{A}:\sharp a\geq 2} \operatorname{Ran}\Omega_a^{\pm}.$$

In plain language, any $u \in \mathcal{H}$ is a superposition of bound states (eigenvectors of $H$) and scattering states with the asymptotics $e^{-itH_a}u_a$, corresponding to the break-up of the system into two or more independent clusters.

The "comparison evolution" must be modified further if the assumptions on the decay rate of the pair potentials are relaxed. Roughly speaking, we say that $V_{ij}$ are *long-range* if $\partial_x^\alpha V(x) = O(\langle x\rangle^{-|\alpha|-\mu_1})$ for all $|\alpha| \leq 2$, where $0 < \mu_1 \leq 1$. (Different authors may use slightly different regularity assumptions.) In particular, Coulomb potentials are long-range.

In the one-body long-range case, the scattering trajectories $e^{-itH}u$ may no longer be approximated by the free evolution $e^{-itH_0}$, since the influence of the potential does not become negligible for large $t$. Instead, one uses an appropriate "modified free evolution" $U_0(t)$, constructed using an (exact or approximate) solution of the corresponding classical Hamilton-Jacobi equation. There are many possible choices of $U_0(t)$; we refer the interested reader to [**DG1**] for a detailed discussion. In this chapter we will use the *Dollard dynamics*, which has the advantage of being both simple and explicit. For $N$-body long-range systems, a similar modification of (6.1.3) is needed.

A somewhat different point of view, leading to the *algebraic scattering theory*, was proposed in [**AGM**], [**M3**]. Its essence is to consider the *asymptotic observables*

$$\text{s-}\lim_{t\to\pm\infty} e^{itH}B(t)e^{-itH},$$

for suitable $B(t)$, as the central objects of scattering theory. Recall that we have already defined and discussed such observables in Chapter 4. In particular, we considered the *asymptotic velocity* observable $P^{\pm}$ defined by:

$$\text{s-}\lim_{t\to\pm\infty} e^{itH}f(\frac{x}{t})e^{-itH} =: f(P^+),\ f \in C_0^\infty(\mathbb{R}^n).$$

Asymptotic velocity was first introduced by Dereziński in [**De3**], and played an important role in the proof of asymptotic completeness for long-range $N$-body systems in [**De1**].

In the algebraic approach to scattering, one does not address directly the question of finding the large-time asymptotics of $e^{-itH}$. Instead, one studies the properties of the spectral measure of asymptotic observables in general, and of $P^{\pm}$ in particular. For instance, asymptotic completeness (when formulated appropriately) is equivalent to the statement that $P^{\pm}$ is unitarily equivalent to $D$. The advantage of this approach is that the asymptotic observables are uniquely defined, whereas the definition of the wave operators depends on the choice of the comparison evolution. Furthermore, one may ask (and sometimes answer) meaningful questions concerning asymptotic observables even in cases when asymptotic completeness need not hold. (See e.g., the *asymptotic absolute continuity* in [**DG1**]). Note also that certain scattering data, such as *scattering cross-sections*, may be recovered directly from the asymptotic velocity (see [**De3**]). The disadvantage is of course that one does not obtain an explicit asymptotics for the evolution.

We now explain in more detail how these concepts apply in our case. We begin with a brief discussion of the "strongly charged systems", where all proper subsystems are assumed to be charged. The arguments of Section 1.4 and Chapter 5 suggest that in this case all scattering channels describe charged clusters of particles escaping to infinity in the direction of the magnetic field. Indeed, recall that in Section 3.3 we proved a Mourre estimate with a conjugate operator $\frac{1}{2}(\langle z, D_z\rangle + \text{h.c.})$ acting only in the $z$ direction. In Section 4.3 we constructed the asymptotic velocity $\zeta^+$ in the direction of the field; combining this with the Mourre estimate just mentioned, we find that $\zeta^+ \neq 0$ for all (non-zero) scattering states (Subsection 6.3.1). This readily yields the existence and asymptotic completeness of the channel wave operators for short-range interactions (Subsection 6.3.2).

In order to extend these results to the case of strongly charged systems with long-range interactions (Subsection 6.3.3), we must deal with one more issue. In the $N$-body long-range case (magnetic or not), the methods described above are not quite sufficient to prove asymptotic completeness. Instead, they show that all scattering states correspond to a break-up of the systems into clusters whose evolution is governed by time-dependent Hamiltonians $H_a + I_a(t,x)$. The time-dependent potential $I_a$ is the long-range "tail" of the intercluster interaction. Our goal is to prove that the different clusters become stable and asymptotically independent as $t \to \infty$, at least as far as their internal structure is concerned. Unfortunately, $I_a(t,x)$ still couples the internal and external motions of the clusters of $a$. We therefore need an additional argument, the outline of which we now describe.

It was realized early on ([**E4**], [**SS4**]) that this problem can be solved by proving an a priori estimate of the order $O(t^\alpha)$, $\alpha < 1$, on the size of the clusters. Namely, such an estimate implies that the time-dependent potential $I_a(t,x)$ may be replaced by $I_a(t,0,x_a)$ (since the difference is integrable in time along the evolution), and the latter potential depends only on the "external" motion of the clusters. Once this is accomplished, a standard argument yields the asymptotic completeness of modified wave operators. (Note that a somewhat different method, based on an estimate on the energy of two-particle clusters, was used by Enss [**E5**] in his work on 3-body systems.)

The a priori estimates just mentioned are actually quite difficult to prove, and it took several years before Dereziński [**De1**] obtained them for long-range $N$-body systems without external fields. For the class of systems currently under consideration, the proof of such estimates consists of two separate, and rather different, arguments concerning the $y$ and $z$ variables respectively. To estimate the size of clusters in the $z$ direction, we will use the results of [**De1**]. The propagation in the directions transversal to the magnetic field was analyzed in Chapter 5; here we will rely in particular on Proposition 5.3.6.

We now turn to the three-dimensional "dispersive" 3-body case, considered in Section 6.4. As explained earlier, neutral pairs may travel transversally to the magnetic field, hence all directions are now available for scattering. A rough outline of the proof of asymptotic completeness is as follows. We first deduce from the minimal velocity estimate of Subsection 4.4.2 that a scattering state describes a break-up of the system into clusters escaping to infinity with non-zero relative velocities. If all clusters are charged, then the same arguments as in the charged case apply. Otherwise, the system splits up into a neutral pair and a third particle. We use the arguments of Chapter 5 to obtain an a priori estimate on the size of the neutral pair, and proceed to construct the wave operators. Assuming that the pair potentials are of Coulomb type, the wave operators corresponding to such "dispersive" channels will be short-range.

Finally, in Sections 6.5 and 6.6 we consider 2-dimensional systems. For strongly charged systems (all proper subsystems are charged), we prove an HVZ-type theorem, which implies that there are in fact no scattering states. The 3-particle dispersive case is similar to its 3-dimensional analogue, but simpler as we do not have to deal with the propagation along the field. The existence and asymptotic completeness of the (short-range) wave operators is established by more or less the same arguments as above.

Asymptotic completeness for a single particle in a constant magnetic field and an external potential vanishing at infinity was proved in [**Ł1**], [**I1**];

for a neutral pair in three dimensions, it was proved in [**Ł2**]. The $N$-body strongly charged case was solved in [**GŁ1**], [**GŁ2**] for short-range and long-range interactions respectively. Strongly charged three-particle systems of total charge 0 were also studied in [**I2**], where a Mourre estimate was proved for suitable reduced Hamiltonians. The scattering theory for "dispersive" three-particle systems (containing at least one neutral pair) was developed in [**GŁ3**]. The results concerning 2-dimensional systems (Sections 6.5 and 6.6) are new. Note also that Jensen-Nakamura [**JN**] studied the spectral theory of a neutral pair in dimension 2, and found conditions on the pair potential under which the spectrum of $H$ is absolutely continuous.

## 6.2. Channel identification operators

DEFINITION 6.2.1. For $a \in \mathcal{A}$, the *channel identification operator* associated to $a$ is

$$\Pi_a := U_a^* \Big( \mathbf{1}_{\mathrm{pp}}(\tilde{H}^{a,\mathrm{c}}) \otimes \int_{Y_a^{\mathrm{n}\prime}} \mathbf{1}_{\mathrm{pp}}(\tilde{H}^{a,\mathrm{n}}(\eta_a^{\mathrm{n}}))\mathrm{d}\eta_a^{\mathrm{n}} \Big).$$

The channel identification operators are essentially projections on the spaces of bound states of the clusters of $a$. In the sequel, we will use them in the definition of the wave operators to separate the parts of the wave function belonging to different channels.

The above definition uses the concept of the channel reducing transformations $U_a$, which, we recall, were defined in Subsection 2.2.2 and are not unique. However, we have the following lemma.

LEMMA 6.2.2. *$\Pi_a$ is an orthogonal projection, independent of the choice of the channel reducing transformation $U_a$.*

PROOF. That $\Pi_a$ is an orthogonal projection is obvious; we need to prove that it does not depend on the choice of $U_a$. We use the notation of Sections 2.2.1, 2.2.2. Let $\Pi_{C_l}$ be the projection on the space of bound states of $H[C_l]$, then

$$\Pi_a = \bigotimes_{l=1}^{\sharp a} \Pi_{C_l}.$$

Applying Lemma 1.4.2 or 1.4.5 (depending on the total charge of $C_l$) to $H[C_l]$, we find that $\Pi_{C_l}$ is independent of the choice of $U_a$. Hence the same is true of $\Pi_a$. □

PROPOSITION 6.2.3. *We have for all $u \in \mathcal{H}$:*

(*i*) $\lim_{R\to+\infty} \sup_{t\in\mathbb{R}} \|F(\frac{|x^a|}{R} \geq 1)\mathrm{e}^{-\mathrm{i}tH_a}\Pi_a u\| = 0$,

(*ii*) $\lim_{R\to+\infty} \sup_{t\in\mathbb{R}} \|F(\frac{|y_a^{\mathrm{c}}|}{R} \geq 1)\mathrm{e}^{-\mathrm{i}tH_a}\Pi_a u\| = 0$.

PROOF. This follows by an argument similar to that in the proof of Lemma 6.2.2: we apply Proposition 1.4.3 or 1.4.6, depending on the charge of $C_i$, to each cluster Hamiltonian $H[C_i]$, $C_i \in a$. □

## 6.3. Three-dimensional charged systems

In this section we prove asymptotic completeness for 3−dimensional "strongly charged" systems. We will always assume that *(V1), (V2), (M1), (Q2)* hold.

Recall that in Section 4.3 we constructed the observable $\zeta^+$, which we interpreted as the asymptotic velocity along the field; we also obtained certain propagation estimates in the $z$ variables for $\mathrm{e}^{-\mathrm{i}tH}u$, where $u$ lies in the range of appropriate spectral projections of $\zeta^+$. We will prove in Subsection 6.3.1 that under the assumptions stated above we have

$$\mathcal{H}_{\mathrm{bound}} = \mathrm{Ran}\,\mathbf{1}_{\{0\}}(\zeta^+).$$

In the following two subsections, we will combine this with the results of Chapters 4 and 5 and deduce asymptotic completeness for the systems under consideration.

Recall that the sets

$$T_a = Z_a \backslash \bigcup_{b \not\leq a} Z_b, \ a \in \mathcal{A}, \tag{6.3.1}$$

were defined in Section 4.3.

**6.3.1. General considerations.** The main result of this subsection is that for "strongly charged" systems the only states with vanishing asymptotic velocity along the field are those in $\mathcal{H}_{\mathrm{bound}}$. The argument is based on the Mourre estimate (Theorem 3.3.2), which states that the conjugate operator $A = \frac{1}{2}(\langle z, D_z\rangle + \langle D_z, z\rangle$ has strictly positive Heisenberg derivative if the energy is localized away from the threshold set; this may be used to localize the propagation in the region of phase space where $\langle A\rangle \geq ct$, and hence $\langle z\rangle \geq c't$, for some $c, c' > 0$ as $t \to \infty$. (A similar abstract argument was explained in detail in Section 4.4).

We must first take care of a technicality. The Mourre estimate in the neutral case ($Q = 0$ in Theorem 3.3.2) was proved for the reduced Hamiltonians $\tilde{H}(\eta_{\mathrm{cm}})$ with a fixed value of the total pseudomomentum. In order to prove a minimal velocity estimate for $\tilde{H} = \int^{\oplus} \tilde{H}(\eta_{\mathrm{cm}}) d\eta_{\mathrm{cm}}$, and hence for $H$, we will need to verify that direct integrals of states with the constant in the Mourre estimate bounded from below by $\epsilon > 0$ are dense in $\mathcal{H}$. This is done in the next two lemmas.

LEMMA 6.3.1. *Assume (V1), (V2), (M1), (Q2), and that* $Q = 0$. *Then the set*

$$\Sigma := \{(\eta_{\mathrm{cm}}, \lambda) \in Y'_{\mathrm{cm}} \times \mathbb{R} \mid \lambda \in \tau_{\mathrm{cm}} \cup \sigma_{\mathrm{pp}}(\tilde{H}(\eta_{\mathrm{cm}}))\}$$

*is closed.*

PROOF. Pick $(\eta_0, \lambda_0) \in Y'_{\rm cm} \times \mathbb{R}$ such that $(\eta_0, \lambda_0) \notin \Sigma$, i.e., $\lambda_0 \notin \tau_{\rm cm} \cup \sigma_{\rm pp}(\tilde{H}(\eta_0))$. By Theorem 3.3.2*(iv)*:

$$[\tilde{H}(\eta_0), \mathrm{i}A] \geq c_0 > 0 \text{ at } \tilde{H}(\eta_0) = \lambda_0.$$

From Proposition 1.3.2 and the explicit form of $[\tilde{H}(\eta_{\rm cm}), \mathrm{i}A]$ it follows that for any $\chi \in C_0^\infty(\mathbb{R})$ the map

$$Y'_{\rm cm} \ni \eta_{\rm cm} \mapsto \chi(\tilde{H}(\eta_{\rm cm}))[\tilde{H}(\eta_{\rm cm}), \mathrm{i}A]\chi(\tilde{H}(\eta_{\rm cm}))$$

is norm continuous. Hence

$$[\tilde{H}(\eta_{\rm cm}), \mathrm{i}A] \geq \frac{1}{2}c_0 \text{ at } \tilde{H}(\eta_{\rm cm}) = \lambda,$$

for $(\eta_{\rm cm}, \lambda)$ in an open neighborhood $U$ of $(\eta_0, \lambda_0)$. By Lemma 3.2.6*(i)*, this implies that $\lambda \notin \sigma_{\rm pp}(\tilde{H}(\eta_{\rm cm}))$ for $(\eta_{\rm cm}, \lambda) \in U$, i.e., that $U \cap \Sigma = \emptyset$. Hence $\Sigma$ is closed. □

LEMMA 6.3.2. *Assume (V1), (V2), (M1), (Q2), and that $Q = 0$. Then the set of vectors $u \in \mathcal{H}$ such that $U_{\rm cm}u = \sum_1^k v_i$, where*

$$v_i = f_i(D_{y_{\rm cm}})\chi_i(\tilde{H})v_i$$

*for some $f_i \in C_0^\infty(Y'_{\rm cm})$, $\chi_i \in C_0^\infty(\mathbb{R})$ satisfying*

$$\operatorname{dist}(\operatorname{supp}\chi_i, \tau_{\rm cm} \cup \sigma_{\rm pp}(\tilde{H}(\eta_{\rm cm}))) > \epsilon, \ \forall \eta_{\rm cm} \in \operatorname{supp} f_i, \tag{6.3.2}$$

*is dense in $\mathcal{H}_{\rm scatt}$.*

PROOF. Let $u \in \mathcal{H}_{\rm scatt}$, $U_{\rm cm}u = v$. Let $\mathrm{d}E$ be the joint spectral measure of the commuting operators $D_{y_{\rm cm}}$ and $\tilde{H}$, and let $\mathrm{d}\mu = (v, \mathrm{d}Ev)$. By Theorem 3.3.2*(iii)*, we have

$$\mathbf{1}_{\rm pp}(\tilde{H}(\eta_{\rm cm})) = \mathbf{1}_{\tau_{\rm cm} \cup \sigma_{\rm pp}(\tilde{H}(\eta_{\rm cm}))}(\tilde{H}(\eta_{\rm cm})).$$

From this and the definition of $\mathcal{H}_{\rm scatt}$ it follows that

$$\mu(\Sigma) = 0.$$

The set of $u \in \mathcal{H}_{\rm scatt}$ such that $\mathrm{d}\mu$ has a compact support $K = K(u)$ is dense in $\mathcal{H}_{\rm scatt}$. For any such $u$, $\Sigma \cap K$ is compact by Lemma 6.3.2, hence for any $\epsilon > 0$ there is a finite collection of rectangles $(M_i)_{1 \leq i \leq p}$ in $Y_{\rm cm} \times \mathbb{R}$ such that:

$$\Sigma \cap K \subset M, \ \mu(M) \leq \epsilon, \tag{6.3.3}$$

where $M = \bigcup_{i=1}^p M_i$. Let $v_\epsilon = \int_{K \setminus M} \mathrm{d}Ev$. Clearly, $v_\epsilon$ may be written as

$$v_\epsilon = \sum_{i=1}^k v_i, \ v_i = \chi_i(D_{y^n_{\rm cm}})f_i(\tilde{H}^{\rm cm})v_i,$$

for some $\chi_i$, $f_i$ as in the lemma. Moreover, by (6.3.3) we have $\|v - v_\epsilon\| \leq \epsilon$. □

We now prove the main result of this subsection.

THEOREM 6.3.3. *Assume (V1), (V2), (M1), (SC), (Q2). Let $\zeta^+$ be the asymptotic velocity for the evolution $\mathrm{e}^{-\mathrm{i}tH}$, defined in Theorem 4.3.2. Then*

$$\mathcal{H}_{\mathrm{bound}} = \mathrm{Ran}\, \mathbf{1}_{\{0\}}(\zeta^+).$$

PROOF. By Proposition 1.4.3 if $Q \neq 0$, and by Proposition 1.4.6 if $Q = 0$, we have $\mathcal{H}_{\mathrm{bound}} \subset \mathbf{1}_{\{0\}}(\zeta^+)\mathcal{H}$. It remains to prove the converse inclusion:

$$\mathcal{H}_{\mathrm{scatt}} \subset \mathbf{1}_{Z\setminus\{0\}}(\zeta^+)\mathcal{H}. \tag{6.3.4}$$

Let $u \in \mathcal{H}_{\mathrm{scatt}}$. Assume first that $Q \neq 0$. By a density argument, we may assume that $u = \chi(H)u$ for $\chi$ supported near $\lambda \in \tau \cup \sigma_{\mathrm{pp}}(H^{\mathrm{cm}})$. By Theorem 3.3.2 we have

$$[H, \mathrm{i}A] > 0 \text{ at } H = \lambda.$$

Following verbatim the proof of the minimal velocity estimate for standard $N-$particle Hamiltonians (see e.g., [**Gr**] or [**DG1**, Prop. 6.6.8]), we prove that there is an $\epsilon_0 > 0$ such that

$$\int_1^{+\infty} \|F(\frac{|z^{\mathrm{cm}}|}{t} \leq \epsilon_0)\chi(H)\mathrm{e}^{-\mathrm{i}tH}u\|^2 \frac{\mathrm{d}t}{t} \leq C\|u\|^2, \tag{6.3.5}$$

$$\mathrm{s-}\lim_{t\to+\infty} F(\frac{|z^{\mathrm{cm}}|}{t} \leq \epsilon_0)\chi(H)\mathrm{e}^{-\mathrm{i}tH} = 0. \tag{6.3.6}$$

Hence $F(|\zeta^+| \leq \epsilon_0)\chi(H^{\mathrm{cm}}) = 0$. Since $u = \chi(H)u$, we obtain that $F(|\zeta^+| \leq \epsilon_0)u = 0$, and that $u \in \mathbf{1}_{Z\setminus\{0\}}(\zeta^+)\mathcal{H}$ as claimed.

Assume now that $Q = 0$. By Lemma 6.3.2, we may assume that $U_{\mathrm{cm}}u = v$, where

$$v = f(D_{y_{\mathrm{cm}}})\chi(\tilde{H}^{\mathrm{cm}})v, \tag{6.3.7}$$

and $f, \chi$ are $C_0^\infty$ functions as in (6.3.2). Moreover, by a covering argument we may assume that $\mathrm{supp}\,\chi \subset [\lambda - \epsilon, \lambda + \epsilon]$, where $\epsilon > 0$ is sufficiently small and

$$\lambda \notin \tau_{\mathrm{cm}} \cup \bigcup_{\eta_{\mathrm{cm}} \in \mathrm{supp}\, f} \sigma_{\mathrm{pp}}(\tilde{H}(\eta_{\mathrm{cm}})).$$

From Theorem 3.3.2*(iv)* we have

$$f(D_{y_{\mathrm{cm}}})[\tilde{H}, \mathrm{i}A]f(D_{y_{\mathrm{cm}}}) > 0 \text{ at } \tilde{H} = \lambda,$$

or, equivalently,

$$f(K_{\mathrm{cm}}^\perp)[H, \mathrm{i}A]f(K_{\mathrm{cm}}^\perp) > 0 \text{ at } H = \lambda.$$

As in the case $Q \neq 0$, we repeat the proof of Proposition 6.6.2 of [**DG1**] and obtain that for $\epsilon_0 > 0$ small enough:

$$\int_1^{+\infty} \|F(\frac{|z^{\mathrm{cm}}|}{t} \leq \epsilon_0)f(K_{\mathrm{cm}}^\perp)\chi(H)\mathrm{e}^{-\mathrm{i}tH}u\|^2 \frac{\mathrm{d}t}{t} \leq C\|u\|^2, \tag{6.3.8}$$

(6.3.9) $$\text{s-}\lim_{t\to+\infty} F(\frac{|z^{\rm cm}|}{t}\leq \epsilon_0)f(K^{\perp}_{\rm cm})\chi(H)\mathrm{e}^{-\mathrm{i}tH}=0.$$

This implies that $F(|\zeta^+|\leq\epsilon_0)f(K^{\perp}_{\rm cm})\chi(H)=0$, hence by (6.3.7) $F(|\zeta^+|\leq\epsilon_0)u=0$. □

**6.3.2. The short-range case.** We first consider the (easier) short-range case: we assume that *(V1)*, *(V2)*, *(M1)*, *(SC)*, *(Q2)* hold, and that $V^{\rm l}_{ij}\equiv 0$ in *(SC)*. By Theorem 6.3.3, all states in $\mathcal{H}_{\rm scatt}$ must separate asymptotically into two or more clusters, the distances between which grow linearly in time. We will see below that if all pair potentials $V_{ij}=V^{\rm s}_{ij}$ are short-range, the interaction between the propagating clusters is integrable in time and therefore asymptotically negligible. Thus the existence and completeness of short-range wave operators will be a direct consequence of Theorems 6.3.3 and 4.3.3.

Recall that $\Pi_a$ and $T_a$ were defined in Definition 6.2.1 and in (6.3.1) respectively.

THEOREM 6.3.4. *Assume (V1), (V2), (M1), (Q2), and (SC) with* $V^{\rm l}_{ij}=0$. *Then:*

*(i) the* short-range wave operators

$$\Omega^+_{{\rm sr},a}:=\text{s-}\lim_{t\to+\infty}\mathrm{e}^{\mathrm{i}tH}\mathrm{e}^{-\mathrm{i}tH_a}\Pi_a,\ \ a\in\mathcal{A},\ \ a\neq a_{\max},$$

*exist and their ranges are mutually orthogonal;*

*(ii) the system is* asymptotically complete*:*

$$\mathcal{H}_{\rm scatt}=\bigoplus_{a\in\mathcal{A},a\neq{\rm cm}}{\rm Ran}\Omega^+_{{\rm sr},a}.$$

*Moreover,*

$${\rm Ran}\Omega^+_{{\rm sr},a}=\mathbf{1}_{T_a}(\zeta^+)\mathcal{H}.$$

PROOF. We will use the notation and results of Section 4.3, in particular of Theorem 4.3.3. Since $V^{\rm l}_{ij}\equiv 0$, we have $U_{a,0}(t)=\mathrm{e}^{-\mathrm{i}tH_a}$. Recall that $\zeta^+_{a,0}$ is the asymptotic velocity for the evolution $U_{a,0}(t)$; let also $\zeta^{+(a)}$ be the asymptotic velocity for the evolution $\mathrm{e}^{-\mathrm{i}tH^a}$, acting on $L^2(X^a\times Y_a)$.

Let $(\zeta^+_{a,0})_a,(\zeta^+_{a,0})^a$ be the projections of $\zeta^+_{a,0}$ on $Z_a,Z^a$. Since $H_a=\frac{1}{2}D^2_{z_a}+H^a$, we have

(6.3.10) $$(\zeta^+_{a,0})^a=\zeta^{+(a)}.$$

Arguing exactly as in the proof of Proposition 1.4.6, we see that

(6.3.11) $$(\zeta^+_{a,0})_a=D_{z_a}.$$

From (6.3.1), (6.3.11), (6.3.10) we have

$$\mathbf{1}_{T_a}(\zeta^+_{a,0})=\mathbf{1}_{T_a}((\zeta^+_{a,0})_a)\mathbf{1}_{\{0\}}((\zeta^+_{a,0})^a)=\mathbf{1}_{T_a}(D_{z_a})\mathbf{1}_{\{0\}}(\zeta^{+(a)}).$$

Since $Z_a \backslash T_a$ has Lebesgue measure 0 in $Z_a$, we have

$$\mathbf{1}_{T_a}(D_{z_a}) = \mathbf{1}.$$

Applying Theorem 6.3.3 to $H^a$ and using Definition 6.2.1, we obtain that

$$\mathbf{1}_{\{0\}}(\zeta^{+(a)}) = \Pi_a, \tag{6.3.12}$$

hence

$$\mathbf{1}_{T_a}(\zeta^+_{a,0}) = \Pi_a. \tag{6.3.13}$$

By Theorem 4.3.3, the limits:

$$\text{s-}\lim_{t\to+\infty} \mathrm{e}^{\mathrm{i}tH}\mathrm{e}^{-\mathrm{i}tH_a}\mathbf{1}_{T_a}(\zeta^+_{a,0}) =: \Gamma^+_{a,0}, \tag{6.3.14}$$

$$\text{s-}\lim_{t\to+\infty} \mathrm{e}^{\mathrm{i}tH_a}\mathrm{e}^{-\mathrm{i}tH}\mathbf{1}_{T_a}(\zeta^+) = \Gamma^{+*}_{a,0} \tag{6.3.15}$$

exist. By (6.3.13), $\Gamma^+_{a,0} = \Omega^+_{\mathrm{sr},a}$, which proves the existence of the latter. Furthermore, by Theorem 4.3.3 again $\Omega^+_{\mathrm{sr},a}$ is a partial isometry from $\mathrm{Ran}\,\Pi_a = \mathrm{Ran}\,\mathbf{1}_{T_a}(\zeta^+_{a,0})$ to $\mathrm{Ran}\,\mathbf{1}_{T_a}(\zeta^+)$; since $T_a$ are disjoint for different $a$, this implies that the spaces $\mathrm{Ran}\,\Omega^+_{\mathrm{sr},a} = \mathrm{Ran}\,\mathbf{1}_{T_a}(\zeta^+)$ are mutually orthogonal. Finally, we have:

$$\bigoplus_{a\neq \mathrm{cm}} \mathrm{Ran}\Omega^+_{\mathrm{sr},a} = \bigoplus_{a\neq \mathrm{cm}} \mathrm{Ran}\mathbf{1}_{T_a}(\zeta^+) = \mathrm{Ran}\mathbf{1}_{Z\backslash\{0\}}(\zeta^+),$$

which, by Theorem 6.3.3, is equal to $\mathcal{H}_{\mathrm{scatt}}$. □

The following corollary of Theorem 6.3.4 and Proposition 6.2.3 states that for the class of systems considered in this section and for any initial condition $u \in \mathcal{H}_{\mathrm{scatt}}$, the "transversal" coordinates of all particles remain bounded uniformly in $t$.

COROLLARY 6.3.5. *Under the same assumptions as in Theorem 6.3.4, for any $u \in \mathcal{H}_{\mathrm{scatt}}$ we have*

$$\lim_{R\to+\infty} \limsup_{t\to+\infty} \|F(\frac{|y|}{R} \geq 1)\mathrm{e}^{-\mathrm{i}tH}u\| = 0.$$

**6.3.3. The long-range case.** In this subsection we assume that *(V1)*, *(V2)*, *(M1)*, *(SC)*, *(Q2)* hold, and that $V^{\mathrm{l}}_{ij} \not\equiv 0$ for at least one pair $(ij)$.

In Theorem 4.3.3 we proved the existence and completeness of partial wave operators intertwining $\mathrm{e}^{-\mathrm{i}tH}$ and $U_{a,0}(t)$. Observe that $U_{a,0}(t)$ couples the internal dynamics within the clusters of $a$ to the relative motion of the clusters via the time-dependent interaction $I_a(t,x)$. As explained in the introduction, the next step in the proof of asymptotic completeness is to undo that coupling. Partial results in this direction were obtained in Propositions 4.3.4 and 5.3.6.

In the short-range case, we were able to use $\mathrm{e}^{-\mathrm{i}tH_a} = \mathrm{e}^{-\mathrm{i}t\frac{1}{2}D_{z_a}^2} \otimes \mathrm{e}^{-\mathrm{i}tH^a}$ as the asymptotic dynamics. For long-range interactions, the external dynamics is not asymptotically free and a modified external evolution must replace $\mathrm{e}^{-\mathrm{i}tD_{z_a}^2}$ in the definition of the wave operators. We will use the Dollard modified dynamics, which we now describe.

Recall that in Section 4.3 we introduced the effective intercluster potentials

$$I_a(t,x) = \sum_{(ij)\not\leq a} F(|x_i - x_j| \geq \frac{t}{\log t}) V_{ij}^{\mathrm{l}}(x_i - x_j).$$

The *Dollard modifiers* are then defined as

$$S_a(t,\zeta_a) := \int_0^t \frac{1}{2}\zeta_a^2 + I_a(s,0,s\zeta_a)\mathrm{d}s.$$

The main result of this section is the following theorem.

THEOREM 6.3.6. *Assume that (V1), (V2), (M1), (SC), (Q2) hold, and that $\mu_{\mathrm{l}} > \sqrt{3}-1$, $\mu_{\mathrm{l}}\mu_{\mathrm{s}} > 1$. Then:*

*(i) the* modified wave operators

$$\text{s-}\lim_{t\to+\infty} \mathrm{e}^{\mathrm{i}tH}\mathrm{e}^{-\mathrm{i}S_a(t,D_{z_a})-\mathrm{i}tH^a}\Pi_a =: \Omega_{\mathrm{lr},a}^+ \tag{6.3.16}$$

*exist and their ranges are mutually orthogonal;*

*(ii) the system is* asymptotically complete*:*

$$\bigoplus_{a\neq \mathrm{cm}} \mathrm{Ran}\,\Omega_{\mathrm{lr},a}^+ = \mathcal{H}_{\mathrm{scatt}}.$$

*Moreover,*

$$\mathrm{Ran}(\Omega_{\mathrm{lr},a}^+) = \mathbf{1}_{T_a}(\zeta^+)\mathcal{H}.$$

As in Section 6.3.2, we also obtain the following corollary.

COROLLARY 6.3.7. *Let $u \in \mathcal{H}_{\mathrm{scatt}}$. Then*

$$\lim_{R\to+\infty}\limsup_{t\to+\infty} \|F(\frac{|y|}{R} \geq 1)\mathrm{e}^{-\mathrm{i}tH}u\| = 0.$$

In the proof of Theorem 6.3.6 we will rely on the following well-known result in two-body scattering theory (see [**DG1**], Section 3.6).

PROPOSITION 6.3.8. *Let $U_{a,z_a}(t)$ be the unitary group on $L^2(Z_a)$ generated by $\frac{1}{2}D_{z_a}^2 + I_a(t,0,z_a)$. Then the limits*

$$\text{s-}\lim_{t\to+\infty} U_{a,z_a}(t)^*\mathrm{e}^{-\mathrm{i}S_a(t,D_{z_a})} =: \omega_a^+,$$

$$\text{s-}\lim_{t\to+\infty} \mathrm{e}^{\mathrm{i}S_a(t,D_{z_a})}U_{a,z_a}(t) = \omega_a^{+*}$$

*exist and are unitary. Let* $p_a^+$ *be the asymptotic velocity for* $U_{a,z_a}$ *on* $L^2(Z_a)$, *then*

$$\omega_a^+ D_{z_a} = p_a^+ \omega_a^+.$$

PROOF OF THEOREM 6.3.6. All of the work involved in proving the theorem has already been carried out; it only remains to collect and assemble the partial results we proved earlier. The details are as follows.

Let $U_{a,z_a}$ and $p_a^+$ be as in Proposition 6.3.8. Recall that in Section 4.3 we introduced the auxiliary dynamics $U_{a,0}(t)$, $U_{a,1}(t)$, generated by $H_a + I_a(t,x)$ and $H_a + I_a(t,y,z_a)$ respectively. We will also use the dynamics $U_{a,2}(t)$ generated by $H_a + I_a(t,0,z_a)$. Note that since $I_a(t,0,z_a)$ commutes with $H^a$, we have

$$U_{a,2}(t) = \mathrm{e}^{-\mathrm{i}tH^a} \otimes U_{a,z_a}(t) \text{ on } \mathcal{H} = L^2(X^a \times Y_a) \otimes L^2(Z_a). \tag{6.3.17}$$

We denote by $\zeta^{(a)+}$ and $\zeta_{a,i}^+$, $i = 1, 2$, the asymptotic velocity operators associated to $\mathrm{e}^{-\mathrm{i}tH^a}$ and $U_{a,i}(t)$ respectively.

By (6.3.17), the projections of $\zeta_{a,2}^+$ on $Z_a$, $Z^a$ are $p_a^+$ and $\zeta^{+(a)}$ respectively, hence

$$\mathbf{1}_{T_a}(\zeta_{a,2}^+) = \mathbf{1}_{T_a(p_a^+)}\mathbf{1}_{\{0\}}(\zeta^{+(a)}).$$

By Proposition 6.3.8, $p_a^+$ is unitarily equivalent to $D_{z_a}$ and, since $Z_a \backslash T_a$ has zero Lebesgue measure in $Z_a$, we have $\mathbf{1}_{T_a}(p_a^+) = \mathbf{1}$. Recall also that $\mathbf{1}_{\{0\}}(\zeta^{+(a)}) = \Pi_a$ (see (6.3.12)). Hence

$$\mathbf{1}_{T_a}(\zeta_{a,2}^+) = \Pi_a. \tag{6.3.18}$$

By Theorem 4.3.5, the limits

$$\begin{aligned} \Gamma_{a,1}^+ &= \mathrm{s-}\lim_{t\to+\infty} U_{a,1}(t)^* \mathrm{e}^{-\mathrm{i}tH} \mathbf{1}_{T_a}(\zeta^+), \\ \Gamma_{a,1}^* &= \mathrm{s-}\lim_{t\to+\infty} \mathrm{e}^{\mathrm{i}tH} U_{a,1}(t) \mathbf{1}_{T_a}(\zeta_{a,1}^+), \end{aligned} \tag{6.3.19}$$

exist. Moreover, $\Gamma_{a,1}^+$ is an isometry between $\operatorname{Ran} \mathbf{1}_{T_a}(\zeta^+)$ and $\operatorname{Ran} \mathbf{1}_{T_a}(\zeta_{a,1}^+)$

Next, applying Proposition 5.3.6 to $U_{a,1}(t), U_{a,2}(t)$ we obtain that the limits

$$\begin{aligned} \Gamma_{a,2}^+ &= \mathrm{s-}\lim_{t\to+\infty} U_{a,2}(t)^* U_{a,1}(t), \\ \Gamma_{a,2}^{+*} &= \mathrm{s-}\lim_{t\to+\infty} U_{a,1}(t)^* U_{a,2}(t) \end{aligned} \tag{6.3.20}$$

exist, and that $\Gamma_{a,2}^+$ is an isometry between $\operatorname{Ran} \mathbf{1}_{T_a}(\zeta_{a,1}^+)$ and $\operatorname{Ran} \mathbf{1}_{T_a}(\zeta_{a,2}^+) = \operatorname{Ran} \Pi_a$.

Finally, from (6.3.17) and Proposition 6.3.8 we obtain that the limits

$$\begin{aligned} \Gamma_{a,3}^+ &= \mathrm{s-}\lim_{t\to+\infty} \mathrm{e}^{\mathrm{i}S_a(t,D_{z_a}) + \mathrm{i}tH^a} U_{a,2}(t), \\ \Gamma_{a,3}^{+*} &= \mathrm{s-}\lim_{t\to+\infty} U_{a,2}^* \mathrm{e}^{-\mathrm{i}S_a(t,D_{z_a}) - \mathrm{i}tH^a} \end{aligned} \tag{6.3.21}$$

exist, and that $\Gamma_{a,3}^+$ is unitary on $\operatorname{Ran} \Pi_a$.

Using the chain rule for wave operators (cf. Subsection 4.1.2), we obtain from (6.3.19), (6.3.20), (6.3.21) that (6.3.16) exists and is an isometry between $\operatorname{Ran}\Pi_a$ and $\operatorname{Ran}\mathbf{1}_{T_a}(\zeta^+)$. Since $\{T_a\}_{a\in\mathcal{A}}$ is a partition of $Z$, we see that $\operatorname{Ran}\Omega^+_{\mathrm{l},a}$ are mutually orthogonal. Moreover, since $\bigcup_{a\neq a_{\max}} T_a = Z\backslash\{0\}$, we have

$$\bigoplus_{a\neq a_{\max}} \operatorname{Ran}\Omega^+_{\mathrm{lr},a} = \mathbf{1}_{Z\backslash\{0\}}(\zeta^+)\mathcal{H} = \mathcal{H}_{\mathrm{scatt}},$$

by Theorem 6.3.3. This completes the proof of the theorem. □

## 6.4. Three-dimensional dispersive systems

We now consider 3-particle systems in $\mathbb{R}^3$ for which *(Q1)* holds, but *(Q2)* fails. Such systems must contain at least one neutral pair (i.e., $q_i + q_j = 0$ for some $i, j$). Throughout this section we will assume that *(V1), (V4), (M1), (S)* hold; recall from Subsection 1.2.3 that this implies *(V2), (SC)*. In the last part of the proof of asymptotic completeness, we will also have to assume *(C)*.

An important step in the proof of asymptotic completeness is the verification that all states in $\mathcal{H}_{\mathrm{scatt}}$ break up into at least two clusters, the distance between which grows linearly in $t$ as $t \to \infty$. In the "strongly charged" case (Sections 6.3.2, 6.3.3), it follows from the Mourre estimate of Section 3.3 that $\operatorname{Ran}\mathbf{1}_{\{0\}}(\zeta^+) = \mathcal{H}_{\mathrm{bound}}$, hence for all states in $\mathcal{H}_{\mathrm{scatt}}$ the clusters separate with non-zero relative velocity along the field. This is mostly because, using a compactness argument in transversal directions, we were able to obtain the Mourre estimate with a conjugate operator acting in the $z$ direction only. In the dispersive case, however, the conjugate operator constructed in Section 3.7 acts in all directions, in order to accommodate the now-permitted motion of neutral clusters across the field. It follows (Section 4.4) that a minimal velocity estimate still holds, but not necessarily in the $z$ direction. We thus need to analyze the hypothetical states in $\mathcal{H}_{\mathrm{scatt}} \cap \operatorname{Ran}\mathbf{1}_{\{0\}}(\zeta^+)$, i.e., scattering states with neutral clusters moving only transversally to the magnetic field.

PROPOSITION 6.4.1. *Let $U_{a,0}$ be the unitary dynamics generated by $H_a + I_a(t,x)$, where $I_a$ was defined in (4.3.2). Let $u \in \mathcal{H}_{\mathrm{scatt}} \cap \operatorname{Ran}\mathbf{1}_{\{0\}}(\zeta^+)$. Then*

$$u = \sum_{a\in\mathcal{A}^{\mathrm{n}},\sharp a=2} u_a,$$

*where $u_a \in \operatorname{Ran}\mathbf{1}_{\{0\}}(\zeta^+)$ and*

$$\lim_{t\to+\infty} U_{a,0}(t)^* \mathrm{e}^{-\mathrm{i}tH} u_a \text{ exists.}$$

PROOF. Let $\mathcal{H}_0 = \mathcal{H}_{\mathrm{scatt}} \cap \mathrm{Ran}\, \mathbf{1}_{\{0\}}(\zeta^+)$. We will construct linear maps $\Gamma_a^+ : \mathcal{H}_0 \to \mathcal{H}_0$, $a \in \mathcal{A}^{\mathrm{n}}$, such that

$$\|\Gamma_a^+\| \leq 1,\ \mathbf{1} = \sum_{a\in\mathcal{A}^{\mathrm{n}}} \Gamma_a^+, \tag{6.4.1}$$

and

$$\lim_{t\to+\infty} U_{a,0}(t)^* \mathrm{e}^{-\mathrm{i}tH}\Gamma_a^+ u \tag{6.4.2}$$

exists. The proposition will then follow by setting $u_a = \Gamma_a^+ u$.

Let $\tau$ be the threshold set defined in (3.7.1). Recall that the operators $H$, $\zeta^+$, $K_{\mathrm{cm}}^\perp$ commute, and that by Theorem 3.7.3*(i)* the set $\sigma_{\mathrm{pp}}(H) \cup \tau$ is closed and discrete. Hence the subset $\mathcal{H}_0'$ consisting of $u \in \mathcal{H}_0$ such that

$$u = \chi^2(H)F^2(K_{\mathrm{cm}}^\perp)u,\ F \in C_0^\infty(Y_{\mathrm{cm}}'),\ \chi \in C_0^\infty(\mathbb{R}\setminus(\tau\cup\sigma_{\mathrm{pp}}(H))), \tag{6.4.3}$$

is dense in $\mathcal{H}_0$. We will first construct $\Gamma_a^+$ on $\mathcal{H}_0'$; by density, we may extend them uniquely to maps $\mathcal{H}_0 \to \mathcal{H}_0$ so that (6.4.1), (6.4.2) continue to hold.

Fix a $u$ satisfying (6.4.3). By Propositions 4.4.10 and 4.2.1, we have for some $C_0, c_0 > 0$:

$$\mathrm{e}^{-\mathrm{i}tH}u = F(\frac{|x^{\mathrm{cm}}|}{t} \geq 2c_0)F(\frac{|x|}{t} \leq C_0)\mathrm{e}^{-\mathrm{i}tH}u + o(1). \tag{6.4.4}$$

Let

$$\mathbf{1} = \sum_{a\in\mathcal{A}} q_a(y^{\mathrm{cm}})$$

be a cylindrical partition of unity on $Y^{\mathrm{cm}}$ as in Proposition 2.3.1, with the constants $\delta_a, \epsilon_a$ small enough (to be chosen later). We may assume that $\mathrm{supp}\, q_{\mathrm{cm}} \subset \{|y^{\mathrm{cm}}| \leq \epsilon_0\}$. Since $N = 3$, we may also assume that for $a \neq a_{\max}$:

$$\mathrm{supp}\, \nabla q_a \subset \{|y^{\mathrm{cm}}| \leq c_0\} \cup Y_{a_{\min},\delta}$$

for some $\delta > 0$, where $Y_{a_{\min},\delta} = \{y \in Y \mid |y_i - y_j| > \delta,\ \forall i,j\}$.

We will prove that

$$\lim_{t\to+\infty} q_a(\frac{y^{\mathrm{cm}}}{t})\mathrm{e}^{-\mathrm{i}tH}u = 0,\ a \in \mathcal{A}^{\mathrm{c}}, \tag{6.4.5}$$

and that

$$\lim_{t\to+\infty} \mathrm{e}^{\mathrm{i}tH}q_a(\frac{y^{\mathrm{cm}}}{t})\mathrm{e}^{-\mathrm{i}tH}u =: \Gamma_a^+ u, \tag{6.4.6}$$

$$\lim_{t\to+\infty} U_{a,0}(t)^* q_a(\frac{y^{\mathrm{cm}}}{t})\mathrm{e}^{-\mathrm{i}tH}u, \tag{6.4.7}$$

exist for $a \in \mathcal{A}^{\mathrm{n}}$. It is then straightforward to verify that $\Gamma_a^+ : \mathcal{H}_0' \to \mathcal{H}_0$, defined by (6.4.6), satisfy (6.4.1)–(6.4.2).

(6.4.5) follows from Theorem 5.3.7 with $F = q_a$ and $\alpha = 1$ if $a \in \mathcal{A}^{\mathrm{c}}$, $\sharp a \geq 2$, and from (6.4.4) if $a = a_{\max}$. It remains to prove that the limits in (6.4.6), (6.4.7) exist. We will only consider (6.4.7), the proof for (6.4.6) being similar.

For $\epsilon > 0$, we fix a cutoff function $F_\epsilon(z^{\mathrm{cm}})$ such that:

- $\operatorname{supp} F_\epsilon \subset \{|z^{\mathrm{cm}}| \leq \epsilon\}$, $F_\epsilon(z^{\mathrm{cm}}) = 1$ for $\{|z^{\mathrm{cm}}| \leq \epsilon/2\}$;
- for each $b \in \mathcal{A}$, there is a neighborhood $U_b$ of $Z_b$ such that $F_\epsilon$ depends only on $z_b$ in $U_b$.

The existence of such $F_\epsilon$ follows from Proposition 5.2.7 with the projections $\pi^{a,q}$ replaced by $\pi^a$. Alternatively, one may use the construction of [**Gr**] or [**DG1**, Definition 5.1.2, Proposition 5.2.8].

Since $u \in \operatorname{Ran} \mathbf{1}_{\{0\}}(\zeta^+)$, we have for any $\epsilon > 0$:

$$\mathrm{e}^{-\mathrm{i}tH}u = F_\epsilon(\frac{z^{\mathrm{cm}}}{t})\mathrm{e}^{-\mathrm{i}tH}u + o(1). \tag{6.4.8}$$

Using also (6.4.3), (6.4.4), we obtain

$$\begin{aligned} & q_a(\tfrac{y^{\mathrm{cm}}}{t})\mathrm{e}^{-\mathrm{i}tH}u \\ = \; & q_a(\tfrac{y^{\mathrm{cm}}}{t})F(\tfrac{|x|}{t} \leq C_0)F_\epsilon(\tfrac{z^{\mathrm{cm}}}{t})\chi^2(H)F^2(K_{\mathrm{cm}}^\perp)\mathrm{e}^{-\mathrm{i}tH}u + o(1). \end{aligned} \tag{6.4.9}$$

From Lemma 2.3.8 we have:

$$\begin{aligned} & [\chi(H), q_a(\tfrac{y^{\mathrm{cm}}}{t})F(\tfrac{|x|}{t} \leq C_0)F_\epsilon(\tfrac{z^{\mathrm{cm}}}{t})] = O(t^{-1}), \\ & (\chi(H) - \chi(H_a))q_a(\tfrac{y^{\mathrm{cm}}}{t}) = o(1). \end{aligned} \tag{6.4.10}$$

Finally, from the definition of $F(K_{\mathrm{cm}}^\perp)$ (see Subsection 1.3.6) and symbolic calculus it follows that

$$[F(K_{\mathrm{cm}}^\perp), q_a(\frac{y^{\mathrm{cm}}}{t})F(\frac{|x|}{t} \leq C_0)F_\epsilon(\frac{z^{\mathrm{cm}}}{t})] = O(t^{-1}). \tag{6.4.11}$$

Combining (6.4.9), (6.4.10), (6.4.11), we find that the limit (6.4.7) is equal to

$$\lim_{t\to+\infty} U_{a,0}(t)^* M(t)\mathrm{e}^{-\mathrm{i}tH}u, \tag{6.4.12}$$

where

$$M(t) = F(K_{\mathrm{cm}}^\perp)\chi(H_a)q_a(\frac{y^{\mathrm{cm}}}{t})F(\frac{|x|}{t} \leq C_0)F_\epsilon(\frac{z^{\mathrm{cm}}}{t})\chi(H)F(K_{\mathrm{cm}}^\perp).$$

It suffices to verify that the assumptions of Lemma 4.1.5*(ii)* are satisfied with $\Phi(t) = M(t)$, $U_1(t) = \mathrm{e}^{-\mathrm{i}tH}$, $U_2(t) = U_{a,0}(t)$. We have

$$\mathrm{D}_{21}M(t) = \partial_t M(t) + \mathrm{i}((H_a + I_a(t,x))M(t) - \mathrm{i}M(t)H$$

$$= I_1 + I_2 + I_3 + I_4,$$

where (using obvious abbreviations):

$$
\begin{aligned}
I_1 &= F(K_{\rm cm}^{\perp})\chi(H_a)(\partial_t(q_a F F_\epsilon) + [H, \mathrm{i} q_a F F_\epsilon])\chi(H)F(K_{\rm cm}^{\perp}),\\
I_2 &= [I_a(t,x), \mathrm{i}F(K_{\rm cm}^{\perp})]\chi(H_a)q_a F F_\epsilon \chi(H)F(K_{\rm cm}^{\perp}),\\
I_3 &= F(K_{\rm cm}^{\perp})[I_a(t,x), \mathrm{i}\chi(H_a)]q_a F F_\epsilon \chi(H)F(K_{\rm cm}^{\perp}),\\
I_4 &= \mathrm{i}F(K_{\rm cm}^{\perp})\chi(H_a)(I_a(t,x) - I_a(x))q_a F F_\epsilon \chi(H)F(K_{\rm cm}^{\perp}).
\end{aligned}
$$

By *(V4)*, we have

$$|\partial_x^\alpha I_a(t,x)| \le C_{\alpha,\epsilon} t^{-\mu_1 - |\alpha| + \epsilon} \ \forall \epsilon > 0, |\alpha| \ge 0.$$

Hence by symbolic calculus:

$$[F(K_{\rm cm}^{\perp}), I_a(t,x)] = O(t^{-1-\mu_1+\epsilon}),$$

and by Lemma 2.3.8:

$$[\chi(H_a), I_a(t,x)] = O(t^{-1-\mu_1+\epsilon}).$$

Finally,

$$(I_a(t,x) - I_a(x))q_a(\frac{y^{\rm cm}}{t}) = -\sum_{(ij)\not\le a} V_{ij}^{\rm s}(x_i - x_j)q_a(\frac{y^{\rm cm}}{t})$$

for $t$ large enough. Hence by Lemma 1.2.2,

$$
\begin{aligned}
&\|\chi(H_a)(I_a(t,x) - I_a(x))q_a(\tfrac{y^{\rm cm}}{t})\|\\
\le\ & C\textstyle\sum_{(ij)\not\le a} \|V_{ij}^{\rm s}(x_i - x_j)q_a(\tfrac{y^{\rm cm}}{t})(D^2+1)^{-\frac{1}{2}}\|\\
\le\ & Ct^{-2},
\end{aligned}
$$

using *(V4)* again. It follows that

$$I_2 + I_3 + I_4 = O(t^{-1-\mu_1+\epsilon}).$$

It remains to consider $I_1$. We write $I_1$ as $I_5 + I_6 + I_7$, where:

$$
\begin{aligned}
I_5(t) =\ & \tfrac{1}{2t}F(K_{\rm cm}^{\perp})\chi(H_a)\Big(\langle \nabla q_a(\tfrac{y^{\rm cm}}{t}), D_{y^{\rm cm}} - \mathbf{A}y^{\rm cm} - \tfrac{y^{\rm cm}}{t}\rangle + \text{h.c.}\Big)\\
&\times F_\epsilon F\chi(H)F(K_{\rm cm}^{\perp}),
\end{aligned}
$$

$$
\begin{aligned}
I_6(t) =\ & \tfrac{1}{2t}F(K_{\rm cm}^{\perp})\chi(H_a)q_a\Big(\langle \nabla F_\epsilon(\tfrac{z^{\rm cm}}{t}), D_{z^{\rm cm}} - \tfrac{z^{\rm cm}}{t}\rangle + \text{h.c.}\Big)\\
&\times F\chi(H)F(K_{\rm cm}^{\perp}),
\end{aligned}
$$

$$
\begin{aligned}
I_7(t) =\ & \tfrac{1}{2t}F(K_{\rm cm}^{\perp})\chi(H_a)q_a F_\epsilon\Big(\langle \nabla F(\tfrac{|x|}{t} \le C_0), D_x - \mathbf{A}x - \tfrac{x}{t}\rangle + \text{h.c.}\Big)\\
&\times \chi(H)F(K_{\rm cm}^{\perp}),
\end{aligned}
$$

We begin with $I_6$. Since $\nabla F_\epsilon$ depends only on $z_b$ in a neighborhood of $Z_b$, and $q_a$ may be chosen to be supported in this neighbourhood, we may write $I_6$ as a sum of terms of the form

$$\frac{1}{2t}F(K_{\mathrm{cm}}^{\perp})\chi(H_a)q_a g_b(\frac{z^{\mathrm{cm}}}{t})(D_{z_b}-\frac{z_b}{t})F\chi(H)(K_{\mathrm{cm}}^{\perp}) =: R_b(t),$$

where $b \in \mathcal{A}$ and $\operatorname{supp} g_b \subset T_b$ for $b \neq a_{\max}$ (recall that $T_a$ were defined in (6.3.1). We pick $\tilde{g}_b$ such that $\tilde{g}_b g_b = g_b$, $\operatorname{supp}\tilde{g}_b \subset T_b$, and rewrite $R_b(t)$ as

$$\frac{1}{2t}F(K_{\mathrm{cm}}^{\perp})\chi(H_a)q_a g_b(\frac{z^{\mathrm{cm}}}{t})(D_{z_b}-\frac{z_b}{t})\tilde{g}_b(\frac{z^{\mathrm{cm}}}{t})\chi(H)F(K_{\mathrm{cm}}^{\perp}) + O(t^{-2}).$$

Hence

$$\begin{aligned}|(u_2, R_b(t)u_1)| \leq\ & \tfrac{1}{t}\||D_{z_b}-\tfrac{z_b}{t}|^{\frac{1}{2}}\tilde{g}_b(\tfrac{z^{\mathrm{cm}}}{t})\chi(H)F(K_{\mathrm{cm}}^{\perp})u_1\| \\ & \times\||D_{z_b}-\tfrac{z_b}{t}|^{\frac{1}{2}}g_b(\tfrac{z^{\mathrm{cm}}}{t})\chi(H)F(K_{\mathrm{cm}}^{\perp})u_2\| \\ & + Ct^{-2}\|u_1\|\|u_2\|.\end{aligned}$$

Applying Proposition 4.3.1 to $\mathrm{e}^{-\mathrm{i}tH}$ and $U_{a,0}(t)$, we find that $I_6$ satisfies (4.1.12).

We now consider $I_7$. We have

$$|(u_2, I_7(t)u_1)| \leq\ \tfrac{C}{t}\|F_1(\tfrac{|x|}{t})\chi(H)F(K_{\mathrm{cm}}^{\perp})u_1\|\|F_2(\tfrac{|x|}{t})\chi(H_a)F(K_{\mathrm{cm}}^{\perp})u_2\|,$$

where $F_i \in C_0^\infty(\mathbb{R})$, $i = 1, 2$ are supported in $\{|\lambda| \geq \frac{C_0}{2}\}$, $F_i \equiv 1$ on $\operatorname{supp}\nabla F$, and $C = \|(D - \mathbf{A}x)\tilde{\chi}(H)\| + \|(D-\mathbf{A}x)\tilde{\chi}(H_a)\|$, where $\tilde{\chi} \in C_0^\infty(\mathbb{R})$ satisfies $\tilde{\chi}\chi = \chi$. Note that $C < \infty$ by Proposition 1.2.1. It now follows from Proposition 4.2.2 that $I_7$ satisfies (4.1.12).

Finally, we consider $I_5$. Recall that $\nabla q_a$ is supported in $\{|y^{\mathrm{cm}}| \leq c_0\} \cup Y_{a_{\min},\delta}$. We pick $\tilde{q}_a \in S^0_{\mathrm{cl}}(X^{\mathrm{cm}})$ such that

$$\tilde{q}_a\nabla q_a = \nabla q_a,\ \operatorname{supp}\tilde{q}_a \subset \{|y^{\mathrm{cm}}| \leq c_0\} \cup Y_{a_{\min},\delta},$$

$\tilde{F}_\epsilon \in C_0^\infty(Z^{\mathrm{cm}})$ such that

$$\tilde{F}_\epsilon F_\epsilon = F_\epsilon,\ \operatorname{supp}\tilde{F}_\epsilon \subset \{|z^{\mathrm{cm}}| \leq \epsilon\},$$

and $\tilde{\chi} \in C_0^\infty(\mathbb{R})$ such that $\tilde{\chi}\chi = \chi$. Then

$$\begin{aligned}& I_5(t) \\ =\ & \tfrac{1}{t}F(K_{\mathrm{cm}}^{\perp})\chi(H_a)\tilde{F}_\epsilon\tilde{q}_a\left(\langle\nabla q_a, D_{y^{\mathrm{cm}}} - \mathrm{A}y^{\mathrm{cm}}\rangle\tilde{\chi}(H_a) - \langle\nabla q_a, \tfrac{y^{\mathrm{cm}}}{t}\rangle\right)FF_\epsilon \\ & \times\tilde{q}_a\tilde{F}_\epsilon\chi(H)F(K_{\mathrm{cm}}^{\perp}) + \text{h.c.} + O(t^{-2}).\end{aligned}$$

Hence for $u_2 \in \mathcal{H}_0, u_1 \in \mathcal{H}_0'$:

$$\begin{aligned}& |(u_2, I_5(t)u_1)| \\ \leq\ & \tfrac{C_0}{t}\|\tilde{q}_a(\tfrac{y^{\mathrm{cm}}}{t})\tilde{F}_\epsilon(\tfrac{z^{\mathrm{cm}}}{t})\chi(H)F(K_{\mathrm{cm}}^{\perp})u_1\|\|\tilde{q}_a(\tfrac{y^{\mathrm{cm}}}{t})\tilde{F}_\epsilon(\tfrac{z^{\mathrm{cm}}}{t})\chi(H_a)F(K_{\mathrm{cm}}^{\perp})u_2\| \\ & + \tfrac{C}{t^2}\|u_1\|\|u_2\|,\end{aligned}$$

where

$$C_0 = \|FF_\epsilon \nabla q_a \tilde{\chi}(H_a)(D_{y^{\mathrm{cm}}} - \mathbf{A}y^{\mathrm{cm}})\| + \|FF_\epsilon \langle \nabla q_a(\frac{y^{\mathrm{cm}}}{t}), \frac{y^{\mathrm{cm}}}{t}\rangle\| < \infty.$$

We write $\tilde{q}_a = \tilde{q}_{a,1} + \tilde{q}_{a,2}$, where $\operatorname{supp} \tilde{q}_{a,1} \subset Y_{a_{\min},\delta}$, $\operatorname{supp} \tilde{q}_{a,2} \subset \{|y^{\mathrm{cm}}| \leq \epsilon_0\}$; then $\operatorname{supp} \tilde{q}_{a,2} F_\epsilon \subset \{|x^{\mathrm{cm}}| \leq \epsilon_0\}$, provided that the $\epsilon \ll 1$ was small enough. Hence we may apply Theorem 5.3.7 (with $\alpha = 1$) to $\tilde{q}_{a,1}$, and Proposition 4.4.10 to $\tilde{q}_{a,2}$, and obtain that $I_5$ satisfies (4.1.12). □

We will now combine Propositions 6.4.1 and 4.3.4, and prove that there are in fact no scattering states in $\mathbf{1}_{\{0\}}(\zeta^+)\mathcal{H}$, provided that $\mu_{\mathrm{l}} > \sqrt{3} - 1$, Unlike for charged systems, this is not simply a consequence of the Mourre estimate. Instead, we follow the following train of thought. By Proposition 6.4.1, any state $u \in \mathcal{H}_{\mathrm{scatt}} \cap \mathbf{1}_{\{0\}}(\zeta^+)\mathcal{H}$ is a superposition of states $u_a$ with asymptotic dynamics generated by $H_a + I_a(t,x)$. Proposition 4.3.4 allows us to replace the $x$ in $I_a$ by $(y, z_a)$. If $I_a(t, y, z_a)$ decays fast enough, one may use a standard 2-body argument (see [**DG1**]) and show that the external component $(\zeta^+)_a$ of the asymptotic velocity is unitarily equivalent to the external momentum $D_{z_a}$. But the spectral measure of $D_{z_a}$ is absolutely continuous, whereupon the conclusion follows.

PROPOSITION 6.4.2. *Assume that* $\mu_{\mathrm{l}} > \sqrt{3} - 1$, $\mu_{\mathrm{l}}\mu_{\mathrm{s}} > 1$. *Then*

$$\mathcal{H}_{\mathrm{bound}} = \mathbf{1}_{\{0\}}(\zeta^+)\mathcal{H}.$$

PROOF. The inclusion $\mathcal{H}_{\mathrm{bound}} \subset \mathbf{1}_{\{0\}}(\zeta^+)\mathcal{H}$ is trivial. It therefore suffices to prove that

$$\mathcal{H}_{\mathrm{scatt}} \cap \mathbf{1}_{\{0\}}(\zeta^+)\mathcal{H} = \{0\}.$$

Let $u \in \mathcal{H}_{\mathrm{scatt}} \cap \mathbf{1}_{\{0\}}(\zeta^+)\mathcal{H}$. By Proposition 6.4.1, we have

$$u = \sum_{a \in \mathcal{A}^{\mathrm{n}}, \sharp a = 2} u_a,$$

where $u_a \in \mathbf{1}_{\{0\}}(\zeta^+)\mathcal{H}$ and

$$\lim_{t \to +\infty} U_{a,0}(t)^* \mathrm{e}^{-\mathrm{i}tH} u_a \text{ exists.}$$

Let $U_{a,1}(t)$ be the evolution generated by $H_a + I_a(t, y, z_a)$. Using Proposition 4.3.4 and the chain rule for the wave operators, we see that the limits

$$\lim_{t \to +\infty} U_{a,1}(t)^* \mathrm{e}^{-\mathrm{i}tH} u_a \tag{6.4.13}$$

exist. Next, we replace $U_{a,1}(t)$ by an appropriate Dollard-type asymptotic evolution. Let

$$S_a(t, y, \zeta_a) = \int_0^t \frac{1}{2}\zeta_a^2 + I_a(s, y, s\zeta_a)\mathrm{d}s,$$

and let $U_{a,D}(t)$ be the unitary dynamics generated by $H_a + S_a(t, y, D_{z_a})$. By Theorem 3.6.2 of [**DG1**], the Dollard-type wave operators

$$\text{s-}\lim_{t\to+\infty} U_{a,D}(t)^* U_{a,1}(t)$$

exist. Combining this with (6.4.13), we see that

$$\lim_{t\to+\infty} U_{a,D}(t)^* \mathrm{e}^{-itH} u_a =: \tilde{u}_a \text{ exists.} \tag{6.4.14}$$

Observe that the dynamics $U_{a,D}(t)$ preserves $D_{z_a}$. Arguing then as in Theorems 3.2.2, 3.6.2 of [**DG1**], we obtain that for $f \in C_0^\infty(Z_a)$:

$$\begin{aligned} & \lim_{t\to+\infty} U_{a,D}(t)^* f(\tfrac{z_a}{t}) U_{a,D}(t) \\ = \; & U_{a,D}(t)^* f(D_{z_a}) U_{a,D}(t) \\ = \; & f(D_{z_a}). \end{aligned} \tag{6.4.15}$$

If $u_a = \mathbf{1}_{\{0\}}(\zeta^+) u_a$, it follows from (6.4.14), (6.4.15) that for each $f \in C_0^\infty(Z_a)$ with $f(0) = 1$ we have

$$f(D_{z_a}) \tilde{u}_a = \tilde{u}_a.$$

But this is possible only if $\tilde{u}_a = 0$, hence $u_a = 0$. □

The next result states that if $(ij)$ is a neutral pair, then $y_i - y_j$ must remain bounded uniformly for all $t$. Roughly speaking, the particles $i$ and $j$ must either remain together or split up. In the former case, $x_i - x_j$ is bounded. Otherwise, the only allowed scattering channels are those in which all clusters are charged; but then the results of Chapter 4 imply that $y_k$ is bounded for all $k$, hence in particular $|y_i - y_j| \le |y_i| + |y_j|$ is bounded. Similar ideas were used in the proof of Proposition 5.3.6 and in Section 3.7.

PROPOSITION 6.4.3. *Let $I(t, x)$ be a real time-dependent potential satisfying (4.1.2), $a \in \mathcal{A}^{\mathrm{n}}$, and let $U_a(t), \tilde{U}_a(t)$ be the unitary propagators generated by $H_a + I(t, x)$ and $H_a + I(t, y_a, z)$, respectively. Assume that $\mu_{\mathrm{l}} > \frac{1}{2}$, $\mu_{\mathrm{l}}\mu_{\mathrm{s}} > 1$. Then the limits*

$$\text{s-}\lim_{t\to+\infty} \tilde{U}_a^*(t) U_a(t) =: \Gamma_a^+, \tag{6.4.16}$$

$$\text{s-}\lim_{t\to+\infty} U_a^*(t) \tilde{U}_a(t) = \Gamma_a^{+*} \tag{6.4.17}$$

*exist and are unitary.*

PROOF. The proof is almost the same as the proof of Proposition 5.3.6: instead of replacing $y$ by 0, one must replace $y^a$ by 0 while treating the coordinates of the third particles as parameters. Here we only indicate the outline of the proof. We will consider only the limit (6.4.16), (6.4.17) being similar.

Let $u \in \mathcal{H}$. By density, we may assume that

$$U_a(t)u = \chi^2(H_a) U_a(t) u + o(1), \; \chi \in C_0^\infty(\mathbb{R}).$$

Let $C_1 = \{i, j\}$ be the neutral pair in $a$. We will apply Theorem 5.3.7 to the Hamiltonian $H[C_1]$ of the neutral pair (see Chapter 2) with the time-dependent perturbations $I(t, x)$ and $I(t, y_a, z)$. We may do so since the cluster decomposition $\{\{i\}, \{j\}\}$ of $H[C_1]$ is strongly charged.

Let $\mathbf{C}_1$ be the center of orbit observable for the pair $C_1$. Pick $\mu_l > \alpha > \max(\frac{1}{2}, \mu_s^{-1})$, and let $F, F_1 \in S^0(Y[C_l])$, $\operatorname{supp} F \subset \{|y_i - y_j| \leq 2\}$, $F \equiv 1$ in $\{|y_i - y_j| \leq 1\}$, $\operatorname{supp} F_1 \subset \{1 \leq |y_i - y_j| \leq 2\}$. Then by Theorem 5.3.7 we have for $U(t) = U_a(t)$ or $\tilde{U}_a(t)$:

$$F(t^{-\alpha}\mathbf{C}_1)U(t)u = U(t)u + o(1), \tag{6.4.18}$$

$$\int_1^{+\infty} \|F_1(t^{-\alpha}\mathbf{C}_1)\chi(H_a)U(t)u\|^2 \frac{\mathrm{d}t}{t} \leq C\|u\|^2. \tag{6.4.19}$$

By (6.4.18), it suffices to show that the limit

$$\lim_{t\to\infty} \tilde{U}_a(t)F(t^{-\alpha}C_1)\chi^2(H_a)U_a(t)u$$

exists. But this is done by following almost verbatim the proof of Proposition 5.3.6. □

We now state the main result of this section.

THEOREM 6.4.4. *Assume that $N = 3$, and that (V1), (V4), (M1), (S), (C) hold with $\mu_l > \sqrt{3} - 1$. Then:*

*i) the* modified wave operators

$$\text{s-}\lim_{t\to+\infty} \mathrm{e}^{\mathrm{i}tH}\mathrm{e}^{-\mathrm{i}S_a(t,D_{z_a})-\mathrm{i}tH^a}\Pi_a =: \Omega^+_{\mathrm{lr},a},\ a \in \mathcal{A}^{\mathrm{c}},$$

$$\text{s-}\lim_{t\to+\infty} \mathrm{e}^{\mathrm{i}tH}\mathrm{e}^{-\mathrm{i}tH_a}\Pi_a =: \Omega^+_{\mathrm{lr},a},\ a \in \mathcal{A}^{\mathrm{n}}$$

*exist and their ranges are mutually orthogonal.*

*ii) the system is asymptotically complete:*

$$\bigoplus_{a\neq \mathrm{cm}} \operatorname{Ran}\Omega^+_{\mathrm{l},a} = \mathcal{H}_{\mathrm{scatt}}.$$

*Moreover,*

$$\operatorname{Ran}(\Omega^+_{\mathrm{l},a}) = \mathbf{1}_{T_a}(\zeta^+)\mathcal{H}.$$

The key property of Coulomb type interactions that is needed here is that if $C_l$ is a neutral cluster, then

$$\text{if } x_i = x_j \text{ for all } i, j \in C_l, \text{ then } \sum_{i\in C_l} V^{\mathrm{l}}_{ik}(x_i - x_k) = 0. \tag{6.4.20}$$

It follows that if $a$ is a cluster decomposition containing a neutral pair, the effective intercluster interaction $I_a(x_a)$ is short-range, even though each $V_{ij}$ may be a long-range potential. Thus, unlike in Subsection 6.3.3, we will not need to modify the free dynamics.

PROOF OF THEOREM 6.4.4. Recall that in Section 6.3.3 we introduced the unitary dynamics $U_{a,2}(t)$, $a \in \mathcal{A}^{\rm c}$, generated by $H_a + I_a(t, 0, z_a)$. For $a \in \mathcal{A}^{\rm n}$, we define $U_{a,2}(t)$ to be the unitary dynamics generated by $H_a + I_a(t, y_a, z_a)$. As before, the asymptotic velocity associated to $U_{a,2}(t)$ is denoted by $\zeta_{a,2}^+$.

Observe that *(V4)* implies *(SC)* with $\mu_{\rm s} = 2$. Since we are assuming that $\mu_{\rm l} > \sqrt{3} - 1$, the condition $\mu_{\rm l}\mu_{\rm s} > 1$ of Propositions 5.3.6 and 6.4.3 is satisfied. Using the chain rule, we may therefore combine Theorem 4.3.5 with Proposition 5.3.6 if $a \in \mathcal{A}^{\rm c}$, and with Proposition 6.4.3 if $a \in \mathcal{A}^{\rm n}$, and obtain that the limits

$$\text{s-}\lim_{t\to+\infty} U_{a,2}(t)^* \mathrm{e}^{-\mathrm{i}tH} \mathbf{1}_{T_a}(\zeta^+) =: \Gamma_{a,2}^+, \tag{6.4.21}$$

$$\text{s-}\lim_{t\to+\infty} \mathrm{e}^{\mathrm{i}tH} U_{a,2}(t) \mathbf{1}_{T_a}(\zeta_{a,2}^+) = \Gamma_{a,2}^{+*}, \tag{6.4.22}$$

exist. Furthermore, $\Gamma_{a,2}^+$ is a partial isometry satisfying

$$\Gamma_{a,2}^{+*}\Gamma_{a,2}^+ = \mathbf{1}_{T_a}(\zeta^+), \ \Gamma_{a,2}^+\Gamma_{a,2}^{+*} = \mathbf{1}_{T_a}(\zeta_{a,2}^+),$$
$$\Gamma_{a,2}^+\zeta^+ = \zeta_{a,2}^+\Gamma_{a,2}^+.$$

We have already proved in (6.3.18) that for $a \in \mathcal{A}^{\rm c}$ we have

$$\mathbf{1}_{T_a}(\zeta_{a,2}^+) = \Pi_a. \tag{6.4.23}$$

If $a \in \mathcal{A}^{\rm n}$, by (6.4.20)) we have $I_a(t, y_a, z_a) = 0$, hence $U_{a,2}(t) = \mathrm{e}^{-\mathrm{i}tH_a}$. In particular, (6.4.23) holds also for $a \in \mathcal{A}^{\rm n}$.

The existence of the wave operators $\Omega_{{\rm lr},a}^+$ follows now from the existence of (6.4.21) and from Proposition 6.3.8. Since $\mathrm{Ran}\Omega_{{\rm lr},a}^+ = \mathrm{Ran}\Gamma_{a,2}^+ = \mathbf{1}_{T_a}(\zeta^+)$, the ranges of $\Omega_{{\rm lr},a}$ are mutually orthogonal. Finally, from Proposition 6.4.2 we have

$$\bigoplus_{a\neq\mathrm{cm}} \mathrm{Ran}\,\Omega_{{\rm l},a}^+ = \mathbf{1}_{Z\setminus\{0\}}(\zeta^+)\mathcal{H} = \mathcal{H}_{\rm scatt}.$$

This completes the proof of the theorem. □

## 6.5. Two-dimensional charged systems

We now turn to the 2-dimensional case. This section concerns $N$-particle systems containing no neutral subsystems; the general 3-particle case is deferred to the next section.

We will prove below that if *(Q1)*, *(Q2)* hold for a 2-dimensional system, then, under fairly weak assumptions on the pair potentials, there are no scattering states. The physical intuition behind this result is, roughly speaking, as follows. A scattering state would describe a break-up of the system into at least two asymptotically independent clusters. Since the system is 2-dimensional, the clusters would have to escape to infinity transversally to the magnetic field. But we saw in Chapter 1 that conservation of the pseudomomentum prohibits charged clusters from doing so.

THEOREM 6.5.1. *Assume that (V1), (V2), (Q1), (Q2) hold. Then:*
*(i) if $Q = 0$, $\mathcal{H}_{\rm scatt} = \{0\}$;*
*(ii) if $Q \neq 0$, $\mathcal{H}_{\rm scatt} = \{0\}$ and $\sigma(H)$ is a closed countable set consisting of eigenvalues of infinite multiplicity.*

Theorem 6.5.1 will be deduced from the following proposition.

PROPOSITION 6.5.2. *Under the assumptions of Theorem 6.5.1, we have:*
*(i) if $Q = 0$, $\sigma_{\rm ess}(\tilde{H}^{\rm cm}(\eta^{\rm n}_{\rm cm}))$ is a closed countable set for $\eta^{\rm n}_{\rm cm} \in Y'_{\rm cm}$;*
*(ii) if $Q \neq 0$, $\sigma_{\rm ess}(\tilde{H}^{\rm cm})$ is a closed countable set.*

PROOF OF THEOREM 6.5.1. By Proposition 6.5.2, $\tilde{H}^{\rm cm}$ has only pure point spectrum if $Q \neq 0$, and $\tilde{H}^{\rm cm}(\eta^{\rm n}_{\rm cm})$ has only pure point spectrum for any $\eta^{\rm n}_{\rm cm}$ if $Q = 0$. Hence in both cases $\mathcal{H}_{\rm scatt} = \{0\}$. Moreover, if $Q \neq 0$, $H$ is unitarily equivalent to $1 \otimes \tilde{H}^{\rm cm}$ acting on $L^2(\tilde{Y}_{\rm cm}) \otimes L^2(\hat{Y}_{\rm cm} \times Y^{\rm cm})$, which has only pure point spectrum of infinite multiplicity. □

PROOF OF PROPOSITION 6.5.2. Consider first the case $Q \neq 0$. We will prove an HVZ-type result:

$$\sigma_{\rm ess}(\tilde{H}^{\rm cm}) \subset \bigcup_{a \in \mathcal{A}, \sharp a \geq 2} \sigma(H_a). \tag{6.5.1}$$

To this end, it suffices to prove the following:

$$\begin{aligned} &\text{if } \chi \in C_0^\infty(\mathbb{R}) \text{ and } \mathrm{supp}\chi \cap \sigma(H_a) = \emptyset \text{ for all } a,\ \sharp a \geq 2, \\ &\text{then } \chi(\tilde{H}^{\rm cm}) \text{ is compact on } L^2(\hat{Y}^{\rm c}_{\rm cm} \times Y^{\rm cm}). \end{aligned} \tag{6.5.2}$$

Let

$$1 = \sum_{a \in \mathcal{A}} q_a^{\rm cm}(y^{\rm cm})$$

be an $N-$body partition of unity on $Y^{\rm cm}$ as in Proposition 2.3.1. By Lemma 2.3.8, we have for $\chi \in C_0^\infty(\mathbb{R})$:

$$\begin{aligned} \chi(H) &= \textstyle\sum_{a \in \mathcal{A}} q_a(\frac{y^{\rm cm}}{R})\chi(H) \\ &= \textstyle\sum_{a \in \mathcal{A}} q_a(\frac{y^{\rm cm}}{R})\chi(H_a) + o(R^0). \end{aligned}$$

If, moreover, $\mathrm{supp}\chi \cap \sigma(H_a) = \emptyset$ for all $a < a_{\max}$, then $\chi(H_a) = 0$ for such $a$'s, so that

$$\chi(H) = q_{\rm cm}(\frac{y^{\rm cm}}{R})\chi(H) + o(R^0).$$

SInce $U_{\rm cm}$ preserves $y^{\rm cm}$, we obtain that

$$\chi(\tilde{H}^{\rm cm}) = q_{\rm cm}(\frac{y^{\rm cm}}{R})\chi(\tilde{H}^{\rm cm}) + o(R^0). \tag{6.5.3}$$

Next, we observe that $q_{\mathrm{cm}}(\frac{y^{\mathrm{cm}}}{R})\chi(\tilde{H}^{\mathrm{cm}})$ is compact on $L^2(\hat{Y}^{\mathrm{c}}_{\mathrm{cm}} \times Y^{\mathrm{cm}})$. Hence, by (6.5.3), $\chi(\tilde{H}^{\mathrm{cm}})$ is compact on $L^2(\hat{Y}^{\mathrm{c}}_{\mathrm{cm}} \times Y^{\mathrm{cm}})$ as a norm limit of compact operators. This ends the proof of (6.5.1).

Since $H_{a_{\min}}$ describes a system of $N$ non-interacting charged particles in the constant magnetic field $\vec{B}$, its spectrum is the set of Landau levels:

$$\sigma(H_{a_{\min}}) = \Big\{ \sum_{i=1}^{N} \frac{|q_i b|}{m_i}(n_i + \frac{1}{2}) : \ n_i \in \mathbb{N} \Big\},$$

which is a closed countable set. We also observe that if the essential spectrum $\sigma_{\mathrm{ess}}(A)$ of an operator $A$ is a closed countable set, then $\sigma(A)$ is also a closed countable set and is pure point. Applying inductively (6.5.1) to all channel Hamiltonians $H_a$, we obtain that $\sigma(H_a)$ is a closed countable set for all $a < a_{\max}$. This proves the proposition if $Q \neq 0$.

Assume now that $Q = 0$, and define $q_a$ and $\chi$ as before. Using again that $U_{\mathrm{cm}}$ preserves $y^{\mathrm{cm}}$, we obtain

$$\chi(\tilde{H}^{\mathrm{cm}}(\eta^{\mathrm{n}}_{\mathrm{cm}})) = q_{\mathrm{cm}}(\frac{y^{\mathrm{cm}}}{R})\chi(\tilde{H}^{\mathrm{cm}}(\eta^{\mathrm{n}}_{\mathrm{cm}})) + o(R^0),$$

which implies that $\chi(\tilde{H}^{\mathrm{cm}}(\eta^{\mathrm{n}}_{\mathrm{cm}}))$ is compact on $L^2(Y^{\mathrm{cm}})$. Hence

$$\sigma_{\mathrm{ess}}(\chi(\tilde{H}^{\mathrm{cm}}(\eta^{\mathrm{n}}_{\mathrm{cm}}))) \subset \bigcup_{a<\mathrm{cm}} \sigma(H_a). \tag{6.5.4}$$

Then we argue as in the case $Q \neq 0$. □

COROLLARY 6.5.3. *Let $a = \{C_1, \ldots, C_k\} \in \mathcal{A}^{\mathrm{cc}}$ be a strongly charged cluster decomposition (i.e., if the particles $i_1, \ldots, i_l$ belong to the same cluster $C_j$ of $a$, then $\sum_{k=1}^{l} q_{i_k} \neq 0$). Then the spectrum of $H_a$ is a closed countable set; in particular, $H_a$ has only pure point spectrum.*

PROOF. To prove the corollary, it suffices to write $H_a$ as the sum of cluster Hamiltonians $H[C_j]$ (Section 2.2), and to apply Proposition 6.5.2*(ii)* and Theorem 6.5.1 to these Hamiltonians. □

## 6.6. Two-dimensional systems with neutral pairs

In this section we continue to consider the 2-dimensional case. We assume that $N = 3$, and that the system contains at least one neutral pair. Unlike in Section 6.5, the scattering theory will now be non-trivial, as a neutral pair is allowed to escape to infinity transversally to the magnetic field. The results and methods of this section will be similar to those of Section 6.4. Throughout this section we assume that *(V1)*, *(V4)*, *(Q1)* hold; in order to prove the existence and completeness of wave operators, we will also assume that the interactions are of Coulomb type. Note, however, that (as in Chapter 3) we will not need the implicit assumption *(S)*, which was required in the 3−dimensional case.

We denote by $I_a(t,y)$ the effective time-dependent potential

$$I_a(t,y) := \sum_{(ij)\nleq a} F(|y_i - y_j| \geq \frac{t}{\log t})V^{\mathrm{l}}_{ij}(y).$$

As before, $U_{a,0}(t), U_{a,1}(t)$ are the unitary dynamics generated by $H_a + I_a(t,y)$ and $H_a + I_a(t,0,y_a)$ respectively.

The first two results are the analogues of Propositions 6.4.1 and 6.4.3.

PROPOSITION 6.6.1. *Let $u \in \mathcal{H}_{\mathrm{scatt}}$. Then*

$$u = \sum_{a\in\mathcal{A}^{\mathrm{n}}, \sharp a=2} u_a,$$

*where*

$$\lim_{t\to+\infty} U_{a,0}(t)^* \mathrm{e}^{-\mathrm{i}tH} u_a \text{ exists.}$$

PROPOSITION 6.6.2. *Assume that $\mu_{\mathrm{l}} > \frac{1}{2}$, $\mu_{\mathrm{l}}\mu_{\mathrm{s}} > 1$. Then the limits*

$$\mathrm{s-}\lim_{t\to+\infty} \tilde{U}^*_{a,1}(t)U_{a,0}(t) =: \Gamma^+_a, \tag{6.6.1}$$

$$\mathrm{s-}\lim_{t\to+\infty} U^*_{a,0}(t)\tilde{U}_{a,1}(t) = \Gamma^{+*}_a \tag{6.6.2}$$

*exist and are unitary.*

The proofs are identical to those of Proposition 6.4.1 and 6.4.3, except that the $z$ variables are omitted and that Proposition 4.4.11 is used instead of Proposition 4.4.10.

THEOREM 6.6.3. *Assume that $N = 3$ and that (V1), (V4), (Q1), (C) hold with $\mu_{\mathrm{l}} > \sqrt{3} - 1$. Then:*

*(i) the wave operators*

$$\mathrm{s-}\lim_{t\to+\infty} \mathrm{e}^{\mathrm{i}tH}\mathrm{e}^{-\mathrm{i}tH_a}\mathbf{1}_{\mathrm{c}}(H_a) =: \Omega^+_a,\ a \in \mathcal{A}^{\mathrm{n}}$$

*exist and their ranges are mutually orthogonal.*

*(ii) the system is asymptotically complete:*

$$\bigoplus_{a\in\mathcal{A}^{\mathrm{n}}} \mathrm{Ran}\,\Omega^+_a = \mathcal{H}_{\mathrm{scatt}}.$$

PROOF. We first prove the existence of $\Omega^+_a$. Let $\tau(H_a)$ be the threshold set for $H_a$, defined in Section 3.8; by Theorem 3.8.3*(ii)*, the set $\sigma_{\mathrm{pp}}(H_a) \cup \tau(H_a)$ is closed and discrete. By density, it therefore suffices to prove that $\Gamma^+_a u$ exists for $u \in \mathcal{H}_{\mathrm{c}}(H_a)$ such that $u = \chi^2(H_a)u$, where $\chi \in C_0^\infty(\mathbb{R})$ is supported near some $\lambda \notin \tau_a$.

Applying the minimal velocity estimate of Proposition 4.4.11, we obtain that for some $\epsilon_0 > 0$

$$\mathrm{e}^{-\mathrm{i}tH_a}u = F(\frac{|y^{\mathrm{cm}}|}{t} \geq \epsilon_0)\chi^2(H_a)\mathrm{e}^{-\mathrm{i}tH_a}u + o(1).$$

Let $\mathbf{C}$ denote the center of orbit observable for the Hamiltonian $H^{a,\mathrm{n}}$ of the neutral pair in $a$ (note that $\mathbf{C}$ commutes with $H_a$). By Proposition 5.2.7, there is an admissible function $F_1 \in S^0(Y^{a,\mathrm{n}})$ such that $\operatorname{supp} F_1 \subset \{|y^a| \leq 2\}$, $F_1 \equiv 1$ in $\{|y^a| \leq 1\}$, $y^{a,\mathrm{n}} \cdot \nabla F_1 \leq 0$. Applying Proposition 5.3.5 to $H^{a,\mathrm{n}}$, we obtain that

$$\mathrm{e}^{-\mathrm{i}tH_a}u = F_1(t^{-\alpha}\mathbf{C})\mathrm{e}^{-\mathrm{i}tH_a}u + o(1),$$

for $\mu_1 > \alpha > \frac{1}{2}$. By Lemma 2.3.8*(ii)*, Proposition 5.2.5, and Lemma 5.1.3, we have

$$\begin{aligned} & F(\tfrac{|y^{\mathrm{cm}}|}{t} \geq \epsilon_0)F_1(t^{-\alpha}\mathbf{C})\chi^2(H_a) \\ = \; & \chi(H_a)F(\tfrac{|y^{\mathrm{cm}}|}{t} \geq \epsilon_0)F_1(t^{-\alpha}\mathbf{C})\chi(H_a) + o(1) \\ = \; & \chi(H_a)F(\tfrac{|y^{\mathrm{cm}}|}{t} \geq \epsilon_0)F(\tfrac{|y^a|}{t^\alpha} \leq c)F_1(t^{-\alpha}\mathbf{C})\chi(H_a) + o(1). \end{aligned}$$

Note that for $t$ large enough we have

$$F(\frac{|y^a|}{t^\alpha} \leq c)F(\frac{|y^a|}{t} \leq \epsilon_0^2) = F(\frac{|y^a|}{t^\alpha} \leq c), \tag{6.6.3}$$

and that, since $N = 3$, the function $F(|y^{\mathrm{cm}}| \geq \epsilon_0)F(|y^a| \leq \epsilon_0^2)$ is supported in $Y_a \backslash \bigcup_{b \not\leq a} Y_b$, provided that $\epsilon_0$ is small enough. Hence Lemma 2.3.8 yields that

$$(\chi(H) - \chi(H_a))F(\frac{|y^{\mathrm{cm}}|}{t} \geq \epsilon_0)F(\frac{|y^a|}{t^\alpha} \leq c) \in o(1).$$

Combining the above estimates, we see that it suffices to prove the existence of

$$\lim_{t \to +\infty} \mathrm{e}^{\mathrm{i}tH}M(t)\mathrm{e}^{-\mathrm{i}tH_a}u, \tag{6.6.4}$$

where

$$M(t) = \chi(H)F(\frac{|y^{\mathrm{cm}}|}{t} \geq \epsilon_0)F_1(t^{-\alpha}\mathbf{C})\chi(H_a).$$

This will be done using Lemma 4.1.5. To this end, we have to estimate

$$\mathrm{D}_{21}M(t) = \partial_t M(t) + \mathrm{i}HM(t) - \mathrm{i}M(t)H_a.$$

The terms in $\mathrm{D}_{21}M(t)$ coming from $\mathrm{D}_{21}F(\frac{|y^{\mathrm{cm}}|}{t} \geq \epsilon_0)$ are estimated as in the proof of Proposition 6.4.1, using the minimal velocity estimates in Proposition 4.4.11. The terms coming from $\mathrm{D}_{21}F_1(t^{-\alpha}\mathbf{C})$ are estimated as in the proof of Proposition 6.4.3. The only remaining term is

$$\chi(H)F(\frac{|y^{\mathrm{cm}}|}{t} \geq \epsilon_0)I_a(y)F_1(t^{-\alpha}\mathbf{C})\chi(H_a).$$

By Lemma 5.1.3 again, this is equal to

$$\chi(H)F(\frac{|y^{\mathrm{cm}}|}{t} \geq \epsilon_0)F(\frac{|y^a|}{t^\alpha} \leq c)I_a(y)F_1(t^{-\alpha}\mathbf{C})\chi(H_a) + O(t^{-2\alpha}).$$

We write $I_a(y) = I_a^{\mathrm{s}}(y) + I_a^{\mathrm{l}}(y)$. By (6.6.3) and *(V4)*,

$$\chi(H)F(\frac{|y^{\mathrm{cm}}|}{t} \geq \epsilon_0)F(\frac{|y^a|}{t^\alpha} \leq c)I_a^{\mathrm{s}}(y) \in O(t^{-2}).$$

Moreover, *(C)* implies that $I_a^{\mathrm{l}}(0, y_a) = 0$, hence

$$\begin{aligned} & F(\tfrac{|y^{\mathrm{cm}}|}{t} \geq \epsilon_0)F(\tfrac{|y^a|}{t^\alpha} \leq c)I_a^{\mathrm{l}}(y) \\ = \; & F(\tfrac{|y^{\mathrm{cm}}|}{t} \geq \epsilon_0)F(\tfrac{|y^a|}{t^\alpha} \leq c)y^a \textstyle\int_0^1 \nabla_{y_a} I_a^{\mathrm{l}}(sy^a, y_a)\mathrm{d}s \\ = \; & O(t^{-1-\mu_{\mathrm{l}}+\alpha}), \end{aligned}$$

where we used that $|y^a| \leq ct^\alpha$ and $|y_a| \geq c't$ on the support of $F(\frac{|y^{\mathrm{cm}}|}{t} \geq \epsilon_0)F(\frac{|y^a|}{t^\alpha} \leq c)$. Thus the assumptions of Lemma 4.1.5 are satisfied, and we obtain the existence of (6.6.4) and of $\Omega_a^+$. It is clear that $\mathrm{Ran}\Omega_a^+$ are mutually orthogonal.

It remains to prove that $\bigoplus_{a \in \mathcal{A}^{\mathrm{n}}} \mathrm{Ran}\Omega_a^+ = \mathcal{H}_{\mathrm{scatt}}$. Let $u \in \mathcal{H}_{\mathrm{scatt}}$. By Proposition 6.6.1, we may write

$$u = \sum_{a \in \mathcal{A}^{\mathrm{n}}, \sharp a = 2} u_a,$$

where

$$\lim_{t \to +\infty} U_{a,0}(t)^* \mathrm{e}^{-\mathrm{i}tH} u_a \text{ exists.}$$

By Proposition 6.6.2, the limit

$$\lim_{t \to +\infty} U_{a,1}(t)^* \mathrm{e}^{-\mathrm{i}tH} u_a \tag{6.6.5}$$

exists. But by *(C)* we have $I_a(t, 0, y_a) = 0$, so that $U_{a,1}(t) = \mathrm{e}^{-itH_a}$. Finally, it is clear that the limit (6.6.5) is in $\mathbf{1}_{\mathrm{c}}(H_a)\mathcal{H}$, hence $u_a \in \mathrm{Ran}\Omega_a^+$. This completes the proof of the theorem. □

CHAPTER 7

# Open Problems

We conclude this book by discussing some of the many open problems in the scattering theory of magnetic $N-$particle Hamiltonians.

**1. Find an Agmon-type framework for magnetic $N-$particle Hamiltonians.** Standard $N-$particle systems are often cast into the framework of *Agmon Hamiltonians* (see eg [**DG1**, Sect. 6.1]). This is an abstract generalization of $N$-body Schrödinger operators which captures the geometry of the $N-$body potentials in a transparent way and allows certain more general situations, such as $k-$particle interactions or particles of infinite mass. A natural question is to find a similar framework for magnetic $N-$particle Hamiltonians to which the results of this book would extend. (Preferably, such a framework should also allow neutral particles.) The challenge here is to reconcile the geometry of the $N$-body potential with the additional effects of the magnetic field.

**2. Remove Hypothesis *(S)* from our proof of the Mourre estimate for 3-particle systems, or check for which $3-$particle systems this hypothesis holds.** An important part of our proof of asymptotic completeness for 3-particle systems was the Mourre estimate for the Hamiltonian of a neutral pair. Recall that the latter is unitarily equivalent to the direct integral

$$H = \int_{\mathbb{R}^d}^{\oplus} H(k)dk, \text{ acting on } L^2(\mathbb{R}^d, \mathrm{d}k; \mathcal{H}'),$$

where $\mathcal{H}' = L^2(\mathbb{R}^d, \mathrm{d}x)$. (Explicit formulas for $H(k)$ are given in (3.5.2), (1.3.35).) In the two-dimensional case $H(k)$ has purely discrete spectrum and $H$ is analytically fibered, hence the Mourre estimate follows directly from the results of [**GN1**] (Proposition 3.6.1). The case $d = 3$ is more difficult, as $H(k)$ may have both eigenvalues and essential spectrum. The eigenvalues of $H(k)$ contribute to the continuous spectrum of the 3-body channel Hamiltonian and the proof of the Mourre estimate must take this into account. The eigenvalues of $H(k)$ below the essential spectrum are handled again by the methods of [**GN1**]. However, it is conceivable that $H(k)$ might also have embedded eigenvalues with irregular behavior (e.g., discontinuous). In our proof of the Mourre estimate in the 3-dimensional dispersive case we assume Hypothesis *(S)*, which states that $H(k)$ does not have such eigenvalues. We expect this to be true for generic pair potentials.

However, it would be interesting to find explicit conditions on the pair interactions under which this assumption holds, or to prove that *(S)* holds for "almost all" (in an appropriate sense) pair potentials. It would be even more interesting (though probably very difficult) to remove the Hypothesis *(S)* altogether.

**3. Prove asymptotic completeness for 3−particle systems with general long-range interactions.** In our proof of asymptotic completeness in the 3-particle dispersive case we had to assume that the pair potentials are of Coulomb type. From the point of view of physics this is not an unreasonable restriction, as the allowed class of potentials includes Coulomb interactions. Nonetheless, it would be more satisfactory to have a similar result without the Coulomb-type restriction, especially because the mathematical question involved seems quite interesting.

The main problem here would be to find the correct modified dynamics for a bounded neutral pair moving away from the third particle and to prove the appropriate propagation estimates. We recall that the dynamics of such a neutral pair is governed by a time-dependent Hamiltonian

$$H(t) = H(D_y) + I(t, x, y),$$

where $y = i\nabla_k$ is the position observable canonically conjugate to $k$, and $I(t, x, y)$ is a time-dependent potential due to the interaction with the third particle. We have

$$|\partial^\alpha_{x,y} I(t, x, y)| = O(t^{-|\alpha|-\mu}),\ 0 < \mu \le 1.$$

The states living in the continuous spectral subspace of $H(k)$ describe a break-up of the pair and can be easily handled by our methods. The difficult part is to understand what happens to the eigenfunctions of $H(k)$ if the time-dependent perturbation $I(t, x, y)$ is added. Assuming that the condition *(S)* holds (see above), the problem is roughly equivalent to studying the scattering theory for a time-dependent Hamiltonian

$$H(t) = M(D_y) + I(t, y)$$

acting on $L^2(\mathbb{R}^d, \mathrm{d}y; \mathbb{C}^n)$, where $M(k)$ and $I(t, y)$ are $n \times n$ Hermitian matrices on $\mathbb{C}^n$. This becomes highly non-trivial if $M(k)$ is allowed to have multiple eigenvalues for some values of $k$. Note that the same model problem appears in the scattering theory for long-range perturbations of a periodic Schödinger Hamiltonian [**GN2**].

**4. Extend the results of Chapter 5 to the case of charged clusters.** We proved in Chapter 5 that a charged cluster having no neutral subclusters cannot escape to infinity in the transversal direction faster than $O(t^\alpha)$ for some $0 < \alpha < 1$. It would be interesting to clarify whether this result extends to arbitrary charged clusters.

**5. Prove a Mourre estimate for $N$−particle systems containing neutral clusters.** Very little is known about the scattering theory for

$N-$particle systems containing neutral clusters for $N \geq 4$. The case $N = 3$, which we considered in Chapter 3, is already rather difficult. The problems arising here are similar to those in the study of the so-called *dispersive Hamiltonians*

$$H = \sum_{i=1}^{N} \omega_i(D_{x_i}) + \sum_{i<j} V_{ij}(x_i - x_j),$$

where $\omega_i(k)$ are differential operators (*dispersion relations*). Indeed, the Hamiltonian of a neutral pair (see above) restricted to certain scattering channels can be reduced to roughly this form. Similar, but much more complicated, Hamiltonians seem to arise for general neutral clusters. Note also that both the transformed Hamiltonians of neutral clusters and dispersive Hamiltonians are translation-invariant but not Gallilei-invariant. A Mourre estimate has been proved for some classes of dispersive systems (see [**De2**], [**G1**]), but these results are not sufficient for our purposes.

**6. Prove asymptotic completeness for general $N-$body systems in constant magnetic fields.** This problem presents even more serious difficulties than Problem 5. In addition to a Mourre estimate, one would need propagation estimates for neutral clusters travelling transversally to the magnetic field. Solving Problem 4 could also be an intermediate step in this direction.

**7. Prove asymptotic completeness for $N-$particle systems with neutral particles allowed.** Our proof of asymptotic completeness, in both the charged and the dispersive case, requires all the particles to be charged (although we erroneously claimed otherwise in [**GŁ1**]). It would therefore be interesting to consider systems in which some of the particles are neutral. This should be an easier question than Problems 5 or 6, since a neutral particle can be seen as a neutral cluster with no internal structure. Recently Adachi [**A1**], [**A2**] proved a Mourre estimate and the asymptotic completeness for $N-$particle systems containing one charged and $N-1$ neutral particles.

This problem, as well as Problems 4–6, should perhaps be considered in connection with Problem 1.

**8. Include an external electric potential.**

Many physical systems, such as atoms or molecules with infinitely heavy nuclei, can be described by a Hamiltonian

$$H_{\mathrm{e}} = H + \sum_{i=1}^{N} V_i(x_i),$$

where $H$ is an $N$-particle Hamiltonian and $V_i$ is an additional external potential vanishing at infinity, describing the interaction of the $i$-th particle with the fixed nuclei. Due to the latter interactions, the operator $H_{\mathrm{e}}$ is not translation-invariant. It may have bound states, corresponding to all particles remaining close to the nuclei. The states in the continuous spectrum correspond to scattering states where some particles of finite mass travel to infinity, possibly forming bounded clusters. It should be possible to extend our results on asymptotic completeness to this case.

**9. Investigate the case when the magnetic field is constant only at infinity.** Instead of considering $N$-particle systems in a *constant* magnetic field $\vec{B}$, one can assume that the magnetic field is given by $\vec{B} = \vec{B}_0 + \vec{B}_1$, where $\vec{B}_0$ is constant and $\vec{B}_1$ vanishes at infinity. There are several results ([**C**], [**BP**], [**E4**], [**LT1**], [**LT2**], [**Ni**], [**NR**]) on asymptotic completeness and related questions for a single particle in a magnetic field vanishing at infinity. It may be possible to handle the case of a magnetic field as above by combining our methods with those of the articles just cited. The results of [**LT1**], [**LT2**], [**E4**] suggest that at least the case of a short-range additional field $\vec{B}_1$ (such that the vector potential $A_1(x) = \frac{1}{2}(\vec{B}_1(x) \times x)$ is bounded by $C|x|^{\frac{1}{2}-\epsilon}$ at infinity) and systems containing no proper neutral subsystems should be similar to that of a constant field.

**10. Study the decay of non-threshold eigenfunctions.** For standard $N-$particle Hamiltonians, it is known [**FH2**] that eigenfunctions corresponding to non-threshold eigenvalues decay exponentially at infinity. While this is not directly related to questions such as asymptotic completeness, it does have non-trivial consequences in scattering theory – for instance, it can be used to obtain a complete description of the joint spectrum of the energy and asymptotic velocity (see [**DG1**]). Also, some of the methods used to prove exponential decay (e.g., those of [**FH2**]) are reminiscent of those of scattering theory, particularly the Mourre estimate.

For one-particle Hamiltonians in a magnetic field, Gaussian decay of the eigenfunctions can be shown in some cases ([**Er**], [**N**], [**So**]). So far nothing is known about the decay of eigenfunctions for $N-$particle Hamiltonians in a constant magnetic field. The center of orbit observable may be a useful tool in proving the decay of eigenfunctions in the directions transversal to the magnetic field.

**11. Consider time-periodic magnetic fields.** Recently there have been some results ([**A3**], [**M**]) on the scattering theory for $N-$particle Hamiltonians in an exterior time-periodic electric field (the so-called AC Stark Hamiltonians). It would be interesting to consider also the case of time-periodic magnetic fields.

CHAPTER 8

# Appendix

We collect here a few technical results in analysis and pseudodifferential calculus that were used in the main text of the book.

We start with a well-known Hardy inequality.

LEMMA A.1. *We have for* $u \in C_0^\infty(\mathbb{R}^n)$*:*

$$\int |x|^{-2\alpha}|u|^2 dx \leq C_\alpha \int |\nabla_x u|^2 + |u|^2 dx,$$

*if* $n = 3$ *and* $0 \leq \alpha \leq 1$*, or if* $n = 2$ *and* $0 \leq \alpha < 1$.

PROOF. We first claim that for $\alpha < n/2$, $\alpha \leq 1$, and $u \in C_0^\infty([0, \infty))$:

$$\begin{aligned} &\int_0^\infty r^{n-1-2\alpha}|u|^2 dr \\ \leq\ & C_\alpha \int_0^\infty r^{n-1}|\partial_r u|^2 dr + C_\alpha \int_0^\infty r^{n-1}|u|^2 dr. \end{aligned} \tag{A.1}$$

To prove (A.1), we pick $\chi \in C_0^\infty(\mathbb{R})$, $\chi \equiv 1$ near 0. Using that $(r^{n-2\alpha}\chi)' = (n-2\alpha)r^{n-1-2\alpha}\chi + r^{n-2\alpha}\chi'$ and integrating by parts, we obtain

$$\begin{aligned} &\int_0^\infty r^{n-1-2\alpha}\chi|u|^2 dr \\ \leq\ & C\int_0^\infty r^{n-2\alpha}\chi|u||\partial_r u| dr + C\int_0^\infty r^{n-2\alpha}|\chi'||u|^2 dr. \end{aligned}$$

By the Cauchy-Schwarz inequality,

$$\begin{aligned} &\int_0^\infty r^{n-1-2\alpha}\chi|u|^2 dr \\ \leq\ & C\left(\int_0^\infty r^{n-1-2\alpha}\chi|u|^2 dr\right)^{\frac12}\left(\int_0^\infty r^{n+1-2\alpha}\chi|\partial_r u|^2 dr\right)^{\frac12} \\ & +C\int_0^{+\infty} r^{n-2\alpha}|\chi'||u|^2 dr \\ \leq\ & \tfrac12 \int_0^\infty r^{n-1-2\alpha}\chi|u|^2 dr + C\int_0^\infty r^{n+1-2\alpha}\chi|\partial_r u|^2 dr \\ & +C\int_0^{+\infty} r^{n-2\alpha}|\chi'||u|^2 dr. \end{aligned}$$

Using also that $r^{n-1-2\alpha}(1-\chi) \leq Cr^{n-1}$, $r^{n-2\alpha}|\chi'| \leq Cr^{n-1}$, $r^{n+1-2\alpha}\chi \leq Cr^{n-1}$ for $\alpha \leq 1$, we obtain (A.1).

Let now $(r, \omega) \in \mathbb{R}^+ \times S^{n-1}$ be the polar coordinates in $\mathbb{R}^n$. We have

$$dx = C_n r^{n-1} dr d\omega,\ \Delta_x = r^{1-n}\partial_r r^{n-1}\partial_r + \frac{1}{r^2}\Delta_\omega,$$

where $\Delta_\omega$ is the Laplace- Beltrami operator on $S^{n-1}$. Hence:

$$\int |x|^{-2\alpha}|u|^2 dx = C_n \int r^{n-1-2\alpha}|u|^2 dr d\omega,$$

$$\int |\nabla_x u|^2 dx \geq \int r^{n-1}|\partial_r u|^2 dr d\omega.$$

This together with (A.1) implies the lemma. □

The following lemma describes a way to split a time-dependent potential into a short-range part and a long-range part with better regularity properties.

LEMMA A.2. *Let $I(s,x)$ satisfy*

$$|\partial_x^\alpha I(s,x)| \leq C s^{-\mu-|\alpha|},\ |\alpha| \leq 2.$$

*Then for any $\delta > 0$ there is a decomposition*

$$I(s,x) = I_1(s,x) + I_2(s,x),$$

*where:*

$$I_1(s,x) = O(s^{2\delta-2-\mu}),$$

$$\partial_x^\beta I_2(s,x) = \begin{cases} O(s^{-(\mu+|\beta|)}), & |\beta| \leq 2, \\ O(s^{-\mu-2-\delta(|\beta|-2)}), & |\beta| \geq 3, \end{cases}$$

PROOF. We pick a cutoff function $j \in C_0^\infty(X)$ such that $\int j(x)dx = 1$, $\int xj(x)dx = 0$, and set

$$I_2(s,x) = \int I(s, x + s^\delta y) j(y) dy.$$

By Taylor's formula, we have $I_1 = I - I_2 = O(s^{2\delta-2-\mu})$. Moreover,

$$\nabla_x^k I_2(x) = (-1)^k s^{-\delta(k-2)} \int \nabla_x^2 I(s, x + t^\delta y) \nabla_y^{k-2} j(y) dy,$$

so that $I_2$ satisfies the estimates of the lemma. □

In Chapter 3 we used the following easy result about real interpolation spaces. We first recall a few definitions (see e.g., [**ABG**] for more details). Let $E \subset F$ be two Banach spaces. Let

$$\kappa(\tau, f) := \inf_{e\in E}(\tau\|e\|_E + \|f-e\|_F),\ 0<\tau<1, f\in F.$$

For $0 < \theta < 1$, the real interpolation space $(E,F)_{\theta,\infty}$ is defined as:

$$(E,F)_{\theta,\infty} := \{f \in F :\ \sup_{0<\theta<1} \tau^{1-\theta}\kappa(\tau, f) < \infty\},$$

with the norm

$$\|f\|_{(E,F)_{\theta,\infty}} = \sup_{0<\theta<1} \tau^{1-\theta}\kappa(\tau, f).$$

LEMMA A.3. *For $m \in \mathbb{N}, \mu \in \mathbb{R}$, let $S^{\mu}_m$ be the Banach space*

$$S^{\mu}_m = \{f \in C^m(X) : \ |\partial_x^{\alpha} f| \le C_{\alpha}\langle x\rangle^{\mu-\alpha}, \ |\alpha| \le m\}.$$

*Then for $0 < \mu_0 < \mu < 1$:*

$$S^{-\mu}_m \subset (S^{-1}_m, S^{-\mu_0}_m)_{\theta,\infty}, \ \theta = \frac{1-\mu}{1-\mu_0}.$$

PROOF. Let $f \in S^{-\mu}_m$. We fix $\chi \in C_0^{\infty}(\mathbb{R})$, $\chi \equiv 1$ near 0 and set $\chi_R(x) = \chi(R^{-1}|x|)$. We have $f = \chi_R f + (1-\chi_R)f$, $\chi_R f \in S^{-1}_m$, $(1-\chi_R)f \in S^{-\mu_0}_m$. By a direct computation:

$$\|\chi_R f\|_{S^{-1}_m} \le CR^{1-\mu}\|f\|_{S^{-\mu}_m},$$
$$\|(1-\chi_R)f\|_{S^{-\mu_0}_m} \le CR^{\mu_0-\mu}\|f\|_{S^{-\mu}_m}.$$

In particular,

$$\kappa(\tau, f) \le \inf_{R\ge 1} C(\tau R^{1-\mu} + R^{\mu_0-\mu}),$$

and the minimum is attained for $R^{\mu_0-1} = \tau$. Hence

$$\kappa(\tau, f) \le C\|f\|_{S^{-\mu}_m}\tau^{(\mu_0-\mu)(\mu_0-1)^{-1}},$$

which proves the lemma. □

In the next two lemmas we state some easy results about pseudodifferential calculus with operator-valued symbols. Recall that $S^{\epsilon}_{\mathrm{cl}}(X)$, $\epsilon \in \mathbb{R}$, denotes the space of functions $f$ such that $\partial_x^{\alpha} f = O(\langle x\rangle^{-\epsilon-|\alpha|})$ for all $\alpha$.

LEMMA A.4. *Let $X = X_1 \times X_2$, and let $P$ be a mapping*

$$X_1' \ni \xi_1 \mapsto P(\xi_1) \in B(L^2(X_2))$$

*satisfying*

$$\sup_{\xi_1} \|\langle x_2\rangle^n \partial_{\xi_1}^{\alpha} P(\xi_1)\|_{B(L^2(X_2))} < \infty, \ \forall n, |\alpha| \in \mathbb{N}. \tag{A.2}$$

*Then:*

*(i) $\langle x_2^n\rangle P(D_1)$ preserves $D(\langle x\rangle^n)$ for all $n \in \mathbb{N}$;*

*(ii) for $q \in S^{\epsilon}_{\mathrm{cl}}(X), \epsilon \in \mathbb{R}$,*

$$[q(x), P(D_1)]\langle x\rangle^{1-\epsilon} \text{ is bounded on } L^2(X);$$

*(iii) let $r_i \in S^0_{\mathrm{cl}}(X)$, $i = 1, 2$, satisfy $r_1 \equiv 0$ on $\operatorname{supp} r_2$, then*

$$r_1(x)P(D_1)r_2(x)\langle x\rangle^2 \text{ is bounded on } L^2(X);$$

*(iv) let $q(t,x) \in C^{\infty}(X)$ satisfy*

$$|\partial_x^{\alpha} q(t,x)| \le C_{\alpha}(t + \langle x\rangle)^{-1-\epsilon}, \ \epsilon > 0,$$

*then*

$$[q(t,x), P(D_1)](\langle x\rangle + t)^{1+\epsilon} \text{ is uniformly bounded.}$$

PROOF. We write:

$$\langle x\rangle^n = \sum_{|j|\leq n} \langle x_1\rangle^{n-|j|} x_2^j q_j(x), \tag{A.3}$$

for $q_j \in S^0_{\rm cl}(X)$. By pseudodifferential calculus, $\langle x_1\rangle^{n-j}\langle x_2\rangle^{j+n}P(D_1)\langle x\rangle^{-n}$ is bounded, which proves *(i)*.

To prove *(ii)*, we write

$$q(x) = q(x_1) + x_2 q_{-1+\epsilon}(x),$$

$$\langle x\rangle^{1-\epsilon} = \langle x_1\rangle^{1-\epsilon} + x_2 q_{-\epsilon}(x).$$

Then:

$$[q(x), P(D_1)]\langle x\rangle^{1-\epsilon} = [q(x_1), P(D_1)]\langle x_1\rangle^{1-\epsilon}$$

$$+[q(x), P(D_1)]x_2 q_{-\epsilon}(x) + [q_{-1+\epsilon}(x), P(D_1)]\langle x\rangle^{1-\epsilon}.$$

The first term on the right-hand side is bounded by pseudodifferential calculus. The remaining two terms are easily seen to be bounded by undoing the commutators, using *(i)* and (A.2).

Next, we prove *(iii)*. By (A.3) and (A.2), it suffices to prove that

$$r_1 P(D_1) r_2 \langle x_1\rangle^2 \text{ is bounded.}$$

We have $r_i(x) = r_i(x_1) + x_2\partial_{x_2} r_i(x_1) + s_i(x)$, $s_i \in S^{-2}_{\rm cl}(X)$, from which the above claim follows by pseudodifferential calculus.

Finally, *(iv)* is proved by the same argument as *(ii)*. We write

$$q(t,x) = q(t,x_1) + x_2 q_{-1-\epsilon}(t,x),$$

and show that

$$[q(t,x), P(D_1)]\langle x\rangle^{1+\epsilon},\ [q(t,x), P(D_1)]t^{1+\epsilon}$$

are both uniformly bounded. □

LEMMA A.5. *Let $X = X_1 \times X_2$ be as in Lemma A.4. Let $B_2$ be an operator acting on $L^2(X_2)$ such that $\langle x_2\rangle^2 B, D^2_{x_2}B$ is bounded. Let $N = n(x, D_x)$ denote a pseudodifferential operator on $L^2(X)$ whose symbol $n(x,\xi)$ is a second order polynomial in $x, \xi$ satisfying $n(x,\xi) \geq x^2 + \xi^2 - c$. Then*

$$[\mathbf{1}\otimes B_2, N](N+C)^{-\frac{1}{2}} \textit{ is bounded on } L^2(X).$$

PROOF. We have:

$$[\mathbf{1}\otimes B_2, N] = \sum_{\alpha_1,\alpha_2,\beta_1,\beta_2} a_{\alpha,\beta} x_1^{\alpha_1} D_{x_1}^{\beta_1} \otimes [B_2, x_2^{\alpha_2} D_{x_2}^{\beta_2}].$$

The only non-zero terms in the above expansion are those where $|\alpha_2| + |\beta_2| \geq 1$, and hence $|\alpha_1| + |\beta_1| \leq 1$. Undoing the commutator $[B_2, x_2^{\alpha_2} D_{x_2}^{\beta_2}]$ and using our assumption on $B_2$, we obtain the lemma. □

# Bibliography


[ABG] W. Amrein, A. Boutet de Monvel, V. Georgescu: *$C_0$-Groups, Commutator Methods and Spectral Theory of N-Body Hamiltonians*, Birkhäuser, Basel-Boston-Berlin, 1996.

[Ag] S. Agmon: *Lectures on exponential decay of solutions of second order elliptic equations*, Princeton University Press, Princeton 1982.

[A1] T. Adachi: Scattering theory for a two-body quantum system in a constant magnetic field, preprint Ecole Polytechnique (2000), to appear in J. Math. Sci. Tokyo Univ.

[A2] T. Adachi: On spectral and scattering theory for $N-$body Schrödinger operators in a constant magnetic field (2000), to appear in Rev. Math. Phys.

[A3] T. Adachi: Scattering theory for N-body quantum systems in a time-periodic electric field (2000), to appear in Funkcialaj Ekvacioj.

[AT] T. Adachi, H. Tamura: Asymptotic completeness for long-range many-particle systems with Stark effect, II, Comm. Math. Phys. 174 (1996), 537–559.

[AHS1] J. Avron, I. Herbst, B.Simon: Schrödinger operators with magnetic fields I: General interactions Duke Math. J. 45(1978), 847–884.

[AHS2] J. Avron, I. Herbst, B.Simon: Schrödinger operators with magnetic fields II: Separation of the center of mass in homogeneous magnetic fields, Ann. Phys. 114 (1978), 431–451.

[AGM] W. O. Amrein, I.V. Georgescu, P.A. Martin: Approche algébrique de la théorie non-relativiste de la diffusion aux canaux multiples, in *Physical Reality and Mathematical Description*, Enz/Mehra eds. (1974), 255-276.

[BFSS] V. Bach, J, Fröhlich, I.M. Sigal, A. Soffer: Positive commutators and the spectrum of Pauli-Fierz Hamiltonians of atoms and molecules, Comm. Math. Phys. 207 (1999), 557–587.

[BG] A. Boutet de Monvel, V. Georgescu: Spectral and scattering theory by the conjugate operator method, Algebra and Analysis 4 (1992), 73–116.

[BGS] A. Boutet de Monvel, V. Georgescu, A. Soffer: $N$-body Hamiltonians with hardcore interactions, Rev. Math. Phys. 6 (1994), 515–596.

[BP] A. Boutet de Monvel, R. Purice: Limiting absorption principle for Schrödinger Hamiltonians with magnetic fields, Comm. Partial Differential Equations 19(1994), 89–117.

[C] M. Combescure: Long range scattering for the two-body Schrödinger equation with "Hörmander-like" potentials, Comm. Partial Differential Equations 11(1986), 123–165.

[Co] J. Cook: Convergence of the Møller wave matrix, J. Math. and Phys. 36 (1957), 82–87.

[CFKS] H.L. Cycon, R. Froese, W. Kirsch, B. Simon: *Schrödinger operators with applications to Quantum Mechanics and Global Geometry*, Texts and Monographs in Physics, Springer 1987.

[DHS] S. DeBièvre, P.D. Hislop, I.M. Sigal: Scattering theory for the wave equation on non-compact manifolds, Rev. Math. Phys. 4(1992), 575–608.

[De1] J. Dereziński: Asymptotic completeness for $N-$particle long -range quantum systems, Ann. Math. 138(1993), 427–476.

[De2] J.Dereziński: The Mourre estimate for dispersive N-body Schrödinger operators, Trans. AMS 317 (1990), 773–798.

[De3] J. Dereziński: Algebraic approach to the $N$-body long range scattering, Rev. Math. Phys. 3(1991), 1–62.

[DG1] J. Dereziński, C. Gérard: *Scattering Theory of Classical and Quantum $N$-Particle Systems*, Texts and Monographs in Physics, Springer-Verlag 1997.

[DG2] J. Dereziński, C. Gérard: Asymptotic completeness in quantum field theory. Massive Pauli-Fierz Hamiltonians, Rev. Math. Phys. 11 (1999), 383–450.

[DG3] J. Dereziński, C. Gérard: Spectral and scattering theory of spatially cut-off $P(\phi)_2$ Hamiltonians, Comm. Math. Phys 213 (2000) 39-125.

[DG4] J. Dereziński, C. Gérard: unpublished.

[Di] J. Dixmier: *Les $C^*-$algèbres et leurs représentations*, Gauthier-Villars, 1964.

[DiS] M. Dimassi, J. Sjöstrand: *Spectral Asymptotics in the Semiclassical Limit*, London Math. Soc. Lect. Notes Series 268, Cambridge Univ. Press 1999.

[Do] J. Dollard: Asymptotic convergence and Coulomb interaction, J. Math.Phys. 5(1964), 729–738.

[DS] N. Dunford, J.T. Schwartz: *Linear Operators*, vol. I–III, John Wiley and Sons 1958-1971 (reprinted 1988).

[E1] V. Enss: A note on Hunziker's theorem, Comm. Math. Phys. 52 (1977), 233–238.

[E2] V. Enss: Asymptotic completeness for quantum-mechanical potential scattering, I. Short-range potentials, Comm. Math. Phys. 61(1978), 285–291.

[E3] V. Enss: Quantum scattering theory for two- and three-body systems with potentials of short and long range, in: Schrödinger Operators, S.Graffi ed., Springer LN Math. 1218, Berlin, 1986.

[E4] V. Enss: Quantum scattering with long-range magnetic fields, Operator Theory: Adv. and Appl. 57(1993), 61–70.

[E5] Enss, V.: Long-range scattering of two- and three-body quantum systems, Journées "Equations aux dérivées partielles", Saint Jean de Monts, Juin 1989, Publications Ecole Polytechnique, Palaiseau 1989

[Er] L.Erdös: Gaussian decay of the magnetic eigenfunctions, Geom. Funct. Anal. 6 (1996), no. 2, 231–248.

[F] L. Faddeev: *Mathematical aspects of the three-body problem in quantum scattering theory*, Publ. of the Steklov Institute, 1963.

[Fe] C. Fefferman: Stability of matter with magnetic fields, *Partial differential equations and their applications*, CRM Proc. Lecture Notes, 12, Amer. Math. Soc., Providence, RI, 1997.

[FH1] R. Froese, I. Herbst: A new proof of the Mourre estimate, Duke Math. J. 49 (1982), 1075–1085.

[FH2] R. Froese, I. Herbst: Exponential bounds and absence of positive eigenvalues for $N-$body Schrödinger operators, Comm. Math. Phys. 87 (1982) 429-447.

[FHi] R. Froese, P. Hislop: Spectral analysis of second-order elliptic operators on noncompact manifolds, Duke J. Math. 58 (1989), 103–129.

[FHP] R. Froese, P. Hislop, P. Perry: A Mourre estimate and related bounds for hyperbolic manifolds with cusps of non-maximal rank, J. Funct. Anal. 98 (1991), 292–310.

[FW1] R. Froese, R. Waxler: Ground state resonances for hydrogen in an intense magnetic field, *Mathematical quantum theory. II. Schrödinger operators* (Vancouver, BC, 1993), 285–292, CRM Proc. Lecture Notes, 8, Amer. Math. Soc., Providence, RI, 1995.

[FW2] R. Froese, R. Waxler: Ground state resonances of a hydrogen atom in an intense magnetic field, Rev. Math. Phys. 7 (1995), no. 3, 311–361.

[FW3] R. Froese, R. Waxler: The spectrum of a hydrogen atom in an intense magnetic field, Rev. Math. Phys. 6 (1994), no. 5, 699–832.

[G1] C. Gérard: The Mourre estimate for regular dispersive systems, Ann. Inst. H. Poincaré 54 (1991), 59–88.

[G2] C. Gérard: Sharp propagation estimates for $N$-particle systems, Duke Math. J. 67(1992), 483–515.

[GG] C. Gérard, V. Georgescu: On the virial theorem in quantum mechanics, Comm. Math. Phys. 208 (1999), 275–281.

[GIS] C. Gérard, H. Isozaki, E. Skibsted: $N$-body resolvent estimates, J. Math. Soc. Japan 48 (1996), 135–160.

[GŁ1] C. Gérard, I. Łaba: Scattering theory for $N-$particle systems in constant magnetic fields, Duke Math. J. 76(1994), 433–465.

[GŁ2] C. Gérard, I. Łaba: Scattering theory for $N-$particle systems in constant magnetic fields, II. Long-range interactions, Comm. Partial Differential Equations 20(1995), 1791–1830.

[GŁ3] C. Gérard, I. Łaba: Scattering theory for $3-$particle systems in constant magnetic fields: dispersive case, Ann. Inst. Fourier (Grenoble), 46, 3(1996), 801–876.

[GN1] C. Gérard, F. Nier: The Mourre theory for analytically fibered operators, J. Funct. Anal. 152(1998), 202–219.

[GN2] C. Gérard, F. Nier: Scattering theory for the perturbations of periodic Schrödinger operators, J. Math. Kyoto Univ. 38 (1998), 595–634.

[GS] G. Gérard, I.M. Sigal: Space-time picture of semiclassical resonances, Comm. Math. Phys. 145 (1992), 281–328.

[Gr] G.M. Graf: Asymptotic completeness for $N-$body short range quantum systems: A new proof, Comm. Math. Phys. 132 (1990), 73–101.

[Gri] M. Griesemer: $N$-body quantum systems with singular interactions, Ann. Inst. H. Poincaré 69 (1998), 135–187.

[HeSj] B. Helffer, J. Sjöstrand: Equation de Schrödinger avec champ magnétique et équation de Harper, Springer Lectures Notes in Physics n° 345 (1989) 118–197.

[HMS1] I. Herbst, J. S. Møller, E. Skibsted: Spectral analysis of $N$-body Stark Hamiltonians, Comm. Math. Phys. 174 (1995), 261–294.

[HMS2] I. Herbst, J. S. Møller, E. Skibsted: Asymptotic completeness for $N$-body Stark Hamiltonians, Comm. Math. Phys. 176 (1996), 509–535.

[Hö] L. Hörmander: *The Analysis of Linear Partial Differential Operators*, vol. III, IV, Springer-Verlag, 1985.

[HüSp] M. Hübner, H. Spohn: Spectral properties of the spin-boson Hamiltonian, Ann. Inst. H. Poincaré 62 (1995), 289–323.

[Hu] W. Hunziker: Cluster properties of multiparticle systems, J. Math. Phys. 6 (1965), 6–10.

[HuSi] W. Hunziker, I.M. Sigal: *The General Theory of N-Body Quantum Systems*, CRM Proceedings and Lecture Notes, Volume 8, 1995.

[HSS] W. Hunziker, I.M. Sigal, A. Soffer: Minimal escape velocities, Comm. Partial Differential Equations 24 (1999), 2279–2295.

[If] A. Iftimovici: Hard-core scattering for $N$-body systems, Math. Phys. Anal. Geom. 1 (1994), 265–313.

[Iv1] V. Ivrii: Heavy atoms in a superstrong magnetic field, *Differential operators and spectral theory*, Amer. Math. Soc. Transl. Ser. 2, 189, Amer. Math. Soc., Providence, RI, 1999.

[Iv2] V. Ivrii: Asymptotics of the ground state energy of heavy molecules in a strong magnetic field. II, Russian J. Math. Phys. 5 (1997), no. 3, 321–354.

[I1] H. Iwashita: On the long-range scattering for one- and two-particle Schrödinger opeators with constant magnetic fields, Tsukuba J. Math. 19(1995), 369–376.

[I2] H. Iwashita: Spectral theory for 3−particle quantum systems with constant magnetic fields, in: *Spectra, scattering theory and related topics*, Kyoto University, Kyoto, 1994 (Japanese).

[Ja] J.M. Jauch: Theory of the scattering operator II: Multichannel scattering, Helv. Phys. Acta 31 (1958), 661–684.

[JMP] A. Jensen, E. Mourre, P. Perry: Multiple commutator estimates and resolvent smoothness in quantum scattering theory, Ann. Inst. H. Poincaré 41 (1984), 208–225.

[JN] A. Jensen, S. Nakamura: The 2D Schrödinger equation for a neutral pair in a constant magnetic field, Ann. Inst. H. Poincaré, Phys. Théor. 67(1997), 487–510.

[Ł1] I. Łaba: Long-range one-particle scattering in a homogeneous magnetic field, Duke Math. J. 70(1993), 283–303.

[Ł2] I. Łaba: Scattering for hydrogen-like systems in a constant magnetic field, Comm. Partial Differential Equations 20(1995), 741–762.

[Ł3] I. Łaba: Multiparticle quantum systems in constant magnetic fields, in: *Multiparticle Quantum Scattering With Applications to Nuclear, Atomic and Molecular Physics*, D. Truhlar and B. Simon, Eds., IMA Volumes in Mathematics and its Applications, Volume 89, Springer-Verlag 1997, 147–215.

[Li] E. Lieb: Large atoms in large magnetic fields, *Méthodes semi-classiques*, Vol. 2 (Nantes, 1991). Astérisque No. 210, (1992), 9, 237–246.

[LSS] E. Lieb, H. Siedentop, J.P. Solovej: Stability of relativistic matter with magnetic fields, Phys. Rev. Lett. 79 (1997), no. 10, 1785–1788.

[Lo] M. Loss: Stability of matter in magnetic fields, *XIIth International Congress of Mathematical Physics 97* Internat. Press, Cambridge, MA, 1999.

[LT1] M. Loss, B. Thaller: Scattering of particles by long-range magnetic fields, Ann. Phys. 176(1987), 159–180.

[LT2] M. Loss, B. Thaller: Short-range scattering in long-range magnetic fields: the relativistic case, J. Diff. Equ. 73(1988), 225–236.

[MeS] M. Merkli, I.M. Sigal: A time-dependent theory of quantum resonances, Comm. Math. Phys. 201 (1999), 549–576.

[Mø] C. Møller: General properties of the characteristic matrix in the theory of elementary particles, Danske Vid. Selsk. Mat-Fys. Medd. 23 (1945), 1–48.

[M] J. S. Møller: Two-body short-range systems in a time- periodic electric field, preprint (1999).

[M1] E. Mourre: Absence of singular continuous spectrum for certain selfadjoint operators, Comm. Math. Phys. 78 (1981), 519–567.

[M2] E. Mourre: Operateurs conjugués et propriétés de propagation, Comm. Math. Phys. 91 (1983), 279–300.

[M3] E. Mourre: Quelques résultats dans la théorie algébrique de la diffusion, Ann. Inst. H. Poincaré Sect. A (N.S.) 18 (1973), 215–226.

[N] S. Nakamura: Gaussian decay estimates for the eigenfunctions of magnetic Schrödinger operators. Comm. Partial Differential Equations 21, (1996), no. 5-6, 993–1006.

[Ni] F. Nicoleau: Théorie de la diffusion pour l'opérateur de Schrödinger en présence d'un champ magnétique, Thèse de Doctorat de l'Université de Nantes, 1991.

[NR] F. Nicoleau, D. Robert: Théorie de la diffusion quantique pour des perturbations à longue et courte portée du champ magnétique, Ann. Fac. Sci. de Toulouse 12(1991), 185–194.

[Pe] P. Perry: *Scattering Theory by the Enss Method.* Harwood Academic Press, London, 1983.

[PSS] P. Perry, I.M. Sigal, B. Simon: Spectral analysis of $N-$body Schrödinger operators, Ann. Math. 114 (1981), 519–567.

[RS] M. Reed, B. Simon: *Methods of Modern Mathematical Physics*, vol. III, *Scattering Theory*, Academic Press.

[RWHG] H. Ruder, G. Wunner, H. Herold, F. Geyer: *Atoms in Strong Magnetic Fields*, Springer-Verlag 1994.

[Se] I. Segal: Transforms for operators and asymptotic automorphims over a locally compact abelian group, Math. Scand. 13 (1963) 31–43.

[Sig1] I.M. Sigal: On the long-range scattering, Duke Math. J. 60(1990), 473–496.

[Sig2] I.M. Sigal: Non-linear wave and Schrödinger equations, I. Instability of Periodic an Quasiperiodic Solutions, Comm. Math. Phys. 153 (1993), 297–320.

[SS1] I.M. Sigal, A. Soffer: The $N-$particle scattering problem: asymptotic completeness for short-range quantum systems, Ann. Math. 125 (1987), 35–108.

[SS2] I.M. Sigal, A. Soffer: Long-range many-body scattering. Asymptotic clustering for Coulomb-type potentials, Invent. Math. 99(1990), 115–143.

[SS3] I.M. Sigal, A. Soffer: Local decay and velocity bounds, preprint, Princeton University, 1988.

[SS4] I.M. Sigal, A. Soffer: Asymptotic completeness for four-body Coulomb systems, Duke Math. J. 71(1993), 243–298.

[SS5] I.M. Sigal, A. Soffer: Asymptotic completeness of $N$ particle long range scattering, J. Amer. Math. Soc. 7(1994), 307–334.

[Sim1] B. Simon: Geometric methods in multiparticle quantum systems, Comm. Math. Phys. 55 (1977), 259–274.

[Sim2] B. Simon: Phase space analysis of simple scattering systems: extensions of some work of Enss, Duke Math. J. 46(1979), 119–168.

[Sk1] E. Skibsted: Propagation estimates for $N$-body Schrödinger operators, Comm. Math. Phys. 142 (1991), 67–98.

[Sk2] E. Skibsted: Absolute spectral continuity for $N$-body Stark Hamiltonians, Ann. Inst. H. Poincaré 61 (1994), 223–243.

[Sk3] E. Skibsted: Asymptotic completeness for particles in combined constant electric and magnetic fields, Duke Math. J. 89 (1997), 307–350.

[Sk4] E. Skibsted: Spectral analysis of $N$-body systems coupled to a bosonic system, Rev. Math. Phys. 10 (1998), 989–1021.

[SW1] A. Soffer, M. Weinstein: Time–dependent resonance theory, Geom. Funct. Anal. 8 (1998), 1086–1128.

[So] V. Sordoni: Gaussian decay for the eigenfunctions of a Schrödinger operator with magnetic field constant at infinity, Comm. Partial Differential Equations 23 (1998), no. 1-2, 223–242.

[VZ] S. A. Vugalter, G. M. Zhislin: Localization of the essential spectrum of the energy operators of $n$-particle quantum systems in a magnetic field, Theor. and Math. Phys. 97(1)(1993), 1171–1185.

[Y] D. Yafaev: Radiation conditions and scattering theory for $N-$particle Hamiltonians, Comm. Math. Phys. 154 (1993) 523-554.

[Yo] K. Yosida: *Functional Analysis*, Grundelehren der mathematischen Wissenschaften, Springer-Verlag Berlin Heidelberg 1980.

# Index